DECISION SUPPORT SYSTEMS
FOR
SUSTAINABLE DEVELOPMENT

A Resource Book of Methods and Applications

T0190200

DECISION SUPPORT SYSTEMS
FOR
SUSTAINABLE DEVELOPMENT

A Resource Book of Methods and Applications

Gregory (Grzegorz) E. Kersten
Zbigniew Mikolajuk
Anthony Gar-On Yeh

Editors

Kluwer Academic Publishers
Boston/Dordrecht/London

Distributors for North, Central and South America:
Kluwer Academic Publishers
101 Philip Drive
Assinippi Park
Norwell, Massachusetts 02061 USA
Telephone (781) 871-6600
Fax (781) 871-6528
E-Mail <kluwer@wkap.com>

Distributors for all other countries:
Kluwer Academic Publishers Group
Distribution Centre
Post Office Box 322
3300 AH Dordrecht, THE NETHERLANDS
Telephone 31 78 6392 392
Fax 31 78 6546 474
E-Mail <orderdept@wkap.nl>

 Electronic Services <http://www.wkap.nl>

Library of Congress Cataloging-in-Publication Data

Decision support systems for sustainable development : a resource book of
    methods and applications / Gregory (Grzegorz) E. Kersten,
Zbigniew Mikolajuk, Anthony Gar-on Yeh, editors.
        p.   cm.
    Includes bibliographical references.
    ISBN 978-1-4419-5094-9   e-ISBN 978-0-306-47542-9
    1. Environmental management.   2. Sustainable development.
I. Kersten, Gregory E.   II. Mikolajuk, Zbigniew.   III. Yeh, Anthony
G. O., 1952-   .
GE300.D43   1999
363.7--dc21   99-37325
                                                              CIP

Copyright © International Development Research Centre 2010

All rights reserved. No part of this publication may be reproduced, stored in a retrieval
system or transmitted in any form or by any means, mechanical, photo-copying,
recording, or otherwise, without the prior written permission from Kluwer Academic
Publishers, 101 Philip Drive, Assinippi Park, Norwell, Massachusetts 02061.

Published in paperback by the
International Development Research Centre
PO Box 8500, Ottawa, ON, Canada K1G 3H9
http://www.idrc.ca/books/

Printed on acid-free paper in the United States of America.

For
Margaret Kersten,
Anna Mikolajuk and
Brenda Yeh
with thanks for
their inspiration
and encouragement.
Thank you!

Contents

IV. DECISION SUPPORT SYSTEMS FOR SUSTAINABLE DEVELOPMENT

PREFACE

There is a growing sense of concern among the public and organizations from all spectra and corners of the world over depletion of renewable resources, which is offset by an awareness of the opportunity to manage them sustainably. This book is a response to that concern. Its objective is to provide resources on methods and techniques that take advantage of this opportunity and are useful to both developing and developed countries. The authors and editors believe in the contributions decision support systems can and should have for the responsible and sustainable use of natural resources and prudent decisions concerning our environment. We hope that the reader will share this belief with us and see the great potential of information technologies and particularly decision support systems for making better and more informed decisions.

The Rio Earth Summit of 1992 was convened to address worldwide interest in sustainable development. *Agenda 21*, prepared for the Summit, and the *Rio Declaration on Environment and Development* provide frameworks for bringing government and nongovernment organizations, businesses, and universities into a joint effort to resolve the issues that prevent the practise of sustainable development. The importance of information and information technologies for decision-making in the area of sustainable development is the topic of Chapter 40 of *Agenda 21*.

The challenge of the Rio Summit has been answered by many initiatives to address the problems of sustainable development and develop methods and techniques to solve them. One such initiative was the *Expert Group Workshop on Decision Support Systems for Sustainable Development,* jointly organized by the International Institute for Software Technology of the United Nations University and the International Development Research Centre (IDRC), Canada. Held in Macau between 26 February and 8 March 1996, its objective was to explore the use of information and computing technologies for decision support systems (DSSs) in the field of sustainable development. The consensus among the participants was that DSSs are necessary for effective planning and management of sustainable development efforts.

The prerequisites for the wise and sustainable use of natural resources are the collection, management, processing, and dissemination of information to extension advisers, planners, administrators, and policy makers who deal with the issues involved in sustainable development. Innovative and efficient information and computing technologies allow for systematic collection and timely sharing of data across regions and countries. The transformation of the available data into information and knowledge is crucial to the creation of a sound environmental impact assessment, for example, or the construction and evaluation of plans, and their efficient implementation and monitoring. The objective of DSS methods and techniques is to integrate data, information, and knowledge to help decision-makers reach well-informed decisions.

For many developing countries, DSSs are new techniques to deal with sustainable development – which is itself a new concept. The objective of this book is to provide an introduction to DSS methods and technologies, as well as a reference on their use in the sustainable planning and management of renewable resources. It contains selected papers presented at the Macau Expert Group Workshop and several solicited papers with innovative DSS applications for developing countries. The applications and case studies presented include the use of DSSs in economic, ecological, and social systems, as well as in the development processes for sustainable development.

Information technology (IT) is the theme of this book; but it also has provided us with a mode of communication and collaboration. Although the contributors, reviewers, and editors live and work in 17 different countries and may have never met, we all became a small virtual community collaborating on this project. Without the Internet, this project could never have come to its fruitful conclusion.

Many organizations and people have contributed to making the publication of this book possible. Our sincere gratitude goes to the chapter authors who contributed their work. Each paper was reviewed by at least two referees. We thank the following external reviewers for their assistance: Tung Bui, Witold Chmielarz, Luigi Iannuzzi, Raimo Hämäläimen, Pekka Korhonen, Anita Lee-Post, Geoff Mallory, Stan Matwin, Wojtek Michalowski, Włodek Ogryczak, Maragareta Soismaa, Abdul Sow, Robert Valentin and Pirkko Walden. Our special thanks go to Tung Bui who agreed to write the book's overview. We appreciate the IDRC support in funding publication of this book and the International Institute for Applied System Analysis (IIASA), Austria, for its help in the organization of the editors' meeting and providing organizational support.

Our thanks go to Garry Folven, senior editor at Kluwer Academic, who made the transition from initial proposal to final product possible, and to Bill Carman, managing editor at IDRC Books, Barbara Hauser, project administrator at IIASA, and Victor Spassov, president of VSES Communications, who helped with the editing of most of the chapters.

Finally, we want to thank our wives and children for their support, encouragement and patience. While we are sincere in advocating computer-based support, we know well that without their support we would not have been able to complete this project.

Gregory E. Kersten
Zbigniew Mikolajuk
Anthony Gar-On Yeh
Ottawa-Singapore-Hong Kong

DECISION SUPPORT SYSTEMS FOR SUSTAINABLE DEVELOPMENT
An overview
Tung X. Bui

The chapters in this book are written out of a common interest in leveraging information technology to support sustainable development. This interest is of special importance as advanced information technology is being unleashed around the globe, transforming the way organizations – private, public or non-governmental alike – plan, implement and control business in a world tightly connected both economically and environmentally.

The use of computer-based decision support systems for sustainable development is however a complex and delicate endeavor. It involves multiple issues embracing economics, ecology, politics, and others (Glatwin *et al.* 1995). Given the very broad nature of decision support technology, the authors of the chapters in this book do not approach their work on the basis of a common methodology or a shared framework to problem solving. Instead, they are interested in suggesting a variety of ways to use decision support technology in the politics of sustainable development.

1. Sustainable development: A decision-making perspective

Sustainable development means different things to different people. Recognizing that reality, it is important to acknowledge that policies to promote sustainable development require more than just wishful thinking and rhetoric discourse. Rather than submerging in definitional diversity and abstract conception, we postulate in this book that sustainable development is a desire for a better world or future. In operational terms, this desire could be interpreted as a goal to satisfy the needs of the present generation without compromising those of future generations. Even more practically, decisions involved in sustainable development include efforts to:
- maximize resource use and energy efficient,
- reduce environmental impacts,
- avoid/improve social impacts,

- promote the use of renewable and green technologies, and
- enforce democratic decision processes through integrative bargaining.

Sustaining development is driven by multiple and conflicting goals and constraints, complex interdependencies among stakeholders and dictated by considerable moral and ethical considerations. Thus, any sustainable development effort could, and probably should, take a unique approach to deal with this complexity by dealing with predictable and unpredictable, near and far outcomes.

If one adopts our Management Science/Operations Research view of sustainable development (SD), lessons learned from modern management and the practice of decision support technology suggest a number of guiding principles for an effective approach to policy making:

- Make SD a part of mainstream thinking and strategic policy and include SD goals and objectives in the strategic plan of all involved parties,
- Establish all-level leadership to champion SD causes and justify decision rationale,
- Consider SD as a business problem and use sound economic principles to clearly define decision issues, capture all opportunities and constraints, search for team consensus and find optimal solutions,
- Leverage decisions with information technology; and
- Treat implementation of SD decisions as a major endeavor requiring careful project management.

As such, the goals and policies of sustainable development appear to be reasonably well defined and prescriptive in nature. They would lend themselves well for decision support technology.

2. Decision support technology and sustainable development

The development and implementation of decision support systems (DSS) require knowledge and understanding of managerial decision making, levels of reasoning and problem solving and roles of managers in organizations (McNurlin and Sprague 1998).

These prerequisites for using computerized decision support already constitute a challenge for those contemplating the use of information technology for sustainable development. The mandate for a better future is uncontestable but formulating this mandate in a tractable manner is non-trivial. This justifies the use of any support methodology – computerized or not – to help authorities involved in sustainable development sort out all the decision variables and parameters, problem solving heuristics, and appreciate the impacts of potential policy actions.

The process of developing a DSS often revolves around five building blocks:

1. *Information resource management.* In software engineering terms, input data are required for decision analysis and resolution; output data are generated and pre-

sented to decision makers for policy making. Effective management of these data constitutes a first major task of any decision support tool.

2. *Model management.* A model is an abstraction of reality whose purpose is to help decision makers focus on the main elements of a problem. Multiple objective optimization under constraints is a classic modeling approach in management science. Qualitative reasoning, expert heuristics, and data mining are alternative methods to formulate decisions. Given a decision problem, the challenge of DSS is to find the best decision method(s) able to suggest a satisfying solution to policy makers.

3. *Interactive problem solving.* Direct interaction between the DSS and its user allows for a more responsive and user-centered view of the problem. A good DSS is one that provides the right information to the right person at the right time with full transparency. In addition, DSS should provide some cognitive feedback to decision makers by helping them comprehend dynamic changes in the underlying assumptions.

4. *Communications and teamwork support.* Decision making, more often than not, involves more than one decision maker and support for communication and coordination is an important dimension of DSS. Support for information exchange, federated organizational memory, group decision and negotiation is an integral component of organizational decision support.

5. *DSS as non-human co-workers.* In a tightly connected networked world, we postulate a working scenario in which humans will team up with computers as co-workers to optimize execution of business decisions (Negroponte 1995). We envision a new social structure that emerges from the interaction of individuals—both humans and non-humans — operating in a goal-oriented environment under rules that place only bounded demands on each individual's information and computational capacity (Bui 1999). In the multi-dimensional context of sustainable development, various DSS, such as those reported in this book, could serve as task-specific aids to policy makers.

The immediate value of using these five building blocks is to help the DSS users improve their decision outcomes. DSS should achieve its support mission by lending a hand to its users: More quality input data are expected to provide a more complete assessment of the problem situation and a richer set of decision alternatives. More sophisticated decision algorithms are expected to help decision makers find solutions that could not have been found otherwise. Expansive real-time trade-off analyses and interactive simulation are expected to provide decision makers with further insights. Communications and group decision support are expected to increase the chance of finding a shared vision and socially equitable solution. Finally, computerized coordinated DSS workflow should seamlessly enhance the integration of sustainable development at a national or global scale.

Perhaps, a more far-reaching value of using DSS is its ability to improve the way decision makers approach the problem, i.e., new insights into the business, better decisions and faster responses to unexpected situations, and most importantly, a changing consciousness about environmental responsibility. Altogether, DSS should help its users become better informed workers in dealing with their information-intensive sustain-centrist tasks.

A unique feature of this book is a balanced mix of concepts, techniques and case studies to substantiate the framework outlined herewith.

3. Book organization: Scopes, contributions and issues

The book is organized in four parts. All in all, they offer a mix of topics including fundamental concepts of decision support, applications and some directions for further research.

In the first chapter, Yeh and Mikolajuk unequivocally remind the readers that sustainable development is rather an evasive, multi-faceted concept. It is not surprising however if one realizes the breadth and depth of sustainable development activities, ranging from long-term economic growth to short-term preservation of resources, from basic research of resource conservation and replenishment to potentially controversial policy making.

As such, defining decision making tasks, and finding ways to support them, require a comprehensive classification of problem types, information requirements, decision making procedures, selection of decision makers involved in the process and, last but not least, the spatial and temporal impacts of sustainable development decisions. Yeh and Mikolajuk provide some examples how information technology – and particularly, that of decision support tools – can help facilitate these elaborate tasks. Implicit in their analysis is the need of identifying the types of decisions in sustainable development and selecting appropriate decision support tools for these decisions.

Kersten provides a comprehensive review of decision making – both from individual and organization perspectives. He proposes a framework bridging basic levels of individuals' cognitive articulation to Herbert Simon's classic organization decision making model (Simon 1960). This framework serves as a foundation for establishing DSS requirements from a systems development standpoint. The requirements for decision support are presented in the context of the key aspects of problem solving and decision making.

Yeh (Chapter 3) exemplifies the contribution of decision support technology in the formulation and implementation of actual sustainable regional development strategy with the use of Geographic Information Systems (GIS). GIS allows use of digital technology to store and retrieve massively large geographic data bases for interactive modeling, simulation and forecasting of urban, regional or environmental plans. Despite the increasing number of successful applications of GIS, Yeh quickly shares the viewpoint of management information systems (MIS) researchers in cautioning the reader that the success or failure of sustainable development depends more on political and managerial leadership than on advanced technology.

The concept papers, when taken together, are expected to enhance the reader's ability to understand the instances in which decision support systems can be deployed. They argue that information technology can help improve decision quality by putting into place computer systems that help gather relevant information, build context-dependent data and knowledge bases and interactive human-computer interaction.

The nine chapters in Part 2 deal with specific cases of the use of management science and decision support technology in dealing with a variety of sustainable development situations.

Modern management science and advanced information technology have been the two main driving disciplines of this book to support decision making in sustainable development. A unique feature of the DSS applications in this book is the explicit effort of modelers to integrate economic efficiency with social constraints and productivity with ecology. As pointed out repeatedly by the contributing authors, a number of requisites are necessary – i.e., sustaincentrism leadership, innovative approach to modeling sustainable development, and better information resource management, if these two disciplines are to support ecologically and socially sustainable development.

The following five factors should be considered when designing DSS for sustainable development:

1. *Decision makers*: Sustainability is a participatory process of creating and fostering a shared vision that commands prudent use of natural and human resources. Decision makers should be solicited beyond the reliance of public authorities. All stakeholders should democratically and pro-actively assume their decision-making responsibilities in taking charge of their fate and that of future generations – in spite of a decision environment prone to faulty assumptions and lacking of incentives for personal integrity.

2. *Decisions*: Sustainable development must be driven by a commitment to ensuring a life-sustaining earth. As such, decision making in sustainable development should embrace all economic, social, political and environmental components to maximize productivity while assuring the long-term viability of natural systems on which all life depends.

3. *DSS Modeling approach*: Modeling sustainable development requires an exhaustive search and gathering of economic and ecological information, comprehensive goal formulation and constraints, and context-dependent knowledge and heuristics for problem solving. Modeling sustainable development implies management of interdependencies between multiple and conflicting goals, a search for solutions that are equitable to current and future generations, and assessment of potential and chronic threats and protection from counterproductive disruption. Full cost accounting of environmental impacts, multiple objective optimization, simulation and forecasting methods and applied artificial intelligence techniques are possible techniques that can be used for sustainable development.

4. *Database requirements.* Quality data are required for successfully putting modeling into practice. Research in database design often cautions the difficulty in setting up data for DSS. Data needed for DSS are typically historical data with extrapolation potential. The data are typically retrieved and combined from multiple sources, characterized by a varying degree of detail and accuracy. Conventional database management systems are not designed to handle these types of requirements effectively. It is the obligation of the SD researchers not to neglect the challenge of putting into place an information management plan for sustainable development.

5. *Visualization and interface requirements:* Decision algorithms should be transparent to policy makers. The conventional wisdom in human-computer interaction feature – what-you-see-is-what-you-get – is an example of how interface design can promote this transparency. Interface controls should be designed to allow DSS users to "navigate" the problem at hand through time (e.g., past experience, current impacts, and future consequences), space (e.g., local, national, and global implications), problem determinants (e.g., decision variables versus constraints, flows versus stocks), and perspectives (e.g., partial versus holistic scope). (Bui and Loebbecke 1996)

Table 1 provides a synoptic view of these case studies along the dimensions described above; with the exception of the visualization and interface requirements as most systems adopt standard graphical user interface technology.

The chapters in Part 2 demonstrate rather conclusively that sustainable development is indeed a tractable problem when formulated appropriately, and such a formulation attempt is more likely to benefit than confuse the policy makers.

Another remark can be made here. Most if not all of the countries covered in the chapters are developing countries. Implementing DSS in developing countries required the followings to be taken into consideration:

- *"Pay-the-price-later" attitude:* Due to the urgent needs of the present, decision authorities are constantly under pressure of having to neglect the future.
- *Relative political and administrative instability:* The political climate in developing countries has been typically unstable, making institutional commitment rather uncertain.
- *Insufficient infrastructure:* Inadequate infrastructure, compounded with inferior technology, tends to hinder information technology innovation and use. This includes the lack of data, resources, leadership and technical competencies.

Interestingly and oftentimes, sustainable development applications are not treated at a global or national level. Instead, they deal with specific projects of economic development at a regional level, such as such as water quality planning for a specific regional river, rural energy development for a village, and land management of a cooperative. The regional scope of the applications suggests that economic solutions devised for, and lessons learned from, a region could be used to inspire solutions for another region having a similar sustainable development problem situation.

The four chapters of Part 3 explore new research attempts to support sustainable development. Again, adopting a building block approach, these chapters focus on recent trends in information technology in the decision support field. In Chapter 13, Ho proposes a rule induction process that seeks to automatically derive knowledge from raw data. Ho's contribution is an example of the current effort by information scientists to deal with information overload, i.e., attempting to discover interesting but non-obvious patterns of information hidden in large and heterogeneous data sets.

Burstein *et al.* (Chapter 14) look at supporting the process of exploring decision alternatives by providing decision makers with some historical knowledge about the decision context. They advocate the implementation of an organizational memory system to provide an information environment conducive to effective organizational

No.	Author(s)	Country	Sustainable development area	Decision to be supported	Decision maker(s)	DSS modeling approach	Database requirements
4	Yeh and Li	China	Land resource management	Search for equitable use of land resource between generations; identify spatial efficiency in land use	Planners and government officials	Remote sensing and GIS modeling to monitor present and future supply and demand of land resources	Physical properties of land; demographic and economic data
5	Gamboa	Philippines	Water supply planning and investment	Water planning; Search for optimal investment in water supply facilities	Local geo-political units: provincial and national governments, funding agencies	Object modeling and knowledge-based systems	Water demand and supply data; funding data
6	Gailly and Installe	Burundi	Rural Development: land increase and intensification of commercial exchanges	Define incentive strategies: technical assistance to farmers and creation of non-agricultural jobs	Public authorities	Multiple criteria optimization; Impact analysis	Agricultural and economic data including weather and export data
7	Lotov, Bourmistrova and Bushenkov	Russia	Water quality planning in large rivers	Efficient strategies in water planning	All stakeholders involved in water quality planning	Feasible Goals Method and Interactive Decision Maps	Environmental data

No.	Author(s)	Country	Sustainable development area	Decision to be supported	Decision maker(s)	DSS modeling approach	Database requirements
8	Pokharel and Chandra-shekar	Nepal	Rural energy planning	Minimize government energy cost; maximize rural employment; maximize energy use efficiency	Planning authorities	Multi-objective programming methods and geographic information system	Energy data
9	Rais, Gameda, Sajjapongse and Bechstedt	South-East Asia	Land management to support sustainability	Identify the sustainability of the land	Planning authorities	Knowledge-based Inference	Qualitative and quantitative data elicited from local experts
10	Costa and Ensslin	Brazil	Textile industry	Sustain industry survivability with tradeoff analysis	Industry experts and owner-managers of textile firms	Cognitive mapping of critical success factors	Socio-economic data
11	Mohan, Kersten, Noronha, Kersten and Cray	N/A	Train decision authorities for sustainable development	Decision and negotiation in technology transfer, cooperation and trade	Planning authorities	Economic model of negotiation with maximization of joint outcomes	Case study data
12	Raefa	Egypt	Information transfer on crop management and water and soil conservation	Site assessment, seedling production, cultivation, agriculture practice management, disorder diagnosis and remediation	Land-use officials, researchers and farmers	Knowledge-based systems	Expert knowledge bases

decision making. They join Ho's work by noting the potential of knowledge management techniques. The latter could be embedded in their proposed framework to help decision experts collapse large amounts of data into "aggregated" trends, correlating patterns cross-sectionally or over time, and uncovering decision knowledge.

In Chapter 15, Hall attempts to elevate the design of DSS to an international level. He proposes a DSS architecture that supports multiple national languages. An innovative aspect of his work is to go beyond simple interface translation. Rather, Hall advocates the design of a culture-dependent human-computer interaction approach that takes into consideration linguistic user interface and cultural factors (e.g., use of currency, use of color display, legal requirements, etc.). Such a system architecture would be an essential first step if the decision making process for sustainable development is to reach the global level.

Hall sets the stage for Chabanyuk and Obvintsev (Chapter 16) to explore the possibility of using state-of-the-art software technology to design complex multi-layered environmental decision and management processes. Setting aside their technical discussion, the authors try to demonstrate that integration of large-scale environmental management could be done via computer system integration.

In Chapter 17, Abel *et al.* use their experience in the design of a DSS to integrate hydrological models, databases and spatial information systems for water quality management in Australia to suggest an approach to linking various exogenous DSS building blocks together. Using a connector/facilitator concept they argue that a federated software architecture would enhance scalability of the DSS as it evolves over time.

Mahadevan *et al.* close Part 3 with a more detailed discussion of the use of data mining and knowledge discovery methods for building DSS (Chapter 18). These new enabling technologies give the old management adage – information is power – a heightened dimension.

As intended, the chapters in Part 3 venture into the frontiers of technology. The issue one would have to raise is how much technology is sufficient – a threshold of engineering complexity beyond which technology could become a burden rather than an aid to policy makers.

It is rather a challenging task to close an edited volume rich in scope and diversity. To a large extent, the two chapters in Part 4 live up to this challenge. In Chapter 19, Hall *et al.*'s chapter incarnates a considerable and collective view of how, and under what conditions, decision support technology can be an integral part of sustainable development. The critical success factor here is the integral aspect of DSS deployment. This is not just about technology. This is about developing the right technology for the right decision problem to be implemented in the right decision environment.

To close the volume, Kersten and Lo (Chapter 20) provide the readers with a classified set of DSS examples. With a twenty-year or so history of using computer technology for decision support, the list they present is only a subset of the significant achievement researchers from the fields of Management Science, Operations Research, Computer Science and Information Systems. Surely, the list of references provided should be seen as an assurance that DSS can deliver. Better yet, it should be viewed as a source of inspiration for researchers and practitioners seeking to creatively exploit information technology for sustainable development.

4. A resource book for a pro-active approach to sustainable development

The chapters in this book embody the important progress made in decision support technology as it applies to sustainable development. As such, they put us in a far better position to track and analyze decisions related to sustainable development. This in turn should improve our ability to help prescribe policy measures, and thereby the attainment of the ultimate challenge of sustainable development.

Each piece of research in this edited book was conducted without reference to the work of others. Nor were the authors asked to have their chapter be integrated in the general framework of using information technology to support sustainable development. This remark is by no means a weakness of the book. On the contrary, the maturation of DSS technology has resulted in a multitude of highly specialized, problem-focused and mission-critical methodologies and applications. All together, they constitute a concerted and pro-active approach to sustainable development. This book should be seen as a unique set of methods and tools that could be deployed in the context of sustainability.

When dealing with large-scale problems involving multiple constituencies, many people find complexity dull. They argue that simple decision making principles and rules are easier to remember and to apply. They also justify that dramatic arguments can be used to convey these principles to the stakeholders of a decision related to sustainable development. This book is an effort to go beyond an account of generic policy formulation to an analytical and computer-supported platform for effective management and policy making. By doing so, it is expected that the impact of decision making in sustainable development tends to be much more profound than one might anticipate, and the use of decision support technology could be instrumental and long lasting. From this standpoint, this book is an invaluable resource of methods and applications for those who wish to exercise their responsibilities for a better world or future.

References

Bui, T. and C. Loebbecke (1996). Supporting Cognitive Feedback using Systems Dynamics: A Demand Model of the Global Systems of Mobile Telecommunication", *Decision Support Systems* 17, 83-98.

Bui, T. (1999). Building Agent-based Corporate Information Systems: An application to Telemedicine, *European Journal of Operations Research* (forthcoming).

Gladwin, T. N, J. T. Kennelly and T. S. Krause, (1995). "Shifting Paradigms for Sustainable Development: Implications for Management Theory and Research", *The Academy of Management Review*, Volume 20, No. 4, pp. 874-885.

McNurlin, B. C. and R. Sprague Jr. (1998). *Information Systems Management in Practice*, Fourth Edition, Prentice Hall, New Jersey.

Negroponte, N. (1995). *Being Digital*, New York: A. A. Knopf.

Simon, H., A. (1960). *The New Science of Management*, New York: Harper and Row.

I SUSTAINABLE DEVELOPMENT AND DECISION MAKING

1 SUSTAINABLE DEVELOPMENT AND DECISION SUPPORT SYSTEMS
Zbigniew Mikolajuk and Anthony Gar-On Yeh

1. Introduction

In this book, we focus on one aspect of the many processes leading to sustainable development – support for decision-making. A vision for sustainable development implies a long-term process, in which decisions on economic, ecological, and community development are based on the best available information and coordinated at the local, regional, and global levels.

Any discussion on technical support for decision-making in different domains should consider a holistic approach to human problem-solving, the concrete environments in which decision support systems (DSSs) will be used, and the acceptance of the system by the user. The ultimate goal is a system under which international organizations, governments, local authorities, and individuals are able to conduct negotiations as well as coordinate and evaluate their own independent decisions.

A discussion on a global system is beyond the scope of this book, which describes decision support technologies and selected applications. The implementation or even the design of a global system to protect our planet is perhaps beyond our present abilities. However, well-defined components like river basin management or health care support systems can and should be both designed and implemented. Such initial modules then can be further developed, integrated, and refined in the future.

2. The concept of sustainable development

No single term or definition can define the concept of sustainable development precisely. There are many operational definitions. The concept is not new: it can be traced back 2 000 years to the time of the ancient Greeks. The idea of sustainability probably appeared first in the Greek vision of 'Ge' or 'Gaia', the goddess of the Earth, the mother figure of natural replenishment (O'Riordan, 1993). Under her guidance,

Greeks practised a system of sustainability under which local governors were re-warded or punished according to whether the fields looked well-tended or neglected. Later, concern about the limited productivity of land and natural resources can be seen in Malthus' essay on population in 1789 and Ricardo's *Principles of Political Economy and Taxation* in 1817. Their concern was that economic growth might be constrained by population growth and limited available resources.

Toward the end of the 19th and the beginning of the 20th century, the prosperity of the Western economy created an optimistic view of the future. Natural resources were no longer seen as posing severe restrictions on economic growth, as new technologies made far more efficient use of both old and new resources. The fragility of this economic growth was revealed, however, by the world oil crisis and economic recession of the 1970s. Neo-Malthusians began to have doubts about unlimited growth, stressing once again the importance of conserving natural resources by setting limits to economic growth.

In April 1968, the Club of Rome gathered to discuss the present and future predicament of the Earth and its finite resources. The results of its deliberations were published in *The Limits to Growth* (Meadows *et al.*, 1972). The book predicted that the limits of growth on Earth would be reached sometime within the next 100 years if the economy continued to expand at the current rate. The sorry state of our finite resources was the result of exponential growth in global population, resource depletion, and industrial pollution. In 1992, 20 years after this controversial book was published, the same authors published its successor, *Beyond the Limits,* which re-examined the situation of the Earth (Meadows *et al.*, 1992).

With new evidence from global data, the book shows that the exponential growth in the global population, economy, resource consumption, and pollution emissions continues unabated. In 1971, they had concluded that the physical limits to human use of materials and energy would be reached within decades. In 1991, after re-running the computer model with newly compiled data and analyzing the latest development pattern, they realized that in spite of improved sustainable development policies throughout the world, the world might well be approaching its limits even faster than they had thought.

Another sustainability concept can be found in the 1960s notion of carrying capacity in resource management. This concept can be described in wildlife management as "the maximum number of animals of a given species and quality that can, in a given ecosystem, survive the least favourable conditions within a stated time period"; in fisheries management as "the maximum biomass of fish that various water bodies can support"; and in recreation management as "the maximum number of people that a recreational site can support without diminishing the recreational experience that attracted people to it in the first place" (Edwards and Fowle, 1955; Dasmann, 1964; DHUD, 1978).

The most recent concept of sustainable development is a modified derivative of the concepts of growth limits and carrying capacity. It not only stresses the importance of resources in setting limits to economic growth but also draws attention to the need to develop methods that emphasize the potential complementarity between economic development and environmental improvement (Markandya and Richardson, 1992).

In the Brundtland report, sustainable development is defined as "development that meets the needs of the present without compromising the ability of future generations to meet their own needs" (Brundtland, 1987). It is also considered to be "a process of social and economic betterment that satisfies the needs and values of all interest groups, while maintaining future options and conserving natural resources and diversity" (IUCN, 1980).

Sustainable development does not mean no development. It means improving methods for resource management in an environment where there is an increasing demand for resources (Levinsohn and Brown, 1991). Sustainable development represents a new synthesis of economic development and environmental protection. Economic activities inevitably are associated with the consumption of natural resources. As populations increase and the pace of economic development quickens, there is a widespread concern about the scarcity of natural resources.

Modern technologies, however, which are improving much faster than ever before, can help alleviate the environmental impacts of economic development. Social and political configurations also can alter the development pattern for achieving long-term economic growth that can be sustained by the environment. Policies can be designed to modify the behaviour of development by regulating its activities. Planning, environmental management, performance standards, fines, taxes, and licences are other economic instruments for controlling development and internalizing its costs.

Important milestones in discussions on sustainable development are the report *Our Common Future,* published in 1987 by the Brundtland Commission; and the document *Agenda 21,* developed for the United Nations Conference on Environment and Development (UNCED), held in Rio de Janeiro in June 1992. The Gaia Hypothesis (Myers, 1985), formally proposed in 1973, is closely related to concepts of sustainable development and provides very interesting philosophical and theoretical insights into planetary life as a whole.

Three vital processes affect the sustainability of development: economic development, ecological development, and community development (Figure 1). Sustainable development brings the three into balance with each other and negotiates among the interest groups and stakeholders involved in the processes. It provides a program of action for community and economic reform, such that economic development will not destroy ecosystems or community systems. There is a growing consensus that sustainable development must be achieved at the local level if it is ever to succeed on a global basis (ICLEI, 1996).

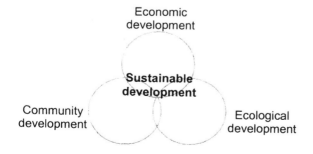

Figure 1. The three interrelated processes of sustainable development.

3. Operational definition of sustainable development

While sustainable development is becoming a widely acknowledged concept (Repetto, 1986; Redclift, 1987; SGSNRM, 1988), there is considerable disagreement on how to implement it. This is because there is no general operational framework for defining sustainable development.

There are more than 70 different definitions of sustainable development, each offering a number of possible modifications to the development process and different reasons for them (Steer and Wade-Gery, 1993). One of these definitions is that development for the future should be based on continuing and renewable processes and not on the exploitation and exhaustion of the principal living resource base (Loening, 1991). Population would be a key factor to be considered here. The implementation of sustainable development can be pursued only if population size and growth are in harmony with the changing productive potential of the ecosystem. The ECCO (enhancement of population carrying capacity options) computer system developed at the Centre for Human Ecology at the University of Edinburgh, is an example of a system that attempts to guide policy-making toward economic development that is more sustainable (Loening, 1991). It tries to answer the question of how we can identify the trade-offs between population growth and standard of living and between intensification of agriculture and soil conservation.

Other definitions of sustainable development focus mainly on sustaining economic development. Sustainable development can be defined as economic development that endures over the long run (Turner, 1993). Some authors, however, think that ecological considerations should be an integral part of economic development. Barbier (1987), for example, argues that sustainable development depends on interactions between three systems: the biological (and other natural resources), the economic, and the social. The goals of sustainable development for the three systems are

- maintenance of genetic diversity, resilience, and biological productivity for the biological system;

- satisfaction of basic needs (poverty reduction), equity enhancement, and increasingly useful goods and services for the economic system; and

- assurance of cultural diversity, institutional sustainability, social justice, and full participation for the social system.

There seems to be no general agreement on the concept of sustainable development. Some people put a high value on a healthy environment, while others prefer high living standards. Income, education, social structure, and ideology are factors that determine the definition of sustainable development in a community. Therefore, some rules need to be designed to guide people toward sustainable development. One such rule might be that a given renewable resource cannot be used at a rate greater than its reproductive rate. Otherwise, it will become totally depleted.

Strict controls on the use of nonrenewable resources are necessary to prevent the possibility of depletion. Substitutions and new technologies can help conserve scarce resources. Abatement measures should be taken to reduce pollution as well. The amount of pollution emissions should not exceed the assimilative capacity of the environment.

Biodiversity in the ecosystem is threatened as more and more species become extinct as a result of human activities. Many species improve the human living environment by generating soil, regulating fresh water supplies, decomposing waste, and cleaning oceans. New genes of great value are being discovered by researchers in many fields. It is thus crucial to maintain a healthy biodiversity by controlling ecosystem balance.

Sustainable development is something every country and every region of the world should work to achieve. Countries differ, however, in their perceptions of sustainable development. There is great difficulty, therefore, in designing universal measurements and indicators of sustainable development because countries also differ in the importance they assign to different components of sustainable development (UNCSD, 1995; Mitchell, 1996). It is here that information technology (IT) can help: IT tools like DSSs can be developed to help local governments and countries to achieve sustainable development once they define their operational indicators.

4. Agenda 21

Agenda 21, a document developed for UNCED, is the result of a comprehensive analysis of what is needed to achieve sustainable development (UNCED, 1993). Its 40 chapters on environmental, economic, social, and organizational issues contain guidelines for developing decision-making processes. Four other key documents were completed at the Rio Conference:

- The Rio Declaration on Environment and Development (The Earth Charter);

- The Statement of Guiding Principles on Forests;

- The Convention on Climate Change; and

- The Convention on Biodiversity.

The purpose of this section on the contents of *Agenda 21* is to help select specific issues for further study on domain-specific decision support and identify the concrete information systems needed in the decision-making processes.

Sustainable development has become an important research issue for many disciplines. Since the Brundtland report, thousands of publications have contributed to a better understanding of the holistic approach to human development and the critical issues of environmental protection, use of natural resources, social development, and well-being. Almost every aspect of life on Earth is represented in this world-scale model of sustainable development.

The *Agenda 21* document presents a framework for processes and actions that will bring the world closer to the concept of sustainable development outlined there. It is based on the principles and recommendations of the Brundtland report and other contributions submitted during the preparatory process by government and nongovernment organizations.

Since the 1992 Rio Conference, much attention has been focused on the political issues of *Agenda 21*, on setting priorities, and on the disagreements surrounding UNCED. At the same time, the implementation of many practical, widely-accepted objectives still remain a subject of scientific discussions and strategic plans. The lack of progress or significant practical steps toward the achievement of sustainable development since 1992 was discussed at the Rio+5 Conference in 1997.

We believe that advances in information and communication technologies in recent years have made it possible to develop practical and effective information systems which can meet the critical needs highlighted in the 40 chapters of *Agenda 21*— improved access to information and support for decision-making. In this and the next chapters we concentrate on the roles DSSs can play in achieving these objectives

Agenda 21 is divided into four sections, each of which addresses one fundamental dimension of sustainable development:

1.   social and economic;

2.   conservation and management of resources for development;

3.   strengthening of major groups; and

4.   means of implementation.

4.1. Section I

The Preamble to *Agenda 21* points out that no nation can achieve sustainable development alone. A global partnership and broad public participation are needed to solve the problems of development. It also recognizes that additional financial resources; the strengthening of public institutions; and well-conceived plans, processes, and implementation strategies will be needed.

In Chapters 2 through 8, on the social and economic dimensions of sustainable development, the authors discuss policies and measures that will have to be designed and implemented to foster trade liberalization, deal with international debts, and create trade and macroeconomic policies that both encourage development and protect the environment.

The document devotes special attention to combating poverty. Some 25% of the people in the developing countries are extremely poor. The creation of new jobs, better education, accessible primary health services, new community-based initiatives, demographic dynamics, and the rehabilitation of degraded resources are considered critical factors for poverty alleviation. *Agenda 21* recognizes that consumption patterns, particularly in developed countries, pose a real threat to the world's ecosystem, and that they must change.

Chapter 6 affirms that the primary health care needs of the world's population must be met if the objectives of sustainable development are to be met. The main issues are the health-care needs of people in rural areas, the control of communicable diseases, the protection of vulnerable groups, urban health problems, and the reduction of health risks from environmental pollution.

Chapter 7 is about the development of human settlement. The proposed program areas indicate the scope and nature of the problems. They include the need to provide shelter to all people and better management of human settlements. The document promotes good managerial practices, especially sustainable land-use planning and management, integrated provision of environmental infrastructure (water, sanitation, drainage, and solid-waste management), sustainable energy and transport systems in human settlements, human settlement planning and management in disaster-prone areas, and sustainable construction-industry activities. Chapter 7 also promotes human resource development and capacity building for human settlements.

Chapter 8 closes the first section and calls for the improvement and fundamental reshaping of decision-making processes at all levels of government, and participation from a broader range of public in decision-making. Improvement of planning and management systems should be based on domestically determined procedures and a holistic approach to economic, social, and environmental issues. The strengthening of national capacity, the development of new methods and tools for data collection, accounting and finance, the strengthening of the legal framework, human resource development, and international cooperation are among the many program initiatives proposed in this chapter.

4.2. Section II

This section focuses on the immediate need to use and manage natural resources wisely so that future generations will be able to meet their survival needs.

Chapter 9 deals with issues related to the Convention on Climate Change. It proposes recommendations on the reduction of atmospheric emissions and the prevention of ozone layer depletion, and calls for international data sharing and research into the modeling and forecasting of atmospheric processes.

Chapters 10 through 16 propose an integrated approach to land-use planning and management. This means that economic and social factors as well as ecological, biological and geological ones are taken into account during the decision-making process. Deforestation is causing serious problems — deterioration of watersheds, soil erosion and loss of land productivity, flooding, loss of genetic diversity, and global warming. Better ways to manage forests, and concrete actions to combat deforestation are outlined in Chapter 11. Related issues of combating desertification and drought are covered in Chapter 12. Developing information systems and sharing knowledge are integral parts of the proposed programs. Knowledge-building, the promotion of management and planning methods, and capacity building at all levels are highlighted in Chapters 13 through 16 by dealing with sustainable agriculture, rural and mountain development, conservation of biological diversity and biotechnology.

Chapters 17 and 18 recognize that oceans and freshwater resources are critical elements of global life. As in the other chapters, the issues of integrated management, environmental protection, sustainable use of resources, stronger cooperation, and knowledge sharing are discussed in the context of concrete program areas.

Chapters 19 through 24 attempt to specify concrete targets and deadlines for handling toxic chemicals, hazardous wastes, solid wastes and sewage, and radioactive wastes. The proposed programs again cover a wide area: information exchange on toxic chemicals and chemical risks, capacity building, training, establishment of government policies, community programs, prevention of illegal international traffic in toxic and dangerous products, research on hazardous waste management, recycling, globally harmonized hazard classification, and compatible labeling systems for chemical materials.

4.3. Section III

This section deals with identifiable groups of people who can contribute to development through education and training and thereby improve their living standards: women, children, indigenous peoples, rural communities, nongovernment organizations (NGOs), local authorities, workers, the community of business and industry, the science and technology community, and farmers. One of the fundamental prerequisites for the achievement of sustainable development is broad public participation in decision-making and access to information.

4.4. Section IV

This section reviews the implementation of sustainable development objectives. Chapters 33 through 39 deal with financial issues; institutional, national, and international arrangements; technology transfer; the role of science; public awareness and training; capacity building; and international legal mechanisms.

Chapter 40 deals with the problem of bridging the gap between developed and developing countries in the availability, quality, coherence, standardization, and accessibility of the data needed to make informed decisions. It sets the following four objectives:

- more cost-effective and relevant data collection and assessment at all levels;

- stronger local, provincial, national, and international capacity to collect and use multisector information in decision-making processes;

- more and stronger local, provincial, national, and international means of ensuring that planning for sustainable development in all sectors is based on timely, reliable, and usable information; and

- relevant, available, and easily accessible information.

Data collection is a necessary but not sufficient condition for better decision-making, however. There is not enough recognition that the capacity building to collect, store, and disseminate information must be accompanied by a concerted effort to develop the theory and practice of DSSs. These and other knowledge-based technologies play a significant role in sustainable development by providing the tools to support the implementation of the Agenda 21 objectives. They make it possible to encode knowledge on the basis of data that are relevant to specific problems, process and aggregate data, and use information to arrive at meaningful conclusions.

5. Technical support for development decisions

5.1. Application domains

DSSs are built for a specific class of problems or a well-defined application domain. *Agenda 21* provides a broad background for the analysis of requirements, scope of development decisions, and specific cases.

A practical and critical barrier to sustainable development is how to make and co-ordinate a wide range of decisions on a local and global level. Technical aids are essential for both information- and judgment-driven decision-making. The interdependence of international, regional, and national development requires a greater exchange of information, sharing of knowledge, and coordination of actions than in the past. Local as well as global factors must be taken into consideration and available academic as well as traditional knowledge must be applied to solve sustainable development problems. DSSs that are well integrated with decision-making processes and valued by decision-makers will contribute to the implementation of the objectives of sustainable development.

Developing resources in a sustainable way requires a holistic approach to human development and the preservation of the natural environment. Decisions on economic, political, or social activities should target concrete problems, but at the same time be based on an analysis of all the implications derived from the paradigm of sustainable development. Many decision-makers are not prepared to handle the complexity of such an analysis. DSSs with a built-in knowledge of the analysis and the capability to investigate and correlate large amounts of data are needed.

A step-by-step approach, international collaboration, determination of priorities, development of adequate methods and technologies, and training are the key factors in this process. A whole range of collaborating DSSs for specific domains will have to be developed and institutionalized to provide effective tools for coordinated actions. Detailed analysis is needed on not only the support requirements but also on how to develop and deploy systems and what their roles are in decision-making.

Since the 1970s, DSS research has concentrated on financial (banking, insurance) and medical systems. In recent years, the following application domains for DSS technology have been investigated from the viewpoint of decision-making for sustainable development:

- land and water management;
- food production and distribution;
- poverty alleviation;
- primary health services;
- public services and administration, governance;
- education;
- pollution control;
- environmental management;
- urban planning and management;
- recovery from a natural disaster;
- population growth control; and
- economic planning.

This is an open list with interrelated and overlapping areas. Useful software applications have been developed for solving many specific problems in these domains. The application areas and references to many specific DSS models and systems are given in Chapter 21. Critical issues are the integration of information systems designed for decision support, standardization, dissemination, institutionalization, and the global networking of systems. The need for multilingual systems and communication (exchange of data and models) between different systems adds to the complexity of system implementation.

Software applications developed for decision support must take into consideration the wide variety of users and their cognitive capabilities. Information, knowledge, and decision-making for development need to be analyzed in the context of human behavior and institutional structures. The main knowledge processor is a human being, not a computer. The computer system merely helps integrate and present the academic and traditional knowledge required for decision-making.

5.2. Decision support systems

There are many definitions for DSSs and their functions in decision-making; several definitions and key functions are discussed in Chapter 2. One should note, however, that different definitions are used in a technology driven market to sell a particular technology. The perspective that we advocate and further elaborate is centered on the user and problem rather than the method and tool. Usually, a combination of different software tools is needed to develop DSSs for sustainable development applications. This is the perspective proposed in Chapter 2 (see Figure 5 in that chapter for a summary).

Policy decisions provide a context and an overall mission statement for operational and resource allocation decisions. For example, one policy decision could be the health improvement of the population in a given region of a country. Such a decision is made as a result of deliberations on government priorities and available resources. In order to achieve the policy goal, operational decisions have to be made at different administrative levels on the allocation of health services, training of medical personnel, supply of fresh water, sanitation, and so forth.

DSS development requires taking into consideration cultural and political factors in addition to technological components. A methodology for the integration of various software technologies, knowledge acquisition tools, and specific user needs is required. The methodological perspective presented in Chapter 2 provides a basis for a comprehensive treatment of decision support and DSS functionalities. Further studies undertaken by interdisciplinary teams of researchers with diverse cultural backgrounds and domain expertise also are needed to focus on the provision of culturally sensitive support methods that allow local knowledge and specific requirements to be incorporated into DSSs.

A background analysis and requirement specification of a support system for decision-making should take into consideration the following factors:

- type of decision problem (policy, operations, resource allocation, etc.);

- domain and scope of the decision problem;

- organizational and structural boundaries;

- decision-making process;

- impact on and synergy with the existing system;

- expected consequences of decision execution;

- profiles of decision-makers (users of the system);

- external constraints and contexts; and

- objectives of a DSS.

The strong interrelation between decisions taken to solve specific problems in different domains introduces additional complexity to the design and use of DSSs. A general model of decision analysis and support for a particular sustainable development domain, such as the management of fresh water resources, for example, may include the format and availability of information, traditional and academic knowledge sources, and interfaces with other systems (land management, legal, market, and others). In addition, it might be required to incorporate the processes of decision-making at those levels of administration that have direct impact on and can contribute to the solution of the problem (household/community, regional, national, and so forth).

Conventional decision-making is aimed at individual problems. Complex geographic or domain interrelationships are not considered because there is a lack of adequate tools to consider different approaches to a given problem.

Decision-making at the level of national governments or large organizations requires different support and tools from those needed for decision-making at the village or single farm level. In either case, the development of a computerized DSS makes economic sense if one or more of the following utility characteristics apply:

- a large amount of data must be collected and processed to produce and analyze decision alternatives (e.g. environmental protection);

- decision-making procedures are applied to many cases within a domain (e.g. immigration application processing) or periodically repeated (e.g. crop management);

- there are many potential users in a given domain (e.g., health services planning);

- it is critical to make top-quality decisions in a short period of time (e.g., recovery from a natural disaster);

- access to the top-level expertise required for decision-making is restricted and expensive (e.g. environmental impact assessment); and

- the possibility of a large number of alternative decisions (e.g., social services planning) with significant and different implications.

The main purpose of a DSS is to help make quality decisions. New methodologies and DSS development tools that include knowledge-based software (see Chapter 12) and knowledge discovery software (see Chapters 13 and 18) are needed to support

DSS developers at each stage of the development cycle. Local knowledge, political culture, and tradition are critical factors in the development and successful institutionalization of a DSS. These issues are discussed at greater length in Chapters 15 and 19.

5.3. DSS technologies

The computer and telecommunication technologies available today allow DSSs to meet demanding functional and user interface requirements. In this section, we deal with some technical topics on the use of specific technologies in DSS development.

A DSS in the domain of sustainable development provides access to and makes use of a variety of databases. In a broad sense, electronic text documents, statistical data, tables of numerical data, digitized maps, drawings, and images are data types processed by a DSS. Domain experts determine which data are needed in the decision-making process, which data are critical for optimal decisions and which sources of data are available.

Formal descriptions of data base structures, data modeling, and the formulation of performance requirements from the point-of-view of decision-making processes are the tasks of DSS developers. For example, a DSS for water quality protection requires all kinds of hydrological, chemical, biological, meteorological, and geographically encoded data on waste water discharge points, manufacturing plants, food processing plants, and water consumption. The DSS developer must identify the appropriate data to show the relationship between water quality and economic activities, municipal water management, and natural phenomena.

The identification of existing data sources and elaboration of procedures and methodologies for collecting data in the field are tasks for the collaborative work of domain specialists and software developers. Data modeling, remote access to data bases, conversion of data, data verification and validation, translation of textual data in a multilingual system, standardization of data formats and access methods, and integration of database management systems with a DSS are technical problems which often have very good solutions. We should consider the adaptation of these solutions in the development of a concrete DSS.

In the real world, decisions are made sometimes despite the fact that complete and accurate data are not available. Data verification modules in a DSS should have built-in knowledge for assessing the validity of the conclusions and the proposed solutions produced by the system in the case of incomplete or unreliable data.

Modeling capability is a major DSS characteristic. Forecasting and decision evaluations are based on computer models. A complex DSS requires a collection of models. Software architecture should include facilities for a model repository, the selection of appropriate models, and the composition of model subsets to solve complex problems. The selection of models in response to a specified problem is a knowledge-based process. An analysis of formal descriptions of decision problems and domain models should generate a scenario for applying a collection of models in a given situation. Take water quality protection, for example: simulation models could help to analyze an emergency situation, like the contamination of a river. A hydrological model of the river and contaminant dispersion would help to determine remedial actions.

Domain knowledge acquisition and encoding are a continuous activity in DSS design and implementation. Decision-making is based on available information and the knowledge of the decision-makers. The acquisition and encoding of that knowledge require specific tools and skills. In the case of government systems, a DSS developer deals with unstructured and distributed knowledge. Legislative documents, policy statements, office procedures and instructions, reports, and programs are sources of knowledge in addition to human expertise and experience. Extracting knowledge from textual documents and integrating it into the document repository with a DSS knowledge base requires software tools such as natural language processors, document management systems, knowledge-acquisition software, and textual information retrieval systems. The analysis of data and textual documents, manipulation of models, generation and evaluation of decision alternatives are knowledge-based procedures.

Knowledge-based software modules (for example, embedded expert systems, rule-based models, and intelligent user interfaces) are components of DSS architecture. Integration of these components and the creation of a consistent knowledge base are the most challenging tasks.

Expert system technology provides facilities for generating an explanation on how and why particular conclusions were drawn. The concept of collaborating expert systems could be useful in solving some problems in a way similar to decision-making by an interdisciplinary group of experts. Expert system technology is useful in addressing the problem of incomplete and unreliable data. A rule-based system with built-in uncertainty factors allows for the analysis of decision-making procedures and evaluation of the probability of results.

The knowledge-based approach to the development of DSSs also involves investigation of the theory and practice of self-learning systems and machine learning.

Support for making optimal decisions to solve a problem is only one step in the decision-making process; determining the consequences of implementing a selected decision is another.

The specification of indicators describing the results of executed decisions, and the collection of pertinent indicator values should constitute a component of DSSs. The domain knowledge base and models can help specify indicators and monitoring procedures. The analysis of indicator values provides feedback for the improvement of the DSS and verification of system software modules. A strict formal analysis supported by software would aid the process of practical implementation of learning algorithms and indicator design.

The design and quality of the user interface are critical if a DSS is to be accepted and used by decision-makers. The complexity of the software and data management has to be understandable to the user, who should be able to concentrate on the decision problem and not be distracted by computer operation-related activities. There are three types of users with different requirements for the interface — developers, domain experts, and decision-makers. The analysis of user profiles and specific domain requirements should encompass the following characteristics and design issues:

- the interactive formulation of a decision problem (structured seminatural language communication between the user and the system),

- the use of graphic tools and images to describe problems,

- case-based dialogue in problem formulation,

- the interactive manipulation of models by means of domain-specific natural language communication,

- the presentation of data in a visual format (GISs, elements of multimedia, and graphics),

- the presentation of decision-making rules and explanation facilities,

- knowledge-based support for user interface features and run-time control (context-sensitive help functions), and

- the ability to adjust to both novice and experienced users.

From the user's point of view, the basic requirements are easy-to-use (intuitive) dialogue facilities and comprehensible responses from the system. A DSS must be attractive to users who do not have the time or willingness to learn computer operations and formal command languages.

The development of DSSs requires the integration of various selected software development tools and methodologies. This should result in specifications for a domain-specific software architecture (DSSA) suitable for development and delivery of a DSS in a particular application domain. Existing system development software, including tools such as DSSA (Tracz, 1995) should be analyzed from the perspective of DSS development. Existing application software modules are building blocks for DSS implementation.

The chapters in this book show how DSSs can be used in the economic, ecological, and social systems as well as development processes of sustainable development. Many of the systems are at the preliminary stage of development. There is a need to identify and further develop DSSs that are useful to the planning and management of sustainable development. No matter how sophisticated and advanced it is, a DSS is useless if it is not being used by decision-makers.

There is a need to persuade government officials and decision-makers to use DSSs in their work. They have to be provided with training so that they can understand and use the technology. Some real-world demonstration projects would be helpful to demonstrate the usefulness of DSSs in the sustainable development field. At the start, such projects do not need to be very sophisticated, but the systems they produce must be operational. Decision-makers should be able to make their own modifications to the parameters and techniques of the DSS to suit the needs of their local communities.

References

Barbier, E.B. (1987). "The Concept Of Sustainable Economic Development", *Environmental Conservation* 14(2), 101-110.

Brundtland (1987). *Our Common Future*, World Commission on Environment and Development, Oxford: Oxford University Press.

Dasmann, R. F. (1964). *Wildlife Management*. New York: Wiley.

De la Court T. (1990). *Beyond Brundtland - Green Development in the 1990s*, New York: New Horizon Press.

DHUD (1978). *Carrying Capacity Action Research: A Case Study in Selective Growth Management*. U.S. Department of Housing and Urban Development, Oahu, Hawaii, Final Report.

Edwards, R. Y. and C.D. Fowle (1955). *Transactions of the 20th North American Wildlife Conference*. Washington, D.C.: Wildlife Management Institute.

ICLEI (1996). *The Local Agenda 21 Planning Guide: An Introduction to Sustainable Development Planning*, Toronto: International Council for Local Environmental Initiatives.

IUCN (1980). *World Conservation Strategy*. Gland, Switzerland: International Union for the Conservation of Nature, Gland, Switzerland, 2.

Levinsohn, A.G. and S. J. Brown (1991). "GIS and Sustainable Development in Natural Resource Management", in Heit, M. and M. Shortreid, (Eds.), *GIS Applications in Natural Resources*, GIS World, Inc. 17-21.

Loening, U.E. (1991). "Introductory Comments: the Challenge for the Future", in A.J. Gilbert and L.C. Braat, (Eds.) *Modelling for Population and Sustainable Development*, London: Routledge, 11-17.

Markandya, A. and J. Richardson (1992). "The Economics of the Environment: An Introduction", in Markandya, A. and Richardson, J. (Eds.), *Environmental Economics*, London: Earthscan, 7-25

Meadows, D.H., D. L. Meadows. and J. Randers (1992). *Beyond the Limits,* London: Earthscan.

Meadows, D.H., D. L. Meadows et al. (1972). *The Limits to Growth*, New York: Universe Books.

Mitchell, G., (1996). "Problems and Fundamentals of Sustainable Development Indicators", *Sustainable Development* 4(1), 1-11.

Myers, N. (Ed.) (1985). *The Gaia Atlas of Planet Management,* London: Gaia Books Ltd.

O'Riordan, T. (1993). "The Politics of Sustainability", in Turner, R.K. (ed.), *Sustainable Environmental Economics and Management:Principles and Practice,* New York: Belhaven Press, 37-69

Redclift, M. (1987). *Sustainable Development: Exploring the Contradictions*, London: Methuen.

Repetto, R. (1986). *World Enough and Time*, New Haven: Yale University Press.

Steer, A. and W. Wade-Gery (1993). "Sustainable Development: Theory and Practice for A Sustainable Future", *Sustainable Development* 1(3), 23-35.

SGSNRM (1988). *Perspectives of Sustainable Development*, Stockholm Group for Studies on Natural Resource Management, Stockholm.

Tracz W. (1995). "DSSA (Domain-Specific Software Architecture) Pedagogical Example", *Software Engineering Notes* 20(3).

Turner, R.K. (1993). "Sustainability: Principles and Practice", in R.K. Turner (ed.), *Sustainable Environmental Economics and Management: Principles and Practice*, New York: Belhaven Press, 3-36.

UNCED (1993). *Agenda 21: The United Nations Programme of Action from Rio*, United Nations Conference on Environment and Development, New York, United Nations.

UNCSD (1995). *Indicators of Sustainable Development: Framework of Methodologies*, United Nations Commission on Sustainable Development, New York: United Nations.

2 DECISION MAKING AND DECISION SUPPORT
Gregory E. Kersten

1. Introduction

Agenda 21, the Brudtland Report and other seminal documents discussed earlier in Chapter 1 provide a number of frameworks for the development and implementation of information systems methods and technologies to aid and support development decisions. There is a great demand for information and tools that facilitate making well informed decisions, assist in evaluating and analyzing their direct results and indirect implications, and provide strong basis for subsequent decisions. The complexity of sustainable development issues requires that all available technologies be used in an integrative and collaborative manner. Subsequent chapters of this book show that there have been many successful developments in the area of information systems and also that individual decision makers still face many challenges on the levels of regional, national and international organizations.

Support for integrated management, environmental protection, and sustainable use of resources requires flexible and content-rich methodologies. Multiple viewpoints, cultures, organizational structures and cognitive efforts need to be considered. Chapter 1 provides a list of various areas of sustainable development issues with those factors and requirements that need to be considered in order to develop and implement a decision support system.

This chapter examines decision making and support from the following four perspectives: information processing, managerial activities, decision problems and human organizations. This broad framework allowed us to identify different aspects and requirements of managerial support and led to a formulation of a set of decision support system design principles. This framework can be applied to position, study and develop a variety of systems, including geographical and spatial decision support systems and also to plan support systems discussed in Chapter 3.

The aim of this discussion is to extend the existing perspectives and taxonomies of decision support systems (DSS) and to position them within the context of information processing activities in both a narrow and broad sense. This leads to a perspective that takes into account recent methodological and technological developments. More importantly, it provides a framework for bridging such key issues as culture, tradition, local knowledge and procedures, multi-dimensionality of decision outcomes and their multiple interpretations, and interactions between the decision makers and their decisions.

The discussion of decision making is based on three levels of reasoning (Kersten and Cray 1996) and Simon's decision phase model (Simon, 1960). The phase model has been recently considered as an obstacle for the evolution of DSS research and practice (Anghern and Jelassi, 1994). In our view an integration of reasoning levels with the phase model provides a sound basis for discussion on DSSs methodologies, functionalities, and most importantly, their integration to the decision process. We argue that the limitations of the model, including the consideration of decision outcomes and knowledge creation, can easily be alleviated.

Several alternative frameworks for DSS development have been proposed. Brooks (1995) suggests that the phase model proposed by Mintzberg et al. (1976), has more relevance to DSS design then the phase model. These two models, however, are not incompatible. The focus of the process model is on support functions and specific tasks as opposed to the main types of activities (phases) in the process. An integration of the decision phases with managerial functions and tasks allows us to enrich the DSS framework.

Four decision making issues are discussed in Section 2. They build upon the cognitive view approach and stress the importance of the concept-to-stimulus focus that positions the decision maker at the center of the decision making and support process. Presentation of the phase model is followed by a discussion of functions and issues in managerial decision making. This provides the basis for a knowledge-based view of decision outcomes.

This discussion aims at the specification of support requirements and support functions that is presented in Section 3. In this section two types of support are also outlined; model-oriented support and data-oriented support.

In Section 4, DSS functional and performance requirements and DSS architecture are discussed. The requirements and design principles for decision support systems that interact with decision makers and aid them in their activities are formulated. Considerations for new system analysis and design methodologies are developed for an architecture centered around knowledge-based methodologies and control components. Support methodologies and functions introduced in the previous sections are then related to different DSS methods.

DSSs can be viewed as active participants in the decision making process. They have to behave, however, according to accepted organizational and managerial norms and rules. To illustrate this point two extreme types of DSS behavior, bureaucratic and entrepreneurial, are introduced in Section 5.

Despite the vast literature on DSSs many gaps remain. There is no unified decision support methodology that integrates the many existing methods, approaches and techniques. This may be due to the evolution of these systems on one hand and multiple technological revolutions, on the other. This chapter does not attempt to provide a unified framework. Rather it complements a decision support ontology proposed by Kersten and Szpakowicz (1994), and attempts to establish linkages between several frameworks and models in an attempt to provide some functional and architectural extensions. We conclude with brief discussion on several other issues relevant to DSS theory and practice.

2. Decision making

The development and implementation of decision support systems (DSS) requires knowledge about, and understanding, of managerial decision making, associated levels of reasoning and problem solving and the roles managers play in organizations. These topics are discussed in this and the following sections and along with their implications for decision support requirements.

2.1. Levels of reasoning

At the most general level problem solving and reasoning about decisions, including managerial decisions, can be articulated at three distinct levels (Kersten and Cray 1996):

1. The *level of needs and values.* Humans experience n and values as a hierarchy that provides a rationale for actions undertaken to solve problems and make decisions. These hierarchies also facilitate the assessment and interpretation of the world and its possible states (Maslow 1954). The decision maker's needs and values have a discriminating effect on her subjective perception of the world; this is one of the key features of the cognitive system. The consideration of the hierarchy of needs as the driving force guiding the decision process has been introduced in decision science literature by Keeney (1992).

2. The *cognitive level.* A cognitive system incorporates needs and values in a structure which is often resistant to change. This system focuses on what needs can be fulfilled, to what extent and how. At the cognitive level, where the world model is formulated and maintained, needs are matched against opportunities and threats. Decision making articulated at the cognitive level involves connecting opportunities for satisfying needs with aspects of the problem and other entities in the world.

3. The *tool and calculation level.* Complex or otherwise difficult decisions require some form of support. Problem elements need to be visualized, computed, compared, analyzed and evaluated with the help of analysts and/or the use of monadic and structural systems. These activities are chosen at the cognitive level and they are carried out at the tool level.

Looking at these three levels and considering the specific activities required to reach a decision, we see that all levels have to be present in a decision making process and the activities have to fit with each other. Further, the cognitive paradigm emphasizes the two-way nature of the human-computer interaction as opposed to the one way channel of classical information systems.

De May (1992) observes that the classical information processing approach tends to operate on the stimulus to concept basis. That is, the process begins with the object or problem domain and activities for the transformation, storage and usage of the sensory input are applied. In contrast, the cognitive view stresses the importance of what is contributed by the user (concept to stimulus) when "knowing" or studying a problem.

The shift to concept-to-stimulus is precisely what is illustrated above by the first and second points, placing the *decision maker at the center of the entire problem analysis and solution synthesis process*. The decision maker has needs and a world model that guides the search for information, the selection of support tools, the construction and evaluation of alternatives, and other problem solving activities. Any active or cognitive system would have to exhibit similar characteristics. It is clear that since these ideas are only recently beginning to appear in the decision analysis community, among practitioners, let alone system developers, the existing decision support systems have not been designed with such attributes.

Most DSS have been primarily conceived and executed primarily at the tool level so that only specified and separate sub-problems are processed. In both behavioral and decision support research, however, there have been successful efforts to build representations of cognitive level perceptions with the use of cognitive mapping techniques (Eden 1988; Eden and Ackerman 1998). While cognitive maps have been used as a mediating technique between managers and consultants they also can be used in a similar capacity in mediating between managers and DSS (an example of the use of cognitive maps in managerial decision making is given in Chapter 10).

2.2. Decision making and decision problems

Support for decision making requires prior knowledge and understanding of the problems, processes and activities leading to a decision. These can be achieved by studying the reasoning activities involved in a choice process, and remembering that they engage human faculties acting within clearly distinguishable phases (Mintzberg et al., 1976; Simon, 1960). The three phases of Simon's model are intelligence, design and choice.

During the *Intelligence* phase information is gathered to understand the problem for which a decision is required, and the necessary assumptions are made explicit. Then, during the *Design* phase, various alternatives are explored through building models and making appropriate calculations to predict the consequences that would arise from each particular alternative. Finally, in the *Choice* phase, a best or satisfactory decision is sought and selected. Often some final verification is undertaken.

Decision theory provides decision makers with a wide range of instruments which can be applied to uncover existing relationships and to help represent, analyze, solve and evaluate a decision problem. The selection and use of a specific method is, however, inherently subjective and guided by the decision maker's preferences expressed in his/her current understanding of the situation. It is often assumed that preferences remain stable, at least for the duration of the choice process, and the selection of a support tool is compatible with these preferences.

This outlook essentially places both choice and reasoning about choice at one level. However, the framing of choice and its impact on a decision process can be articulated by the decision maker at the cognitive level (which also includes needs and values), as well as the tool level.

The multi-faceted character of the cognitive level can be manifested by the hierarchy of the decision maker's objectives and constraints. Objectives provide a rationale for some actions the decision maker undertakes to solve problems and make decisions. Constraints provide the boundaries for choice and distinguish feasible decision alternatives from infeasible ones.

Analysis of the problem should link a decision opportunity to the ability to realize objectives (Heylinghen, 1992). This involves the recognition of the type of problem, the definition and interplay of its components and its relation to earlier experience. Thus, decision making articulated at the cognitive level includes connecting objectives' achievement with problem recognition and definition, and with the situational constraints within which a decision has to be made. The building blocks of the decision problem are considered, major difficulties or obstacles in determining a solution are specified, and the relationships between possible decision outcomes and needs are determined. Choices concerning problem solving strategies and methods that can support them are also made at this level.

While it is hard to conceive of a decision that can be made without being articulated at a cognitive level, many routine and simple decisions are made without the use of the calculation level, where different mechanisms are involved to reduce the mass of information to be processed, or to reduce the complexity of a problem. Support is provides at this level by most DSSs. It can be directed to any of the three phases of Simon's model, as indicated in Fig. 1.

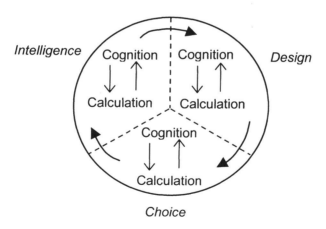

Figure 1. Simon's phase model and levels of articulation

The activities at the calculation level may involve the use of complementary representations to determine the set of feasible alternatives, to investigate the preference structure of the decision maker, and to conduct what-if and sensitivity analyses. All levels of problem articulation may be present in each phase. The importance and the scope of each of the levels, although not indicated in Fig. 1, are likely to vary between the decision phases. They depend on a number of factors, including the decision topic, its complexity and its familiarity to the decision maker, and the context of the decision process. The implication of this for decision support is that both the level at which the need for support is generated and the type of support required are likely to change from phase to phase of the decision process.

2.3. Managerial decision making

Managers are not able to use many DSSs because their design is not compatible with the nature of managerial work nor with approaches to problem solving and decision making managers normally employ. To illustrate this point we give a brief overview of the main aspects of managerial problem solving and decision making and their implications for support.

1. Managerial decisions are always made within an organization and context including the organization's own culture, routines and operating procedures. Support systems need to allow for the influence of cultural and other traits, and must fit the organizational structure. While DSS contribute to changes in the way organizations operate, such changes are inevitably evolutionary in nature so systems that require drastic changes will typically be rejected.

2. Decision problems evolve and their structure changes. Managers devote significant efforts to control organizational which leads to the redefinition of decisions and decision processes. Support has to be well defined in the sense that DSS users need to know not only how the output can be used but also how the system fits in the process and structure.

3. Decision making is a process involving multiple participants and requires conflict resolution. It also involves multiple information sources. This implies that DSS, including group and multi-participant DSS, rarely are the sole source of information. Further, the system should be able to interact with other information systems and make use of historical and current data from multiple sources.

4. Decision makers have different perspectives and use multiple interpretations in creating problem representations and solving problems. The *analytic perspective* allows one to identify and represent sub-problems and apply algorithms to solve them. The *holistic perspective* encourages one to view the overall problem and focus on its few assumptions, issues or implications. A manager can focus only on a few aspects at a time; seeing the big picture, he may choose what he needs or wants to consider in more detail, cycling between analytic and holistic perspectives. Decision analytic methods and models allow for problem decomposition and analysis. Data aggregation and visualization techniques, maps and visual simulation can be used for overall problem and solution assessment.

5. Managers consider and attempt to integrate both qualitative and quantitative aspects of a problem. Problem decomposition and the analysis of problem elements provide only partial understanding of the problem and its implications. Integration of symbolic representations, knowledge processing with quantitative models and preference elicitation schemes partially address this issue.

6. Managers formulate some decision alternatives only superficially and discard them before they are fully articulated. For other alternatives they use partial

quantitative analysis, while still other options may be fully considered. Some systems can be used to assess a problem's importance, others promote a general understanding of the problem and yet other systems perform detailed analysis. The same problem may be considered by different systems at different levels of detail.

7. Decision making does not end with the choice of an alternative. The decision-theoretic concept of outcomes can be used in simple, repetitive, and already well structured decisions. In other cases, managers and executives cannot rely only on precise numerical estimates and expected values that describe outcomes because a decision problem may describe a unique situation with significant qualitative aspects and because decision makers can influence outcomes through decision implementation and control.

8. The assumption that decisions can be made solely on the basis of their utility, i.e., "decision quality in the rational perspective", is unrealistic. Decisions lead to other decisions and opportunities so it is often impossible to assess a priori utilities for these sunsequent decisions. Flexibility that allows the use of unexpected opportunities or adaptation to unforeseen setbacks is an important decision criterion. Other decision criteria, including, fairness, equity, political acceptability, and power implications, may also take precedence over utility. The ability of the decision maker to shape subsequent decisions may also be an important criterion.

9. Decision makers do not conform to one choice paradigm (e.g., rationality, bounded rationality, garbage can, politics and power). Systems must allow for the use of multiple paradigms which are complemented by cognitive and descriptive models, and heuristics. They should also facilitate, not restrict, the role of insight and intuition of decision makers.

10. Making decisions involves the determination of what will happen if a decision is selected, how it can be implemented, what may happen when it is implemented, why this may happen, what happens next, and who may obstruct the implementation and why. What-if and sensitivity analyses, scenario generation and management techniques are important DSS tools. It is no less important, however, to accept that the outcomes of a DSS are only a partial outcome of decision making.

2.4. Decision outcomes

The main aspects of decision making outlined above provide a setting in which a decision is made and a context in which support is provided and utilized. They provide a broader perspective for the decision phase model and indicate that there are results of the process other then a decision selected in the choice phase. Decisions are undertaken to maintain the organization's competitiveness, the productivity levels of the farm, improvement in the regional health care system or the reliability of a water supply system. They set the stage for future decisions. They also allow individuals and organizations to enhance and increase their problem solving and decision making knowledge. In this perspective, called the *knowledge-based view* (Holsapple and

Whinston, 1996), decision making is a creation of descriptive and/or procedural knowledge.

The knowledge view of decision processes positions decision support as an activity oriented toward the manufacturing of new knowledge. This knowledge needs to be stored and maintained because it is used in future problem solving and decision making activities. The outcomes are represented by a set of values on the decision variables and a prescription for their achievement. They also include modification of the context caused by the decision making process and decision implementation, and learning and expertise gained from the process. In this view decision support is considered as both a part of organization-wide information support and knowledge management efforts.

3. Requirements and support functions

The discussion about decision makers, decision process, context and outcomes allows us to specify requirements for decision support and to describe specific functions that may be expected from support systems.

3.1. Fundamental requirements for decision support

There are two fundamental requirements of decision makers that any support system needs to address: *simplicity* and *consistency* (Hill et al., 1982, p. 62-66). Simplicity is needed in selecting and organizing information. Human beings, whether operating as individuals or in groups, can access only a limited amount of information at a time. As Simon (1960) argues, decision makers bound rationality to derive structured, simplified depiction of the decision. Thus, the presentation of a problem within the DSS must be driven by the cognitive capabilities of the decision maker in order to provide information that is critical to the solution of a problem. At the same time the DSS should perform as detailed and comprehensive a computation as possible with the results communicated concisely and succinctly. All additional queries from the decision maker should be treated in a similar manner unless the decision maker wants to enter into a discourse with the system and understand its reasoning. In short, the DSSs must present a simplified version of the problem to the decision maker while maintaining its underlying complexity. At the same time it has to provide easy access to every mechanism used to allow for the verification of the mechanisms' role in the support process.

Consistency in decision making and support can be considered in three dimensions.

The first dimension includes *internal consistency* of data and representations and consistent application of procedures for representation construction and solution derivation. This is an obvious requirement and a prerequisite for simplicity.

The second facet deals with *needs-outcomes consistency*. The relationship between the decision maker's needs and outcomes is the cornerstone of decision analysis. Typically, it involves preference elicitation, alternatives' comparison or determination of a measure of decision quality. While there are numerous approaches for the specification and formal representation of needs, their explanatory power has been questioned (Tversky and Kahneman, 1981).

The third facet of consistency involves the relationship between various decision problems belonging to the same class. This is *inter-decisional consistency* and it reflects the history and expertise of the decision maker and the requirement to maintain consistency among needs and values that are part of the decision maker's personal context (Hill et al., 1982; Newcomb, 1953). With the shift of DSS support to the cognitive level it becomes more important for the system to exhibit inter-decisional consistency, taking into account needs in the decision maker's requests and requirements. Inter-decisional consistency can be achieved when a support system is viewed as a part of a larger organizational information technology infrastructure and its use and results are recorded in the organizational memory (knowledge).

3.2. Methodologies

The functional classification of support for decision making should be interpreted within the frameworks of Simon, Mintzberg, and others. This classification should also be positioned within the general taxonomy of information systems proposed by De May (1992). The purpose is to view decision support within a broader class of information systems that include human decision makers. Following De May we will define three groups of support methodologies.

1. *Monadic methodologies* involve pattern recognition with the use of simple or complex sets of procedures. There is no input analysis, as preprocessed inputs are compared to pre-defined templates. Monadic processing cannot deal with ambiguous inputs and cannot resolve conflicts. Query languages, text retrieval, graphs and data visualization are used to aid decision makers who have to define the scope and limitation of the presentations.

2. *Structural methodologies* involve analysis of specific aspects of information input and decisions as to its further processing. The input is decomposed and a structural analysis of its attributes is performed. Structural analysis allows the resolution of some ambiguities embedded in the input. It can also cope with the conflicting signals. Formal models (e.g., financial, econometric, optimization), are often used to support this function.

3. *Contextual methodologies* involve dealing with the ambiguities not resolved by the structural analysis. There are several possible interpretations of the input and no one can be chosen on the basis of the input alone. The only opportunity to determine proper interpretation is through an assessment of the context in which the inputs were produced and/or received. The key feature of contextual processing is that the inputs and their attributes are analyzed in a broader context outside their domain. Cognitive maps, goal seeking functions and sensitivity analyses can be used to support this function.

Monadic systems provide a variety of presentations of data and models with the use of different media, including text, tables and graphics. Data visualization techniques are successfully used to provide users with information about the organization, potential problems, and the environment. These techniques together with drill-down, drill-up, data slicing and other methods for problem recognition are used in the intelligence phase. In the design phase presentation methods can be used for model construction. Iconic presentations of model primitives allow users to build and view models. Similarly, decisions can be viewed, compared and evaluated separately and in a broader context.

While monadic systems facilitate data, model and solution analyses, structural support systems can conduct these analyses. On-line analytic processing, time series and other statistical methods are used to inform users about potential problems and areas of concern. Regression and log-linear models and modeling environments test model validity and consistency. Data mining methods are used to establish relationships among data and construct representations.

Contextual functions require the application of knowledge to interpret data, modeling efforts, and the implications of alternative decisions. Knowledge-based systems and learning methods are used to resolve ambiguities and to interpret and synthesize data and information. Software agents can represent decision makers and undertake specific decision tasks (Franklin and Graesser 1996).

A general approach to decision support requires that the cognitive predilections of the decision maker acting as the link between the needs fulfillment and circumstantial interpretation provide both a template for support and a limitation for its utilization. A decision maker is unlikely to use a DSS that is incompatible with his/her enacted environment either in terms of decision elements or decision processes. This does not imply that the decision support must be molded to the requirements of a specific decision maker but that it should be flexible enough to match a decision maker's cognitive and reasoning abilities. The support functions positioned in the context of Simon's model and De May's methodologies are presented in Table 1.

Table 1. Decision making phases, support methodologies and support functions

	Intelligence	Design	Choice
Monadic	Presentation	Presentation	Presentation
Structural	Analysis Verification Representation	Analysis Representation Verification	Analysis Representation
Contextual	Interpretation Synthesis	Interpretation Critique Verification	Interpretation Synthesis

3.3. Model-oriented support

Traditionally, decision support has been model-oriented. A decision opportunity or a problem identified in the intelligence phase led to the selection of a modeling technique. A model was constructed in the design phase and used to determine decision alternatives. To obtain alternatives model parameters were calculated with the use of data stored in databases and the user's input. This type of support is centered on user-model interactions

The sequence of key activities in model-oriented support is presented in Fig. 2. The interaction between the decision problem (opportunity) and the model describes activities of model selection and fitting. The data is used inasmuch as it is required to obtain model parameters.

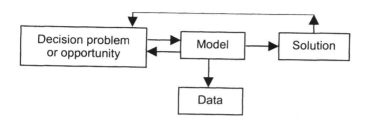

Figure 2. Model-oriented support sequence

Many of the applications presented in Part 2 use model-oriented support (Chapters 5, through 10). Models are used to obtain optimal or efficient decision alternatives, to search through the set of feasible solutions for alternatives with specific characteristics, to conduct sensitivity analyses, and so on.

3.4. Data-oriented support

Model-oriented support assumes that models exist prior to decision making activities. Developed by researchers and analysts, they are embedded in a support system. In contrast, in data-oriented support no model as given, rather it is constructed from the analysis of available data. Data mining and knowledge discovery techniques (see Chapters 13 and 18) are used to extract knowledge and formulate models. This approach depends on the availability of large datasets, often stored in a data warehouse. The sequence of key activities in this type of support is given in Fig. 3.

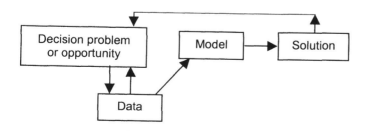

Figure 3. Data-driven support sequence

The realization of a decision problem or opportunity may initiate model construction. This type of support, however, may be initiated by the user but it may also originate with the the system itself. A routine analysis conducted, for example, by on-line analytic processing may indicate the existence of a problem and invoke an extensive data analysis with statistical and other data mining techniques leading to model construction and problem formulation. Only at this stage will the decision maker may enter the process of finding a solution to the problem.

We expect that in future the two types of support will merge. Models generated by data mining techniques will be tested and validated. They can be used to determine parameters for other models or be included in models of an organization, its unit or of a complex activity.

4. Decision support systems

Decision makers' requirements are met with different types of information systems. Management information systems, database management systems (DBMS), on-line analytic processing (OLAP) are just a few examples of systems that provide information used in decision making. It has been suggested that DSS address some or even all of the key requirements.

4.1. Definitions

Decision support systems (DSS) are computer-based systems used to assist and aid decision makers in their decision making processes. Because of the continuously growing number of different types of computer-based systems it is important to distinguish among them and position DSSs within the family of information systems used by decision makers. Little (1970), in one of the earliest works on computer-based decision support, proposed that a DSS be "a model-based set of procedures for processing data and judgments to assist a manager in his decision making".

From the inception of DSSs, it has become clear that they aid and assist decision makers but do not replace them. This feature distinguishes a DSS from other IS. Some IS replace decision makers in well structured, routine and recurring decisions; others are used to verify, record or extract data. Keen and Scott-Morton (1978) note that DSS play a different role and propose the following definition:

> "Decision support systems couple the intellectual resources of individuals with the capabilities of computers to improve the quality of decisions. It is a computer-based support for management decision makers who deal with semi-structured problems."

Moore and Chang (1980), define a DSS in terms of its features and use. They view a DSS as a system that is extendable, capable of supporting ad hoc analysis and decision modeling, oriented towards future planning, and of being used at irregular, unplanned intervals.

Bonczek, Holsapple and Whinston (1980), define DSS in terms of its components. A generic DSS consists of a language system for communication between the user and the DSS, a knowledge system containing problem domain knowledge consisting of data and procedures, and a problem processing system consisting of programs capable of solving decision problems.

The difficulties with defining DSSs were recognized already at the early stage of their introduction. Sprague and Carlson (1982) note that some definitions are so restrictive that only a few existing systems satisfy them, while other definitions are broad so that they include almost all computer systems. Systems for extracting, summarizing and displaying data are also viewed as DSSs (McNurlin and Sprague, 1993). This led Naylor (1982) to observe that "... it seems that virtually every computer hardware and software firm in the industry refers to its products as DSS". This statement is even more justified today as DSSs have gained much popularity and software companies use it as a marketing attribution that indicates their product's innovative character and ability to solve complex managerial problems.

It is clear that DSSs are used to support decision processes as do management information systems (MIS), database management systems (DBMS), on-line analytic processing (OLAP), and also some knowledge-based systems (KBS). All these systems may support decision makers on-line and in an interactive mode so that this feature does not distinguish DSS from other systems.

The main difference between DSS and other information systems lies in the model component: formal quantitative models are an integral part of a DSS (Emery 1987; Bell 1992). These models, for example, statistical, simulation, logic and optimization models, are used to represent the decision problem; their solutions are decision alternatives. However, the model need not be defined a priori but may be constructed during the decision making process. Following the two orientations in decision support introduced in Section 3, we distinguish two types of DSS: *data-oriented* DSS and *model-oriented* DSS.

Dhar and Stein (1997) distinguish between data-driven and model-driven DSS. However, they consider simple aggregation models, query and reporting tools as DSS data-driven systems. We view these systems as a modern version of MIS rather than DSS. While they are helpful in managerial work, including decision making, they are not specifically designed for oriented toward providing support. The inclusion of SQL, OLAP, and other tools used to access and present data would lead to lack of distinction among information systems as most, if not all, provide data that can be used to make a decision. In our view, data-oriented DSS are used to construct models and obtain knowledge about decision problems rather than simply condensing and summarizing large amounts of data (Dhar and Stein 1997, pp. 30-51)

Beulens and Van Nunen (1988) reiterate that a DSS enables managers to use data and models related to an entity (object) of interest to solve semi-structured and unstructured problems with which they are faced. This view allows us to incorporate some of the functions of DBMS and MIS in a DSS. It also emphasizes that a DBMS is an important component of a DSS which also needs reporting capabilities. This is because data used to determine the parameters of a decision model needs to be analyzed and verified. A decision maker requires facilities to extract and view data describing an entity or an object to be able to verify and possibly modify the parameters. While DBMS and MIS are used to provide information about past and present, a DSS is used to determine decisions that will be implemented and will produce outcomes in the future. Thus, the decision maker may need to use models to extrapolate data and obtain a description of the future state of an entity or of an object of interest.

4.2. Functional and performance requirements

From the viewpoints and definitions presented above, we can determine specific characteristics and properties required of a DSS. A support system is used to analyze a problem, determine alternative decisions and select one. Historical data and models are used to generate and evaluate forecasts and decision alternatives. In other words, DSS are future oriented and used to determine the future course of actions in dynamic environments.

Future states and situations can be predicted and assessed only with uncertainty. This requires generating several alternatives or scenarios and imposes a requirement to evaluate the consequences of the alternatives, simulate future situations and provide answers to 'what-if' questions. Alternatively, the user may require a 'goal-seeking' capability, in which the system searches for decision alternatives that can satisfy certain criteria. Sensitivity analysis is also needed to determine the impact of parameter changes on solutions. This approach requires that the robustness of models and procedures be analyzed. This, in turn, imposes a requirement that the solution procedures be efficient and able to generate necessary decision alternatives and to perform sensitivity analysis.

DSS systems often require user involvement in the construction of problem representation and model verification. They also require direct user involvement in the analysis of the decision problem, evaluation of decision outcomes and preference specification. These activities involve subjective judgments and, therefore, a DSS should focus on effective support and not on automatic selection. A DSS, to be effective, needs to be flexible and adaptable to changes in the decision making process and in user requirements.

A DSS participates in the decision process which is controlled and coordinated by the user. Active user involvement requires that a DSS be user friendly and cooperative. This is important in case of episodic or fragmented usage. User involvement also requires that a DSS be well integrated in the decision making process. The implementation of a DSS heavily affects the decision making organization and its procedures and, therefore, needs to be flexible and adaptable to changes in the decision making process.

Active user involvement in problem specification and solution also requires a DSS to support decision processes which embrace qualitative as well as quantitative aspects. The quantitative aspect of the problem may be well structured; it is its qualitative nature (including comparison, evaluation and choice) that makes the problem semi- or unstructured. Decision alternatives need to be judged by both quantitative and qualitative criteria. The use of qualitative and subjective criteria may mean that a satisficing solution is selected rather than an optimal one. This imposes a requirement on the problem solvers and procedures so that the user is able to inspect optimal (efficient) and non-optimal (non-efficient) solutions.

If a DSS is to aid and assist decision makers it must support one or more phases of the decision making process. The required information is provided by quantitative models and data that describe the entity (object) of interest in order to help identify problems or to generate, evaluate, and compare decision alternatives.

A DSS is used to generate, analyze and compare decision alternatives. Data and the parameters that are used to determine them need to be stored separately. We call the database containing alternative solutions and other information a solution base.

The evaluation of decision alternatives or scenarios requires their comparison which a DSS should facilitate. The user-system interface should provide facilities for model- and report-generation, and allow for multiple modes of information display.

We said that models are an important component of a DSS. Many systems are built "around" only one type of model (for example systems that incorporate only linear programming models). If a system is designed to support more than one phase of the decision process including the analysis of both data and solutions then a class of models may be required rather than just one. If the problem is semi-structured, only a part of the problem can be captured by one model and complementary models may be required. In these situations multiple models need to be integrated into the system in such a way that they can interact with each other.

4.3. Difficulties with DSS development and use

The problems associated with building and using DSS partly result from the fact that some of the requirements discussed above are ignored or not fully met. Other problems include insufficient access to databases external to the DSS. This may be due to technical problems such as unknown data models, data models of different types or interfacing difficulties. Those problems may also be caused by badly designed data models and the lack of consideration for systems' interactions.

Many users of DSS have experienced difficulties in learning how to use a system properly and effectively. Beulens and Van Nunen (1988) list the following reasons for these difficulties:

1. the procedures to be used in the system have little in common with procedures or systems that users normally employ;

2. it is difficult to know the interdependencies of the functions provided by the system;

3. it is difficult to keep track of the consequences of a DSS function usage with respect to decision scenaria and the integrity of the database;

4. there are applications that require extensive knowledge of a specific problem domain or technical knowledge (for example optimization or forecasting models);

5. users have to deal with several databases and models, each with different data models and resulting translation problems; and

6. users may have to work on several decision scenarios at the same time. As a consequence they have to keep track of what they have done for each of them.

Model building for semi- and unstructured problems is difficult and requires considerable expertise. Therefore, they are often constructed by system analysts and MS/OR specialists and not by the managers who are the direct users of DSSs. This, however, may defeat the very purpose of the system since it may not meet the managers' requirements, and therefore, is unlikely to participate directly in the decision process. In this case, and also when the decision maker is a direct user, the model may be considered as a black box. A friendly interface (e.g., a graphical user interface and extensive reporting capabilities) does not alleviate the difficulty in the user's understanding of the model assumptions and the relationships between data, models and solutions even though it may make the system easier to access..

4.4. Generic DSS Architecture

The specifications which the DSS needs to satisfy lead to recommendations with respect to the technical/architecture and functional aspects of the DSS. As we show in Fig. 4, the DSS has an architecture comprising of data components and software components. The data components include data, models and solution databases. It also requires a bank with dialog primitives that are used to compose interfaces.

The software components comprise a database management system (DBMS) capable of handling the databases, a model base management system (MBMS) to handle the model base and a solution (scenario) management system. DBMS and MBMS need a large array of data management and representation functions to manipulate the different databases as well as the model base. The solution management component needs a number of functions to generate and evaluate the decision alternatives and scenarios. The dialog system controls the display of information and the system's interaction with the user.

The traditional DSS "dialog-data-model" architecture (Sprague and Watson 1997), limits communication between the system's components and also the components' communication with external systems in ways independent of any pre-specified control mechanism (Kersten and Noronha 1999). A major problem with this architecture is that it does not have a plug-and-play philosophy.

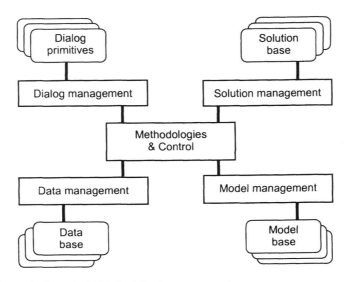

Figure 4. Generic DSS Architecture

A DSS system should be viewed and implemented as a collection of loosely-associated smart objects. Its design must always be open-ended and amenable to change. Many DSS systems, however, are weak in providing a methodology for change. Kersten and Noronha (1999) view system change as an upgrade in the attributes or behaviors of *individual* objects (occasionally, modules), independent of other objects or modules, rather than one synchronized change across the whole system. They propose a *methodologies and control* (M&C) component that allows for:

1. the situation-dependent use of different data analysis, modeling and visualization support methodologies (defined by available data, meeting specified constraints or requested by the user);

2. content rich communication between the user and other systems;

3. use of specialized models for dialog, solution and communication components, in addition to those used in the model management component; and

4. expandability in directions not necessarily envisaged by the developers.

The M&C component comprises knowledge about other DSS components, relationships among them, and their use. This component is used to activate other DSS components depending on the actions undertaken by the user and the system, the system's state and the actions it is to perform. Hence M&C allows for context-sensitive DSS behavior. It also enables the system to intervene actively in the process based on an analysis of the state of the "decision-making arena" (Angehrn 1993).

4.5. DSS methods

Following Alter's (1980) classification DSSs have been categorized as: file drawers, data analysis, information analysis, accounting models, representational models, optimization models and suggestion models. Significant advances in computing and communication technologies, including methods for the development of problem representation, problem solving techniques, and computer and communication technologies, have rendered this classification outdated. Nevertheless, it is still being used to describe the DSS functions and methods (see, e.g., Sauter, 1997; Turban and Aronson, 1998).

Apart from being inadequate from a technological perspective, Alter's classification does not allow for the analysis of DSSs from an information provider perspective. This view implies that six corresponding general support categories should be distinguished (Zachary 1986).

1. *Presentation methods* are computer-graphic, voice and text processing tools to present data and information in a meaningful and suitable form. They are used to make data, information and knowledge used and generated by DSS accessible to the decision maker in terms of the decision maker's own mental representation of the decision process and context.

2. *Information management methods* are used to store, organize, retrieve and summarize data, information and knowledge providing timely and relevant information extending decision makers' ability to access it.

3. *Process modeling methods* are quantitative and qualitative models of real-world processes, and techniques to provide predictions of, and scenarios for, these processes at future points in time and under different conditions.

4. *Choice modeling methods* are used to select and combine decision attributes, define objectives and goals, determine preferences and trade-offs among objectives and utility or value functions. They are also used to determine preference consistency and remove cognitive biases by the consistent use of preferences and trade-offs.

5. *Automated analysis and reasoning methods* are mathematical and logical tools used to automate fully or partially analytical and reasoning activities. They are used to organize the decision process, provide expert knowledge and compensate for situational constraints limiting the unaided decision maker.

6. *Judgement refinement methods* are used to guide decision makers in their efforts, identify and remove systematic inconsistencies and biases that arise from human cognitive limitations. They include aids to structure decision problems, estimate probability distributions, analyze risk and check for consistency of the decision maker's reasoning.

The role of the methods listed above is to organize the technology base and give a framework for the system's functional specification. To establish close linkages between the different aspects of managerial decision making and organizational and individual needs and to position methods and aids in the overall system's architecture, each of the functions needs to be further defined and translated into detailed system requirements.

The six types of DSS methods include techniques and tools that can be used to provide specific decision support functions. In Fig. 5 we identify technologies that can be used to provide the required functionalities. This lists of technologies are not exhaustive but they indicate types that can be used in each category. New technologies are being developed and some existing ones cannot be neatly assigned to a single category (e.g., cognitive mapping and visual simulation).

5. Systems and organizations

We have mentioned that DSSs, in serving their users, interact and co-operate with them. Typically, the aspects of integration and co-operation are considered when enlarging the concept of information system so that it includes both the user and the tool in a human-machine system. If we consider cognizant systems, both human and artificial, then this enlargement seems too limited. Almost every complex entity can be designated a system, but there are some features of systems comprising intelligent entities that differentiate them from others. These systems are *organizations* and we attempt here to analyze the roles of DSS from the organizational point of view. We do not posit that a user and DSS form an organization; rather that an organizational perspective, and especially a cultural one, can highlight the requirements and modes of interaction and co-operation between users and DSS.

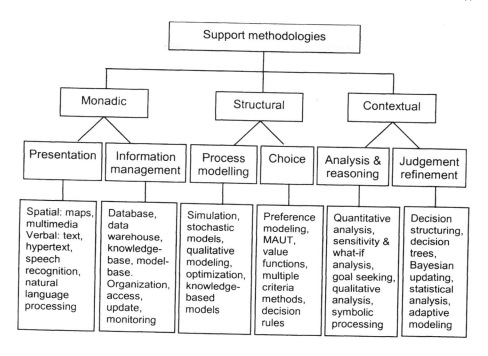

Figure 5. Information system, support categories and technologies

Organizations social units intentionally constructed to achieve a specific set of goals (Etzioni 1968). For our purposes we allow also other, than human, entities to participate in an organization and limit discussion to an organization consisting of the decision maker, i.e., the user and the DSS.

5.1. Bureaucratic DSS

Bureaucratic Weberian organization (Mintzberg's machine bureaucracy) assumes clear-cut specialization of tasks, a hierarchy of authority, accountability, decisions made according to rules, standard procedures and efficiency. The role of the lower-level participants is to implement efficiently decisions of the higher-level authorities. These decisions are not verified, analyzed or evaluated at the lower level but taken at face value. There are clear, universalistic and impartial implementation rules for how these decisions are implemented. The intelligence of the lower-level is thus limited to the ability to use these rules efficiently.

This is also the role of many DSSs; they are used to formulate and solve specific and highly specialized problems with an *a priori* defined sequence of procedures. These are monadic and structural DSSs. They cannot be used for any variation of the problem unless such a variation was earlier implemented as an option or the user transforms the problem so that it fits the system (e.g., a system cannot solve a mini-mization problem if it has been designed to solve only maximization problems). The system does not interpret or evaluate the input; it can only use its procedures or fail. Similarly, it does not make attempts to search for additional data if it cannot solve the problem as stated.

Monadic and structural DSSs, like bureaucratic organizations, perform well in well defined and predictable situations. They solve problems efficiently and provide solutions but only for clearly stated and clearly understood problems. Their provision of support for solving ill-structured problems lies in their ability to determine solutions for different parameter values through sensitivity and what-if analyses. A problem may be ambiguous or ill-structured only for the user, who has to present it to the system as a well-structured one.

This type of DSS may have led (Naylor 1982) to the claim that there was nothing new about DSS that does not already exist in MS/OR and MIS. Interestingly, there is almost no distinction in the DSS literature regarding systems that can be used only by specialists in OR/MS, decision analysis, finance, marketing—users who prepare analyses for the decision makers, and systems which can support decision makers. Perhaps, as Finlay and Martin (Finlay and Martin 1988, p. 530) state in their review, "chauffeured use of DSS is the one and only feasible way forward for most organizations" because decision makers are unable to communicate their needs to the computer. This situation has been also shown in empirical studies, for example, a survey of 580 DSSs in Canada (including file drawer systems and MIS) showed that less than 15% of the users were decision makers themselves (Huff et al. 1984).

5.2. Entrepreneurial DSS

Entrepreneurial organizations (also called organic), are small, highly active entities without a clear and inflexible division of labor. They are focused on effective, but not necessarily efficient, use of resources for fast-breaking production developments in a dynamic and volatile environment. In these organizations the tasks are undertaken not according to hierarchy but expertise and current workload. Members of entrepreneurial organizations are often encouraged to be attentive to the environment, to search for problems and opportunities, and to perform various tasks and solve a variety of problems. In these organizations the exchange of information required for problem solving does not follow routine and fixed paths. Rather, individuals who may have relevant information or are able to provide expertise are co-opted and involved in the decision process.

If a DSS is to be used in such an organization then it has to be an active participant in the decision process. Such a system has a significant amount of autonomy in determining alternative actions and issues that the manager might have overlooked. This requires two independent processes, one directed by the user and the other under the DSS's own control (Rao, Sridhar et al. 1994). In the user/system organization there are parallels to decision making by a group of specialists. They cooperate with each other to define a collective problem and to process problem structures and solutions.

It is obvious that such a DSS has to be contextual if not cognitive. This is because it has to be able to represent, analyze and solve the whole problem, not only some of its elements. While it is not necessary for the system to consider all the perspectives and models of choice it must be able to incorporate the user's perspectives in its processing activities. It must participate in the manager's construction of the cognitive perception of the problem. Monadic and structural DSS do not need to have this feature because the user's perspective is irrelevant or it is fixed and represented with one of the system's constructs (e.g., the objective function). Here, there are multiple perspectives and they may change requiring the system to accommodate variety and change.

Contextual and cognitive DSS, like individuals in entrepreneurial organizations, must have multiple capabilities. They need to be able to incorporate the user's needs and values in their processing, interact with the user on her cognitive level, present the problem in a meaningful way and use specialized (monadic and structural) tools. In an entrepreneurial organization communication encompasses all three levels of managerial problem solving. Therefore, the system needs to have similar capabilities.

6. Conclusions

The theory of decision support appears to be playing a catch-up with the real-life applications and with developments of computer and communication technologies. This is not to say that there is lack of different DSS frameworks and architectures. Our attempt was to select and organize views and models for studying support issues and to formulate prescriptions for DSS design and development.

The cognitive framework provided the basis for further discussion. This is in line with the knowledge-based perspective proposed by Holsapple and Whinston (1996). While they propose a somewhat different DSS architecture than presented in Section 4, we view these two architecture as complementary since both stress the importance of integrating a knowledge-based component.

There are many important issues not discussed here. The effect of culture in the international use of DSSs is significant especially if technologies are modified and systems adapted to specific local environments. Some of the architectural requirements posed by software internationalization are discussed in Chapter 15. Others include the impact of social traditions and values on the selection of decision variables and the relationships between models and cultures (Evans et al., 1989).

DSSs can be used by individuals and groups and in different settings. In Chapter 20 references are given to team and group decision support, negotiated and intra- and inter-organizational support. References to numerous real-life applications in several key areas are also given, complementing the theoretical focus of this chapter. The references clearly show that the DSS area is very active; plethora of systems has been developed in the nineties and used to support almost every type of decision problem.

Acknowledgements

I thank David Cray for helpful comments. This work has been supported by the Social Science and Humanities Research Council of Canada and the Natural Sciences and Engineering Research Council of Canada.

References

Alter, S. L. (1980). *Decision Support Systems. Current Practices and Continuing Challeneges*, Reading, MA: Addison-Wesley.

Angehrn, A. A. (1993). "Computers that Criticize You: Stimulus-based Decision Support Systems", *Interfaces* 23(3), 3-16.

Angehrn, A. A. and T. Jelassi (1993). "DSS Research and Practice", *Decision Support Systems*, 12, 267-275.

Bell, P. C. (1992). "Decision Support Systems: Past, Present and Prospects", *Revue des systèmes de décision* 1(2-3), 126-137.

Beulens, A. J. and J. A. van Nunen (1988). "The Use of Expert System Technology in DSS", *Decision Support Systems*, 4(4): 421-431.

Bonczek, H., C. W. Holsapple and A. Whinston (1980). "Evolving Roles of Models in Decision Support Systems", *Decision Sciences*, 11(2), 337-356.

Brooks, C. H. P. (1995). "A Framework for DSS Development", in P. Gray (Ed.), *Decision Support and Executive Information Systems*, Englewood Cliffs, NJ: Prentice Hall, 27-44.

De May, M. (1992). *The Cognitive Paradigm*. Chicago, The University of Chicago Press.

Dhar, V. and R. Stein (1997). *Intelligent Decision Support. The Science of Knowledge Work*. Upper Saddle River, NJ, Prentice Hall.

Eden, C. (1988). "Cognitive Mapping: A Review", *European Journal of Operational Research* 36(1).

Eden, E. and F. Ackerman (1998). *Making Strategy, The Journey of Strategic Management*. London: Sage.

Emery, J. C. (1987). *Management Information Systems. The Critical Strategic Resource*. New York: Oxford Univ. Press.

Etzioni, A. (1964). *The Active Society*, New York: Free Press.

Evans, W. A., K. C. Hau and D. Sculli (1989). "A Cross-cultural Comparison of Managerial Styles", *Journal of Management Development*, 8(1), 5-13.

Finlay, P. and C. Martin (1988). "The State of Decision Support Systems: A Review", *Omega*, 17(6), 525-531.

Franklin, S. and A. Graesser (1996). Is it an Agent, or just a Program?: A Taxonomy for Autonomous Agents,. *Proceedings of the Third International Workshop on Agent Theories, Architectures, and Language*, Springer-Verlag.

Heylighen, F. (1992). "A Cognitive-systemic Reconstruction of Maslow's Theory of Self-actualization", *Behavioral Science*, 37, 39-58.

Hill, P.H. *et al.*, (1982), *Making Decisions. A Multidisciplinary Introduction*, Reading, MA: Addison-Wesley.

Holsapple, C. W. and A. B. Whinston (1996). *Decision Support Systems. A Knowledge-based Perspective*, New York: West.

Huff, S. L., S. Rivard, A. Grindlay and I. P. Suttie (1984). "An Empirical Study of Decision Support Systems", *INFOR*, 21(1), 21-39.

Kampke, T. (1988). "About Assessing and Evaluating Uncertain Inferences Within the Theory of Evidence", *Decision Support Systems*, 4(4), 433-439.

Keen, P. G. W. and M. S. Scott-Morton (1989). *Decision Support Systems: An Organizational Perspective*, Reading, MA, Addison-Wesley.

Keeney, R.L. (1992). *Value-focussed Thinking. A Path to Creative Decision Making*, Cambridge, MA: Harvard University Press.

Kersten, G. E. and D. Cray (1996). "Perspectives on Representation and Analysis of Negotiations: Towards Cognitive Support Systems", *Group Decision and Negotiation*, 5(4-6), 433-468.

Kersten, G. E. and S. J. Noronha (1999), "WWW-based Negotiation Support: Design, Implementation, and Use", *Decision Support Systems*: (to appear).

Kersten and S. Szpakowicz (1994). "Decision Making and Decision Aiding. Defining the Process, Its Representations, and Support", *Group Decision and Negotiation*, 3(2), 237-261.

Little, J. D. C. (1970), "Models and Managers: The Concept of a Decision Calculus", *Management Science*, 16(8), 35-43.

Maslow, A.H. (1954), *Motivation and Personality*, New York: Harper and Row.

McNurlin, B. C. and Sprague, R. H., Jr. (1993). *Information Systems Management in Practice*, Engelwood Cliffs, NJ: Prentice Hall.

Mintzberg, H., D. Raisingham and A. Theoret (1976). "The Structure of 'Unstructured' Decision Processes", *Administrative Science Quarterly*, 21, 246-275.

Moore, J. H. and M. G. Chang (1980). "Design of Decision Support Systems", *Data Base* 12(1-2).

Naylor, T. H. (1982). "Decision Support Systems or Whatever Happened to MIS?" *Interfaces*, 12(4).

Newcomb, T.M. (1953). "An Approach to the Study of Communicative Acts", *Psychological Review*, Vol. 60, 394-404.

Rao, H. R., V. S. Jacob et al. (1992). "Hemispheric Specialization, Cognitive Differences, and Their Implications for the Design of Decision Support Systems Responses" *MIS Quarterly*, 16(2), 145-154.

Tversky, A. and D. Kahneman, (1981). "The Framing of Decisions and the Psychology of Choice", *Science*, 211, 453-458.

Sauter, V. (1997). *Decision Support Systems*, New York: Wiley.

Simon, H. A. (1960). *The New Science of Management Decision*, New York: Harper & Row.

Sprague, R. H., Jr. and E. D. Carlson (1982). *Building Effective Decision Support Systems*, Engelwood Cliffs, NJ: Prentice Hall.

Sprague, R. H. and H. J. Watson (1997). *Decision Support for Management*. Upper Saddle River, NJ: Prentice Hall.

Turban, E. and Aronson, J.E. (1998). *Decision Support Systems and Intelligent Systems*. Upper Saddle River, NJ: Prentice-Hall, Fifth Edition.

Zachary, W. (1986). "A Cognitively Based Functional Taxonomy of Decision Support Techniques", *Human-Computer Interaction* 2: 25-63.

3 DECISION SUPPORT WITH GEOGRAPHIC INFORMATION SYSTEMS

Anthony Gar-On Yeh

1. Introduction

Geographic information systems (GISs) are a set of computer-based information technologies capable of integrating data from various sources to provide information necessary for effective decision-making in urban and regional planning (Han and Kim, 1989). GISs serve both as a tool box and a database for urban and regional planning (Figure 1). As a tool box, GISs allow planners to perform spatial analysis by using their geoprocessing or cartographic modeling functions such as data retrieval, map overlay, connectivity, and buffer (Berry, 1987; Tomlin and Dana, 1990).

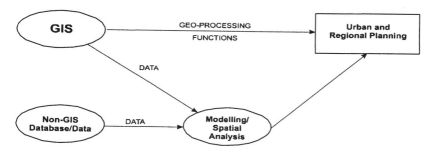

Figure 1. GISs in urban and regional planning.

Of all the geoprocessing functions, map overlay is probably the most useful planning tool. There is a long tradition of using map overlays in land suitability analysis (McHarg, 1960; Steinitz et al., 1976; Hopkins, 1977). Planners also can extract data from the GIS database and input them to other modeling and analysis programs together with data from other databases or conduct special surveys for making planning decisions. GISs have much to offer in regional planning (Marble and Amundson, 1988; Levine and Landis, 1989). They have been used in information retrieval, development control, mapping, site selection, land-use planning (Gross et al., 1987), land-suitability analysis (McDonald and Brown, 1984; Lyle and Stutz, 1987), and programing and monitoring (Yeh, 1990). Recently, works have been published on the application of GISs in environmental modeling and planning (Goodchild et al., 1992; Schuller, 1992).

2. Geographic, spatial, and planning decision support systems

The increase in speed and data storage and the decrease in the price of workstations and microcomputers and their peripherals have made GISs an affordable technology, even in developing countries. GISs can start at an affordable price with a few networked workstations and microcomputers. Additionals can be added incrementally when the system expands without incurring large overhead costs. GIS techniques are quite well developed and operational. Although improvements can still be made, existing techniques are adequate for most urban and regional planning tasks. It is only a matter of putting GISs in operational environments where they can help formulate and implement actual sustainable regional development strategies.

Decision support systems (DSSs) were developed in the late 1960s and early 1970s as a response to the shortcomings of management information systems (MISs). These could neither support analytical modeling capabilities adequately nor facilitate the decision-maker's interaction with the solution process. The DSS concept then was expanded to include the development of spatial decision support systems (SDSSs) (Densham, 1991). An SDSS comprises four main components, as shown in Figure 2.

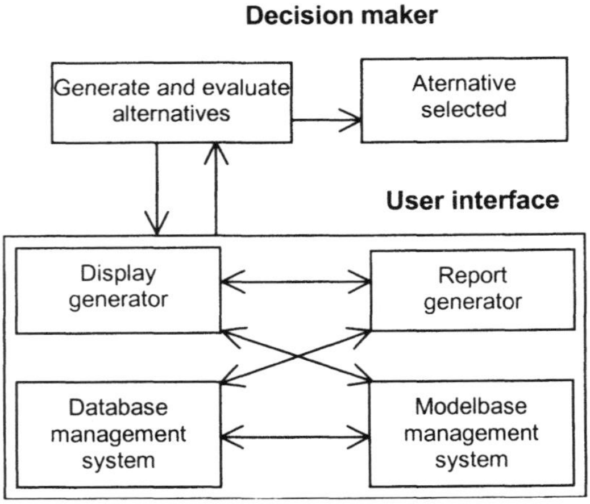

Figure 2. Architecture of a spatial decision support system (SDSS).
(Source: Armstrong et al., 1986.)

An SDSS helps decision-makers choose a location by using the system to generate and evaluate alternatives, such as the optimal location of a service centre, for example (Densham and Rushton, 1988; Armstrong and Densham, 1990). The lack of analysis functions in the past meant that GISs were not considered as part of SDSSs at all. Instead, GISs were used to generate and store spatial data, which then were used as inputs for the analytical models whose results, in turn, were displayed by the GISs.

There is a great deal of published research on the use of GISs in the visualization of the results of the analytical models (Armstrong *et al.*, 1992; Batty, 1992, 1994). Advances also have been made in incorporating analytical models into GISs (Batty and Xie, 1994). For example, the location-allocation model has been incorporated as a standard function in the latest version of ARC/INFO (7.0). Indeed, the distinction between GISs and SDSSs is expected to continue to diminish.

A parallel development in the planning field is the concept of planning support systems (PSSs), first advocated by Harris (1989). This is a combination of computer-based methods and models in an integrated system designed for the planner. Not only does it serve as a DSS for decision-makers, but it also provides the tools, models, and information used by planners (Klosterman, 1995). The PSS consists of a whole suite of related information technologies such as GISs, spreadsheets, models, and databases with different applications for various planning stages (Harris and Batty, 1993; Batty, 1995). It is illustrated in Figure 3.

GISs are an important component of a PSS because of their geoprocessing, graphic display, database, and modeling capabilities. However, a PSS cannot rely on GISs alone. It must also include the full range of the planner's traditional tools for economic as well as demographic analysis and forecasting, environmental modeling, transportation planning, and land-use modeling (Klosterman, 1995). It should also include technologies such as expert systems, decision support aids such as multicriteria decision analysis, hypermedia systems, and group decision support systems.

3. GISs in sustainable-development planning

3.1. Resource inventory

To develop finite resources sustainably, several different conditions must be met. First, the supply and quality of major consumables and inputs to our daily lives and economic production – air, water, energy, food, raw materials, land, and the natural environment – must be secured. Land is important because it is the source of our energy, food, and other raw materials, and also the habitat of our flora and fauna. Like the other resources, however, it is a scarce commodity. Some of the destruction of these natural resources is irreversible – like the conversion of agricultural land into urban land, for example. Land use that is unsuitable for development can harm both the natural environment and human life. For example, land reclamation for housing developments can destroy wetlands that are valuable natural habitats for some wildlife species and lead to their extinction. It can also cause loss of property or even human life, if there were flooding. We therefore must make sure land is used properly.

In China, for example, the 1978 economic reforms have caused a wild rush of land development and led to many land-related problems. One is the encroachment on agricultural land: for example, government statistics show that arable land in the Pearl River Delta has shrunk from 1 044.7 thousand ha in 1980 to 898.2 thousand ha in 1991, or 14% in 10 years. The actual loss of farmland may be far greater. It has depleted the food production capacity of a region that was once one of China's most important agricultural regions. Clearcutting of hills and land reclamation on a massive scale along rivers and estuaries have led to the destruction of natural habitats and an ecological imbalance in the region. Unplanned clearcuts of hills will eventually cause soil erosion and landslides. There is thus an urgent need to formulate and implement a sustainable development strategy for this region.

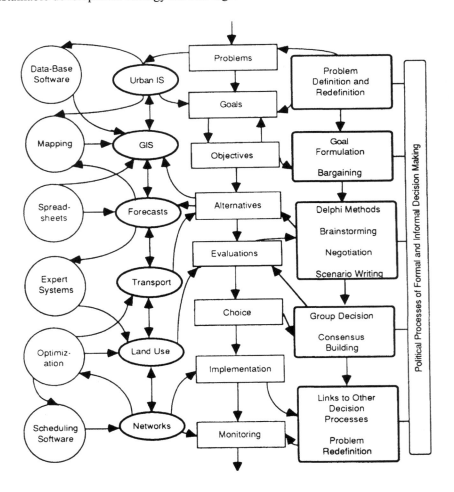

Figure 3. The planning process supported by a formal computation desktop PSS (Batty, 1995).

It is difficult to set operational rules and criteria for the concept of sustainable development, because it has broad meanings and ambiguous boundaries. However, if the focus is on its environment and land components, a detailed planning procedure that favours sustainable development, rather than economic, is easier to establish. The various stages in the formulation and implementation of a sustainable regional development strategy are listed in the left column of Figure 4: determination of objectives, resource inventory, analysis of existing situation, modeling and projection, development of planning options, selection of planning options, plan implementation, and plan evaluation, monitoring, and feedback (Yeh and Li, 1994).

GISs and remote sensing are environment-and land-related technologies. They are therefore particularly useful for dealing with the environment and land components of a sustainable development strategy. However, they provide only some of the data and techniques needed in sustainable development planning. They must be used with other databases and models at the various planning stages. The specific applications of GISs and remote sensing at different stages of the planning process are shown in Figure 4.

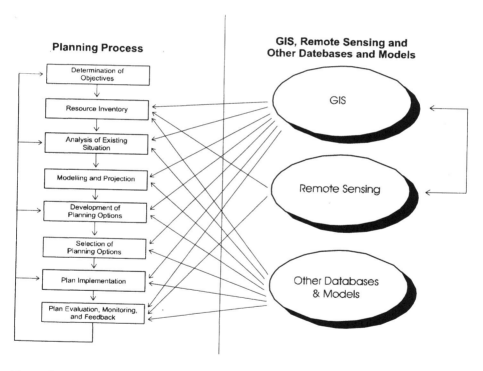

Figure 4. Integration of GIS, databases and models into the planning process of sustainable development.

The integration of GISs with remote sensing can save time in collecting regional land-use and environmental information. Remote sensing also can help in the inventory of land use and natural resources. GISs can help store, manipulate, and analyze physical, social, and economic data; produce land-suitability maps; and estimate the carrying capacity of a region. The GIS environmental data can be used to build land-suitability maps (McDonald and Brown, 1984; Lyle and Stutz, 1987). The carrying capacity of a region can be estimated by means of the land-suitability map and environmental information. Environmentally sensitive areas can be identified by remote sensing and other environmental information (Yeh and Li, 1996).

3.2. Analysis of existing situation

GISs can help to identify areas of conflict between land development and the environment by overlaying existing land-development maps on land-suitability maps. GISs also can estimate the magnitude of encroachment by economic development into the environment and the extent of land degradation simply by overlaying land-use change maps over land-suitability maps. Finally, decision-makers can help assess the pace and impact of economic development on the environment by using GISs to analyze economic growth and urban expansion.

3.3. Modeling and projection

Future population and economic growth need to be projected. Spatial modeling of their likely distribution in space needs to be analyzed and evaluated to estimate the impact of the existing population and economic growth on the environment. GISs can forecast the future impact of economic development on the environment if the current trend continues. This can be done by projecting the future demand for land resources from population growth and economic activities, modeling the spatial distribution of this demand, and using GISs to identify areas of conflict by map overlay analysis. With the socioeconomic and environmental data stored in GISs, environmental-planning models can be developed to identify areas of environmental concern or conflict (Schuller, 1992). GISs can model different development scenarios. They can show the modeling results in graphic form, which makes them accessible to decision-makers. Planners can use the information to formulate different options to guide future development so that it avoids conflict.

3.4. Development of planning options

Land suitability maps are very useful in the development of planning options. They can be used to identify solution space for future sustainable development. The use of spatial-optimization models with GISs (Chuvieco, 1993) can help develop planning options that try to maximize or minimize certain objective functions. The simulation of different scenarios of development with GISs can help in the development of various planning options (Landis, 1995).

3.5. Selection of planning options

By overlaying development options on land-capacity and land-suitability maps, decision-makers can compare and assess the impacts of various planning options on the environment. The use of GISs with multicriteria decision analysis can also help in the selection of planning options (Carver, 1991; Eastman *et al.*, 1993). The integration of spatial and nonspatial models with GISs can help them evaluate different planning scenarios (Despotakis *et al.*, 1993).

3.6. Plan implementation

GISs also can be used for the implementation of sustainable regional-development strategy. They can produce environmental impact assessments of proposed projects to evaluate and minimize the effect of development on the environment. Remedial measures then can be recommended to alleviate the negative impact.

3.7. Plan evaluation, monitoring, and feedback

When used together with remote sensing, GISs can help monitor the environment; for example, they can track land-use changes (Yeh and Li, 1996, 1997). Decision-makers can determine whether land development is following the land-use plan of the region by overlaying the land-development map produced from the analysis of remote sensing images on the land-use plan. GISs also can be used to evaluate the impact of development on the environment and decide whether adjustments to the plan are needed.

4. Sustainable development in developing countries

The most important constraint to the use of GISs in developing countries is the lack of operational definitions and indicators of sustainable development. Another is the lack of qualified personnel to develop and implement sustainable development strategy. As a result, GISs cannot be used effectively for the planning and implementation of sustainable development even if they are available. Other constraints deal with the use of GISs in developing countries generally (Yeh, 1991).

Despite the rapid growth in the use of GISs since the early 1990s, their use for urban and regional planning in the developing countries is still underdeveloped. Nevertheless, some have built more comprehensive GISs and used them more than others. In the late 1980s, a few large-scale GIS projects were being planned or developed, most notably the land information systems in Hong Kong and Singapore. The new millenium is expected to see a large increase in the installation of GISs in the developing countries of Asia.

For the most part, developed countries use general purpose, management-oriented GISs, with many different users and purposes, and regular, routine data updates. In developing countries, however, most GISs are created to serve the needs of particular development projects carried out by government departments and research institutes to explore their usefulness for planning and management. Except for GISs developed for mapping purposes, most of them have no long-term objective of integration with other GISs or databases.

More GISs are developed at the regional level because they are easier and less expensive to develop than at the national or urban levels. GISs also need a reliable telecommunication network that is not always available in many developing countries. They are still experimenting with their setting up, management, and applications and are not ready yet to develop national systems.

The development of urban GISs needs a relatively large system to store the huge volume of large-scale maps and parcel data. Urban GISs are used more frequently in the newly industrialized economies (NIE) because of the availability of funding, good maps, and land records. Most cities in developing countries, however, have difficulty in finding funds to acquire these large systems. Even if funding is not a problem, they may not have the base maps and data necessary to set up such systems. They may also need to devise a geocoding system for linking parcel data with graphic data. For example, in the metro Manila physical survey (MMAPS) project, a basic mapping program was carried out to look into the availability and standard of base maps before the development of a computerized GIS could be considered (Cruz, 1989).

Regional GISs, on the other hand, are easier to develop because data is more readily available from remote sensing. Depending on the resolution required by the system, regional systems also can be managed by a microcomputer if small-scale maps and remote sensing images are stored. Again, this makes them more affordable than national and urban GISs that often require large computer systems. Finally, one reason regional systems may have grown is that GIS knowledge is mainly confined to remote sensing experts who are eager to find applications for their data and techniques.

Most regional GISs are used for resource inventory, resource management, and land-capability evaluation. They are thus used for resource and environmental management rather than for planning. For the most part, they store data on the physical environment, such as topography, soil, vegetation, land use, and hydrology, which can be obtained easily from maps and remote sensing. Socioeconomic data are generally not available. Many systems are experimental and designed by researchers with little input from practitioners. Very often they are one-off systems that will be used only for the research project, without a built-in institutional arrangement to maintain the system after the project is completed. The end product is often a series of research reports but no plans. There are few attempts to apply GISs to creating planning scenarios, allocating regional investment, and evaluating development proposals.

More urban GISs are nevertheless being developed. They already help with the management and planning for the rapidly growing metropolises of Bangkok, Guangzhou, Hong Kong, Manila, and Singapore. They are also often used to improve cadastral mapping, land taxation, and land titling (Angus-Leppan, 1989). There will be an increasing trend to develop urban GISs in the large cities of developing countries because of their urgent need for management of their scarce land resources and rapidly growing populations.

Funding for GISs in developing countries is usually provided by international funding agencies such as the World Bank and the Asian Development Bank, with matching funds from local governments. The better-off NIEs, however, like Hong Kong and Singapore, use mainly their own funding.

As noted above, the use of GISs has expanded significantly as a result of advances in computer technology. Many systems are microcomputer-based and therefore much more affordable to developing countries. Lower-cost workstations and computer networks have also contributed to make large-scale GISs more affordable. Thus a small system consisting of a network of workstations and microcomputers can be set up first, and later expanded into a much larger system. The Hong Kong land information system is based on this incremental approach.

4.1. Data

The lack of available data is one of the major hindrances in the use of GISs in developing countries (Yeh, 1992). GISs are information systems that need volumes of graphic and textual data to function. But if there is no data, there can be no application. As noted above, the most readily available data are from remote sensing, but these are limited mainly to land cover information. Socioeconomic data, which are vital to urban and regional planning, are generally unavailable and often limited to census data.

Socioeconomic data require field surveys, which are expensive and time-consuming. In a study of information systems for planning in Ghana, Akom (1982) found that there was a lack of up-to-date and reliable data because some departments and institutions were slow to acquire it, while coordination of the flow and exchange of information between others was weak. Other inhibiting factors included the lack of financial resources, trained personnel to collect data, and modern and efficient data-processing equipment. The relatively low price of microcomputers can help to alleviate the hardware constraints in electronic data-processing facilities: but the main obstacle is still the lack of government recognition of the need for statistical information and scant interest in mobilizing the resources to collect it.

Base maps, which are more essential to GISs than textual data, are often outdated or simply unavailable. They are compiled by a variety of agencies, to different standards of accuracy and map scales, which make them difficult to integrate into the system. Furthermore, there is also a general lack of standardized geocoding systems, which makes it difficult to link textual data with graphic data. It is very difficult to start GISs without the necessary maps, geocoding system, and textual information, as well as a system and an institutional framework to link them together.

If the availability of data is a problem, so is their quality. Stale-dated data are of limited use for planning. There is usually no institutional arrangement to determine, coordinate, and monitor the frequency of data updating by different departments. Often, a fairly large amount of planning data is collected by other agencies over which the planning agency has little control (Batty, 1990). There is also a general lack of organizational procedure for verifying the quality of collected data.

There has been a misconception that hardware, software, and human resources are the major constraints to the spread of GIS technology for urban and regional planning in developing countries. Yet the greatest hindrance to the effective use of GISs and remote sensing is the lack of up-to-date information. There is therefore an immediate need in most developing countries to establish an institutional framework to collect and update the required data regularly.

4.2. Planning systems

Planning systems in developing countries have not advanced very much when compared to GISs. The skills of planners and the planning system may not be ready to use the data and functions provided by GISs. Planners should know how to use these systems in conjunction with any new planning techniques so they can diagnose potential problems and assess the desirability of alternative plans. Most planners in developing countries are not yet aware of the benefits and potential applications of GISs. Furthermore, little effort has been spent on transforming available data into information for making planning decisions.

4.3. Manpower and training

The rapid growth of GISs has meant a general shortage of personnel skilled in their use – even in developed countries. This shortage is more serious in developing countries both in absolute numbers and in relative terms. Given the number of cities and regions that could benefit from GISs, this shortage constitutes a major problem.

The problem is that few receive GIS training because of the lack of expertise and a shortage of funds. GIS development in developed countries is often supported by teaching and research in the universities, while the reverse is true of developing countries. Their government agencies often buy GISs first, with funding from international banks; a few years later, GISs are installed and courses at the universities and other tertiary institutions. GIS training is provided by software companies, who either send their staff to short on-site courses or send operators and users to be trained at their headquarters. Any other GIS training programs in developing countries are usually taught at departments of surveying, remote sensing, and geography. There is a general lack of GIS courses in planning schools. Yet to use GISs most effectively, planners must be trained not so much in the operation of the system, but in how to use its data and functions in different processes of planning and plan evaluation.

There are five major groups of users who need training: policy makers, planners, programers, technicians, and educators. Courses should be organized to meet the specific needs of the different users. Policy makers should be made aware of the uses and limitations of GISs. They should be informed of developing trends in GISs and the kind of resources needed to make them work. Planners in the field should have a general understanding of data, models, and relational data structures, as well as the use of GIS functions at different stages of urban and regional planning. Programers need a higher level of technological competence. They need to be trained to manage the system and to develop application modules to meet local needs. Technicians need to be trained for data collection and entry, and to recognize the likely types of errors. Educators should be kept informed of the latest developments in GISs. Finally, universities and other higher-education institutions should invest more in GIS training and research to develop local expertise.

There is a general lack of GIS personnel but also of planners in the developing countries. For example, Suzhou, a medium-sized city in China, has a population of 500 000 but only seven or eight trained planners. It is important, therefore, to train both planners and GIS personnel and to do it soon.

4.4. Maintenance of GIS

Most GIS hardware and software are imported from developed countries. It takes a long time to repair computer hardware when the necessary components are not available locally. It is also difficult to consult software companies when problems arise. Most of the services and expertise are concentrated in the main cities, which makes hardware and software maintenance more difficult for GISs located elsewhere.

Very little funding is available to maintain the systems. Very often, GISs cannot operate fully because one or two terminals and peripherals are out of order and the agency responsible does not have the funds to repair them. Institutions often have neither funding nor adequate procedures to update the data. Planning requires up-to-date information; the system is useless if its data are not updated. The development of GISs should be considered as a continuous process and not just a one-time project. The maintenance of hardware, software, and data for GISs cannot be neglected .

4.5. Software development

Software for large-scale systems is purchased mainly from the developed countries. It is expensive and consumes much foreign currency, which is often in short supply. There is a general lack of locally-developed software. Attempts have been made to use low-cost commercial software to perform GIS tasks. The most popular are the combinations of commercial CAD packages such as AutoCAD with commercial database packages such as dBASE III (Yapa, 1988, 1989) and INFORMIX (Cheng *et al.*, 1989). These systems, although limited in their GIS functions, make GISs available to departments and agencies with little funding. However, they still need to be purchased from developed countries.

Software developments in developing countries are fragmented and involve only one or two researchers. They do not have the personnel or institutional setup to develop and maintain software the way developed countries can. Researchers in different parts of the country or region therefore should pool their human and systems resources to develop a package with good documentation, manuals, and support, much like the commercial packages of the developed countries. This type of cooperation might help develop local expertise in GISs and save enormous sums of money in the long run.

Language is one of the major problems with the introduction of information technology in developing countries (Calhoun *et al.*, 1987). Most of the imported GIS programs and manuals are written in English, while most of the users, particularly the decision-makers and local planners, have a limited understanding of English. User-friendly application programs with instructions or pull-down menus written in local languages need to be developed.

4.6. Leadership and organization

The impact of leadership and organization on encouraging the effective use and introduction of computers is very well documented (Danziger *et al.*, 1982). Bardon *et al.*

(1984) observed that a few key individuals interested in computers can become instrumental in the initial acquisition of equipment and guide its applications. Their leadership plays a critical role in many information projects by setting clear goals and objectives, winning acceptance by other information system users, and ensuring commitment to their achievement (Edralin, 1989).

Another critical function of leadership is coordinating the different departments that share the information system. The technical skill and interest of the head of the planning department plays an important role in the decision to purchase computing equipment (Standerfer and Rider, 1983). The head of a planning department needs a good comprehension and appreciation of computer applications before he or she can do battle with the administration to acquire them. As GISs are rather new in developing countries, many department heads may not know about their usefulness and so are slow to demand them for their departments. They need to be exposed to the benefits of GISs before they can take up a leadership role in promoting them.

Calhoun *et al.* (1987) encountered organizational problems such as a lack of prior computer use, little support from the government, poor physical infrastructure, no technological familiarity, and an unfamiliar language in the design and implementation of a microcomputer-based system for the Sudanese Planning Ministry. There was a mismatch in organizational expectations between the suppliers and the receiving organization. The Sudanese example demonstrates the need to develop computer awareness at all levels in the organization if the chances of success are to be maximized.

As noted above, it is important for the different agencies to agree on the type of data that their GISs should maintain. Through geo-relational database design, some of the problems in data sharing can be resolved by the overlaying of geographic data. However, it is necessary to identify the users, their requirements, the types of urban and regional plans produced, and the smallest geographic units for urban and regional planning. Most of the data required for planning analysis come from other departments. There is thus a need to identify the flow and management of information in different departments and to cooperate interdepartmentally in data management and sharing.

GIS projects are very often initiated by international assistance agencies, although they rarely take the organizational setting and personal motivations involved into account. There is evidence that large investments have been made to acquire GIS technology, but less evidence that the systems are functioning satisfactorily and contributing to national development efforts. Moreover, problems arise often in the transfer of expertise as well as the cost of maintenance when international assistance leaves the project (Batty, 1990).

4.7. Standardization

Many developing countries lack standardized data; yet both data and maps must be standardized if they are to be shared. GISs are developed by different agencies at different levels. Data may be stored in different formats because of differences in the systems used. To some extent, this problem has been resolved by the development of a GIS software program that enables data transfer between different software packages. However, these conversions are not always one hundred percent effective, and a great deal of work remains to be done after data are transferred.

The more fundamental problems with data sharing, however, have to do with the accuracy of data, the scale used in digitizing, the type of data-classification system used, and the frequency of data update. The accuracy of the data will affect the quality of the data to be input into the planning process. The scale used in digitizing will affect the resolution of the geographic data collected. It is difficult to perform map overlay analysis if the maps and images entered into various GISs are at different scales and resolutions. Problems of alignment and silver polygons will occur. Very often, it is difficult to share data among government departments because of differences in the systems used to classify geographic information. There is also no standardized geocoding system. This makes the integration of map parcel data with their parcel boundaries difficult.

In the United States, the National Committee for Digital Cartographic Data Standards (NCDCDS) and the Federal Interagency Coordination Committee on Digital Cartography (FICCDC) are trying to establish standards to ensure compatibility among digital spatial data gathered by different agencies (Digital Cartographic Data Standards Task Force, 1988). A similar effort is being made in the United Kingdom (Rhind, 1986) and Canada (Zarzycki, 1984). Chinese officials have made efforts to establish a national standard for the geographic coordinate system for GIS, a classification system for resource and environmental information and the boundaries of administrative, natural, and drainage areas (He *et al.*, 1987). Although it will take much time and effort to set up the data standard for GISs, it is an important and worthwhile task that will greatly facilitate future data sharing and integration of different GISs within a country.

4.8. Overdominance by GIS technocrats

GISs are developed and used by departments such as agriculture, survey and mapping, forestry, and remote sensing, which in the past have used computer-assisted cartography or remote sensing in their work. Most of the staff are trained in remote sensing, computer science, surveying, and computer-assisted cartography, although very few have a planning background. They are more interested in GIS research, such as developing faster algorithms and better database designs, than in the practical aspects of GISs. As a result, the needs of planners often are not recognized in the system design of GISs and their products cannot be incorporated into planning systems.

GIS research and training also tend to concentrate mainly on geography and surveying. There are very few planning schools that offer GIS training in their curriculum. There is thus a need to popularize GISs and disseminate information about them to different parts of the country and to planning professionals.

5. Conclusion and recommendations

There are many definitions of sustainable development: indeed, its definition varies from country to country. But before we can use GISs and DSSs to formulate and implement policies and plans to manage our resources sustainably, we need clear goals and indicators for sustainable development.

First, we must remember that sustainable development is much more a political issue than a technical one. Sustainable development goals and the choice of planning options need to be determined by the community. Technology can assist the community in making decisions, but it cannot substitute for the decision-making process. Because of their database as well as graphic and tabular report generation capacity, GISs are important components of sustainable development DSSs. They are particularly useful for its land and environment components. However, they have to be integrated with population, land use, environmental, water, air, food, and energy models if they are to be useful to the planning and implementation of sustainable development policies and plans.

The types of models to be integrated with GISs depend on understanding the process of formulating and implementing these policies and plans. There is a need to identify and develop GIS techniques and decision support models that are useful for the purpose. Also, no matter how sophisticated and advanced it is, a decision support system is useless if it is not being used by decision-makers. Studies repeatedly show that staff and organizational factors are more important than technology in successful applications of GISs (Campbell, 1994).

There are three sets of conditions that are important in the effective implementation of GISs: 1) an information management strategy which identifies the needs of users and takes into account the resources at the disposal of the organization; 2) commitment to and participation in the implementation of any form of information technology by individuals at all levels of the organization; and 3) a high degree of organizational and environmental stability (Masser and Campbell, 1991; Campbell, 1994). Finally, GISs that are most likely to be used are ones that can deal with identifiable problems. Complex applications are less likely to be used than simple ones.

The use of GISs is often limited to researchers and technocrats in developing countries. Very often, they are not used by government departments which actually are involved in sustainable development policy. So there is a need to educate and persuade government officials and decision-makers to use GISs and DSSs in their work. Training should be provided to government officials and decision-makers so that they can understand and use the technology. In order for them to be used in the real world, GIS and DSS experts must start to work with people from other disciplines in the planning process. One way to begin is to do a pilot project which applies GISs and DSSs to different stages of the sustainable development planning process. This can help to demonstrate their usefulness in the planning process.

Most of the GIS techniques that are needed at various stages of the planning process are quite well developed and operational. There is a need to compile these techniques, which are scattered throughout different sources, into an operational workbook that can be used easily by planners. The workbook can be developed from the pilot project so that the experience and techniques developed can be transferred easily to government officials and decision-makers, who can then apply them to other regions. United Nations agencies already have begun to provide training manuals to planners and decision-makers in developing countries. Notable examples include the United Nations Centre for Regional Development (UNCRD) (Batty et al., 1995) and the United Nations Institute for Training and Research (UNITAR) (Eastman *et al.*, 1993).

Workbooks and training courses on GISs and DSSs in sustainable development are needed as well. Most of the workbooks and training materials are published in English and will need to be translated into local languages because most planners and decision-makers in countries like China do not understand English. A network of researchers and practitioners involved in the use of GISs and DSSs in sustainable development also would be useful to disseminate information and experience on the latest research and applications.

Again, the major hindrance to the use of GISs is the availability of data. A strategy needs to be developed to collect and maintain the necessary data (Yeh, 1992). Without data, GISs would be useless even with the best of hardware, software, and human resources. If data are not available, a data-collection program should be started immediately.

The conflict between economic development and environmental protection is most severe in developing countries. Economic development most often is given priority over environmental considerations. There is a need, therefore, to develop planning support systems to help developing countries guide their development towards sustainable development. Even if they have the will to formulate and implement sustainable development policies and plans, they may lack the personnel to do so. China, for example, has had a great boom in economic development since the adoption of economic reform in 1978. The number of cities has nearly tripled – from around 200 in 1980 to 570 in 1993.

Because of the need for planning and development control, the number of planners rose from 6 000 to 37 000 over the same time period. However, many have not been well trained in urban planning and environmental management. Still, by 1995, the number of Chinese universities offering planning-related courses and degrees had risen to 17 (Yeh, 1995). With an annual output of 500 graduates by one estimate, the total number of graduates produced between 1980 and 1993 was 8 500. This means that at least 22 500 planners currently on the job do not have formal training in their discipline; and this situation may not be unique in developing countries.

It will therefore be difficult for developing countries to formulate and implement a sustainable development strategy even if they have one. A PSS with standardized models and procedures would help. Such systems do not need to be very sophisticated, but they have to be operational. Local planners and decision-makers could make their own modifications to the parameters and models. Such PSSs would mean that sustainable development concepts could be implemented more quickly in developing countries, where they are needed most, and in the process improve the sustainable development process throughout the world.

References

Akom, A.A. (1982). "Information System for Planning in Ghana", *Ekistics*, 292, 73-77.
Angus-Leppan, P.V. (1989). "The Thailand Land Titling Project: First Steps in Parcel-based LIS", *International Journal of Geographical Information Systems* 3(1), 59-68.

Armstrong, M.P. and P.J. Densham (1990). "Database Organization Strategies for Spatial Decision Support Systems", *International Journal of Geographical Information Systems*,4(1), 3-20.

Armstrong, M.P., P.J. Densham, P. Lolonis, and G. Rushton (1992). "Cartographic Displays to Support Locational Decision Making", *Cartography and Geographic Information Systems* 19(3), 154-164.

Bardon, K.S., C.J. Elliott, and N. Stothers (1984). "Computer Applications in Local Authority Planning Departments 1984: A Review", Department of Planning and Landscape, City of Birmingham Polytechnic, Birmingham.

Batty, M. (1990). "Information Systems for Planning in Developing Countries", in: *Information Systems and Technology for Urban and Regional Planning in Developing Countries: A Review of UNCRD's Research Project*, Vol. 2. Nagoya, Japan: United Nations Centre for Regional Development.

Batty, M. (1992). "Urban Modeling in Computer-Graphic and Geographic Information System Environments", *Environment and Planning B* 19, 663-688.

Batty, M. (1994). "Using GIS for Visual Simulation Modeling", *GIS World* 10, 46-48.

Batty, M. (1995). "Planning Support Systems and the New Logic of Computation", *Regional Development Dialogue* 16(1), 1-17.

Batty, M., D.F. Marble, and A.G.O. Yeh (1995). *Training Manual on Geographic Information Systems in Local/Regional Planning*. Nagoya: United Nations Centre for Regional Development (UNCRD).

Batty, M. and Y. Xie (1994). "Urban Analysis in A GIS Environment: Population Density Modeling Using ARC/INFO", in A.S. Fotheringham and P. Rogerson (Eds.) *Spatial Analysis and GIS*. London: Talyor and Francis, 189-219.

Berry, J.K. (1987). "Fundamental Operations in Computer-Assisted Map Analysis", *International Journal of Geographical Information Systems* 1(2), 119-136.

Calhoun, C., W. Drummond, and D. Whittington (1987). "Computerised Information Management in A System-Poor Environment: Lessons From the Design and Implementation of A Computer System for the Sudanese Planning Ministry", *Third World Planning Review* 9(4), 361-379.

Campbell, H. (1994). "How Effective Are GIS in Practice? A Case Study of British Local Government", *International Journal of Geographical Information Systems* 8(3), 309-325.

Carver, S.J. (1991). "Integrating Multi-Criteria Evaluation with Geographical Information Systems", *International Journal of Geographical Information Systems* 5(3), 321-339.

Cheng, B.Z., X.D. Song, and C.Q. Lin (1989). "The Study of Microcomputer-Based Urban Planning and Management Information System", *Proceedings of International Conference on Computers in Urban Planning and Urban Management*, 22-25 August, 1989. Hong Kong: Centre of Urban Studies and Urban Planning, University of Hong Kong, 119-121.

Chuvieco, E. (1993). "Integration of Linear Programming and GIS for Land-Use Modelling", *International Journal of Geographical Information Systems* 7(1), 71-83.

Cruz, E. T. (1989). "The application of metro Manila physical survey project to urban and regional planning, land use and zoning", in *International Conference on Geographic Information Systems: Applications for Urban and Regional Planning*, United Nations Centre for Regional Development (UNCRD), Ciloto, Indonesia.

Danziger, J.N., W.H.Dutton, R. King, and K.L. Kraemer (1982). *Computers and Politics: High Technology in American Local Government*. New York: Columbia University Press.

Densham, P.J. (1991). "Spatial Decision Support Systems", in D.J. Maguire, M.F. Goodchild, and D.W. Rhind, (eds.) *Geographical Information Systems: Principles and Applications* 1, 403-412.

Densham, P. and G. Rushton (1988). "Decision Support System for Location Planning", in R.G. Golledge and H. Timmermans (eds.) *Behavioral Modelling in Geography and Planning*. Beckenham: Croom Helm, 56-90.

Despotakis, V.K., M. Giaoutzi, and P. Nijkamp (1993). "Dynamic GIS Models for Regional Sustainable Development", in M.M. Fisher and P. Nijkamp (eds.), *Geographic Information Systems, Spatial Modelling, and Policy Evaluation.* Berlin: Springer-verlag, 235-261.

Digital Cartographic Data Standards Task Force (1988). "Special Issue on the Proposed Standard for Digital Cartographic Data", *The American Cartographer* 15(1), 9-140.

Eastman, J.R., P.A.K. Kyem et al. (1993). *GIS and Decision Making.* Worcester, M.A: Clark Labs for Cartographic Technology and Geographic Analysis.

Edralin, J. (1989). "Implementing Information Systems/Technology in Local/Regional Planning: A Review of Critical Success Factors", *Integrating Information Systems/Technology in Local/Regional Planning.* Nagoya, Japan: United Nations Centre for Regional Development, 37-46.

Goodchild, M., B. Parks, and L. Staeyert (1992). *Environmental Modeling with GIS,* New York: Oxford University Press.

Gross, M., J.G. Fabos, C.L. Tracy, and M. Waltuch (1987). "Computer-assisted Land Use Planning: A Study of the Connecticut Greenway", *Land Use Planning* 4(1), 31-41.

Han, S.Y. and T. J. Kim (1989). "Can Expert Systems Help with Planning?", *Journal of the American Planning Association* 55(3), 296-308.

Harris, B. (1989). "Beyond Geographic Information Systems: Computers and the Planning Professional", *Journal of the American Planning Association* 55, 85-92.

Harris, B. and M. Batty (1993). "Locational Models, Geographic Information, and Planning Support Systems", *Journal of Planning Education and Research* 12, 184-198.

He, J., H. Zhao, B. Li, and T. Jiang (1987). "Research on the Standardization and Normalization of GIS in China", *Asian Geographer* 6(2), 119-133.

Hopkins, L.D. (1977). "Methods for Generating Land Suitability Maps: A Comparative Evaluation", *Journal of the American Planning Association* 44(4), 386-400.

Klosterman, R.E. (1995). "Planning Support Systems", in R. Wyatt and H. Hossain (eds.), *Proceedings of the Fourth International Conference on Computers in Urban Planning and Urban Management,* Melbourne 7(1), 19-35.

Landis, J.D. (1995). "Imagining Land Use Futures: Applying the California Urban Futures Model", *Journal of the American Planning Association* 61(4), 438-457.

Levine, J. and J.D. Landis (1989). "Geographic Information Systems for Local Planning", *Journal of the American Planning Association* 55(2), 209-220.

Lyle, J. and F.P. Stutz (1987). "Computerized Land Use Suitability Mapping", in W.J. Ripple, (ed.) *Geographic Information Systems for Resource Management: A Compendium.* Falls Church, VA: American Society for Photogrammetry and Remote Sensing and American Congress on Surveying and Mapping, 66-76.

Marble, D.F. and S.E. Amundson (1988). "Microcomputer Based Geographic Information Systems and Their Role In Urban and Regional Planning", *Environment and Planning B* 15(3), 305-324.

McDonald, G.T. and A.L. Brown (1984). "The Land Suitability Approach to Strategic Land-Use Planning in Urban Fringe Areas", *Landscape Planning* 11, 125-150.

McHarg, I. (1960). *Design with nature.* Doubleday, New York.

Rhind, D.W. (1986). "Remote Sensing, Digital Mapping, and Geographic Information Systems: the Creation of National Policy in the United Kingdom", *Evironment and Planning C: Government and Policy* 4(1), 91-102.

Schuller, J. (1992). "GIS Applications in Environmental Planning and Assessment", *Computers, Environment, and Urban Systems* 16, 337-353.

Standerfer, N.R. and J. Rider (1983). "The Politics of Automating a Planning Office", *Planning* 49(6), 18-21.

Steinitz, C., P. Parker, and L. Jordan (1976). "Hand-Drawn Overlays: Their History and Prospective Uses", *Landscape Architecture* 66, 444-55.

Tomlin, C. D. and C. Dana (1990). *Geographic Information Systems and Cartographic Modeling*. Englewood Cliffs, NJ: Prentice Hall.

Yapa, L.S. (1988). "Computer-aided Regional Planning: A Study in Rural Sri Lanka", *Environment and Planning B* 15(3), 285-304.

Yapa, L.S. (1989). "Low-cost Map Overlay Analysis Using Computer-Aided Design", *Environment and Planning B* 16(4), 367-498.

Yeh, A.G.O. (1990). "A land Information System in the Programming and Monitoring of New Town Development", *Environment and Planning B* 17(4), 375-384.

Yeh, A.G.O. (1991). "The Development and Applications of Geographic Information Systems for Urban and Regional Planning in the Developing Countries", *International Journal of Geographical Information System* 5(1), 5-27.

Yeh, A.G.O. (1992). "Data Issues in Geographic Information System Development in the Developing Countries", Paper presented at the *27th International Geographical Congress*, 9-14 August 1992, Washington, D.C., U.S.A.

Yeh, A.G.O. (1995). "The Goals and Directions of Urban Planning Education in China", *Chengshi Guihua (City Planning Review)* 111, 9-10 (in Chinese).

Yeh, A.G.O. and X. Li (1994). "Land Use Analysis and Evaluation Using GIS and Remote Sensing in the Formulation and Implementation of Sustainable Regional Development Strategy", *Proceeding of Symposium on Space Technology and Application for Sustainable Development*. United Nations Economic and Social Commission for Asia and Pacific (UNESCAP) and State Science and Technology Commission of China, 19-21 September, Beijing, 2-4-1 to 2-4-14.

Yeh, A.G.O. and X. Li (1996). "Urban Growth Management in the Pearl River Delta. An Integrated Remote Sensing and GIS Approach", *The ITC Journal* 1996-1, 77-86.

Yeh, A.G.O. and X. Li (1997). "An Integrated Remote Sensing and GIS Approach in the Monitoring and Evaluation of Rapid Urban Growth for Sustainable Development in the Pearl Rive Delta, China", *International Planning Studies* 2(2), 193-210.

Zarzycki, J.M. (1984). "Standards for Digital Topographic Data: the Canadian Experience", *Computers, Environment and Urban Systems* 9(2/3), 209-215.

II APPLICATIONS AND CASE STUDIES

4 DECISION SUPPORT FOR SUSTAINABLE LAND DEVELOPMENT
A Case Study in Dongguan
Anthony Gar-On YEH and Xia LI

1. Introduction

China's adoption of economic reform and an "open door" policy in 1978 has meant a whirlwind of economic development: an annual growth rate as high as 9% between 1980 and 1990, and 13% in 1993. Indeed, economic development has replaced ideological and political debate as the main agenda. Along with rapid economic development, however, has come rapid urbanization, especially in the Pearl River Delta next to Hong Kong and the coastal provinces. Widespread urban sprawl has taken over valuable agricultural land. The resultant loss of food production capacity will erode the ability of these areas to sustain their development.

Rational use of land resources is only possible if urban expansion in China is guided by the concept of sustainable development. Sustainable development is emerging as a new approach to the dilemma between the needs of current versus future generations, as well as between economic development and resource conservation. It drew the attention of many of the world's scientists and governments at the Rio United Nations Conference on Environment and Development in June 1992. Shortly after the Conference, officials from China's State Planning and State Science and Technology Commissions organized an effort of all the government agencies involved to formulate and implement national strategies for sustainable development — China's Agenda 21 (ACCA, 1993). One of the main concerns for China is that rapid encroachment of urban development on valuable agricultural land may keep the country from achieving its sustainable development objectives.

The conversion of agricultural land to urban use is inevitable for many developing countries. However, there is a need to minimize the impacts of agricultural land loss through sustainable land allocation: that is, to create a balance between land development and conservation, so that the needs of both present and future generations can be met. In the conservation of land resources, two types of land-use patterns may be considered unsustainable: development of unused agricultural land and excessively rapid development.

Information is important for making good decisions. However, information on land use and land-use change for most cities in China is often inaccurate, inconsistent, and incomplete. There are often distortions of information, because local governments are unwilling to report the precise area of agricultural land lost to urban development to the central government. For example, Dongguan is a fast-growing city with a significant but under-reported agricultural land loss in recent years. In addition, technologies and systems to monitor rapid land-use change are in short supply.

Before implementing a land-use plan, therefore, the government must do an inventory of land resources and set up a plan to monitor land-use change on an ongoing basis. These are expensive: but this problem can be resolved in part through the use of remote sensing. Land-use types can be mapped by a classification of remote-sensing images; any land-use change would be detected by the comparison of a time series of remote-sensing images. Information obtained from remote sensing and geographic information systems (GISs) databases can thus be used for land-use modeling to support decision-making for sustainable land development. The following sections attempt to provide an operational model for sustainable land development through the integration of remote sensing and GISs.

2. Sustainable land-development model

Sustainability in land use can be defined as land use that satisfies the needs of the current generation yet keeps land-use options for the needs of future generations. Land-use planning should allow future generations to be at least as well off as ours. Three operating principles should guide planning for sustainable development of land:

1.   to maintain an equitable distribution of land resources between generations;

2.   to consume as little cultivated land as possible while maintaining a reasonable rate of economic growth; and

3.   to select land of less importance to agriculture first if any cultivated land is to be sacrificed to economic growth.

These principles can be developed into a sustainable land development model that deals with the interaction between land demand and land supply. The model is implemented through the integration of remote sensing and GISs. Land demand is estimated through population projection. The criterion of equity in land consumption per capita, discussed in the following section, can be used to help allocate land resources fairly to each generation. An inventory of the land area available for urban development is created by remote sensing. The suitability of the land supply is computed through the use of GISs. Once the amount of land needed for each time period is determined and the suitability of the land supply is known, GIS modeling can be used to match land supply to land demand in a sustainable way (Figure 1).

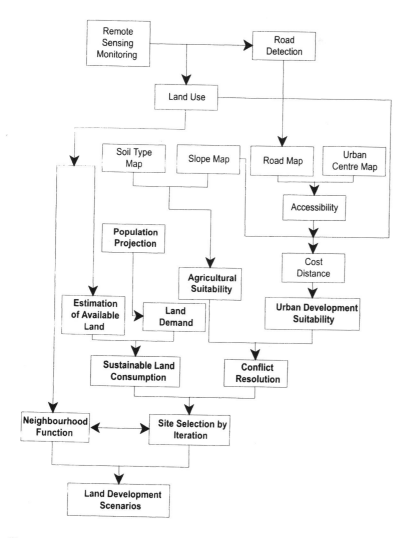

Figure 1. Sustainable land development modeling through the use of remote sensing and GIS.

2.1. Demand for land

Economic and population growth in the cities creates demand for more urban land. This demand has to be met. But it is important to differentiate between present and future demand. The availability of a fair share of the land resource to future generations cannot be compromised to satisfy present land demand. The principle of equitable distribution between generations should determine whether a particular land use is fair to both present and future generations.

Economic and urban growth in Southern China is booming. As a result, a significant amount of agricultural land in the Pearl River Delta has been taken over by urban development (Xu, 1990). This loss seemed inevitable: every year, more land was needed for urban development to accommodate the growth of both population and economy. There was a need for urban planners to manage the demand for land in order to reduce waste of the land resource and minimize the environmental problems that it entails.

2.2. Land supply and suitability

Land supply is defined by two factors: area and quality. The total land supply is fixed, because the area of a state or a city usually cannot change. Also, once agricultural land has been taken over by city blocks, it cannot be reconverted into farmland. As a result, the availability of land for development is decreasing. The quality of land is determined through land assessments which result in maps that show its suitability for various types of land use. The geographical location and physical properties of the land determine whether a particular parcel is better suited for agricultural production or urban development.

2.2.1. Geographical location. Distance from an urban center often determines the value of a parcel of land. In the calculation of that distance, however, it is more appropriate to use network rather than Euclidean distance, because the availability of transportation will have an effect on 'real' distance.

Platt (1972) stresses the importance of location as a determinant of land use. The pattern of spatial distribution of land uses captured the attention of many early urban geographers. For example, Von Thumen (1826) used location theory to explain the spatial distribution of agricultural activities (Chisholm, 1964; Platt, 1972). He suggested that their spatial distribution was determined by distance from a market outlet and transportation costs.

2.2.2. Physical characteristics. The physical properties of land include soil quality, water supply, topography and size.

- *Soil quality.* Agricultural production depends on the quality of the land, which is determined by its soil type. Some soils may not be suitable for growing particular crops. All else being equal, however, soil fertility determines agricultural yield.

- *Water supply.* The availability of water for irrigation is a crucial factor in suitability for agricultural production. The construction of irrigation systems can resolve water shortage problems and improve production capability.

- *Topography.* Topographical maps are necessary for the evaluation of land suitability, especially in areas with complicated geomorphological features. Uneven relief deters both urban development and agricultural activities. Topography ratings are used to evaluate the suitability of land for various uses.

- *Size.* The size of a parcel of land is a parameter that needs to be considered for land development. Also, planners need to aggregate land uses so fragmentation of land uses can be avoided.

2.2.3. Land suitability. There are two major types of land suitability — urban development and agricultural. While both types are determined by topography, urban development is more affected by location and the availability of transportation. A land parcel is suitable for urban development if it is close to an urban center or connected to it by a major highway. Agricultural suitability is more affected by soil type as well as topography: crops grow best on flat, fertile land, which is therefore the most productive. There are often conflicts between urban development and agricultural suitability, because a land parcel can be suitable to both types of land use. There is a need to resolve such conflicts.

2.3. Equitable distribution of land between generations

The following sections apply the concept of equitable inter-generational land sharing to sustainable land-use planning.

2.3.1. Per capita equity formula. To prevent urban sprawl, China has established a quota system to control per-capita land use in its through its land-use planning. Per capita land-use values for Chinese cities may range from 60 to 120 m^2 (Table 1). A very precise population growth estimate is thus required to implement this land quota control system.

Table 1. Urban land use quotas for Chinese cities (in m^2).

Class	Land use per capita
1	60.1 – 75.0
2	75.1 – 90.0
3	90.1 – 105.0
4	105.1 – 120.0

Source: China's urban land-use classification and land-use planning standards (1992), Ministry of Construction, China.

The population growth in a region is expected to stabilize once its population approaches or reaches the maximum permitted within its class. Of course there are many factors that can keep population growth below the allowed maximum. These can include water supply, availability of housing, means of transportation, pollution, and availability of land. The task for Chinese urban planners in the coming years is thus to match land use quotas to population growth. A simple way to establish land-use equity between generations is to ensure that land use per capita remains constant over time. If the total area of land available for development is Q_n, and the total additional population for a given period of time is P_{na}, under the established per capita equitable distribution principle, per capita land use should remain at Q_n/P_{na}. Thus land use Q_t in period t should be

$$Q_t = P_{ta}Q_n/P_{na},$$

where Q_t is the planned land use and P_{ta} is the projected additional population for period t.

2.3.2. Tietenberg's equity formula. In environmental economics, demand and supply curves are assumed to be influenced by price. Thus if the quantity demanded Q_d falls, the quantity supplied Q_s increases as price P rises. The search for efficient resource allocation begins with a determination of the marginal benefit and marginal cost functions. Under ideal conditions, the demand function for a commodity is equal to the marginal benefit function for that commodity, while the supply function is equal to the marginal cost function for that commodity (Common, 1988). The total benefit is simply the sum of the marginal benefits, while total cost is the sum of the marginal costs (Figure 2). The most efficient allocation of a resource is to maximize the net benefit, which is equal to the total benefit minus total cost. It is easy to derive that the net benefit is maximized when the marginal benefit is equal to the marginal cost.

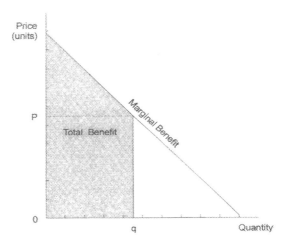

Figure 2. Relationship between marginal benefit and resource use.

The intersection of the marginal benefit (demand) and the marginal cost (supply) yields the output Q_0, the point at which the net benefit is maximized (Figure 3). However, the conditions for efficient allocation can be violated by externalities, which are very common but, usually not accounted for by the market. For example, user A produces wastes, which are externalities that are harmful to user B. Because the market is imperfect, user A does not need to take responsibility (i.e., pay for) for such externalities. A similar situation exists in agricultural land use.

Usually, a land user (for example, an urban developer) needs to compensate the (rural) seller only for the direct loss of income from agricultural production. That loss can be covered easily by the profit from the conversion, because the income from urban development of the land is much greater than the income from its agricultural use. Also, the buyer of the land does not need to compensate anyone for the broader impacts caused by urban land development, which may include pollution, landscape damage, reduction in the food supply and soil erosion. This unfair neglect of external costs tends to lead to excessive use of land resources.

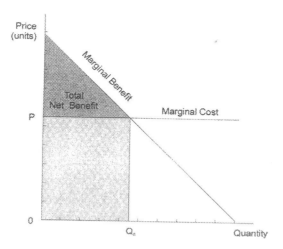

Figure 3. Maximizing the total net benefit.

Tietenberg (1992) uses the theory of environmental economics to propose a dynamic efficient-allocation model which can be used to allocate land resources efficiently and maintain land-use distribution equity between generations. The model allows the allocation of a depletable, nonrecyclable resource under dynamic-efficiency criteria. He suggests that to achieve dynamic efficiency, a society must balance the current and subsequent uses of the resource by maximizing the present value of the net benefit derived from its use. The time value of money is incorporated in the model. For example, a dollar invested today at 10% interest rate will yield $1.10 a year from now. Thus, the present value of a net benefit B_n received n years later is:

$$B_n/(1 + r)^n$$

where r is the interest rate.

The question is how to allocate the amount of the finite resource to different time periods so that the total net benefit can be maximized. Tietenberg assumes that the demand curve for a depletable resource is linear and stable over time. Thus, the inverse demand curve in year t can be written as

$$P_t = a - bq_t,$$

and the total benefit TB derived from extracting an amount of resource q_t over t years is the integral of the function:

$$TB = \int (a - bq_t)dq_t = aq_t - bq_t^2/2.$$

The marginal cost of extracting that resource is further assumed to be a constant c. The total cost TC of extracting the amount q_t, is

$$TC = cq_t.$$

The dynamic efficient allocation of a resource over n years, therefore, should satisfy the maximization condition

$$(a - bq_t - c)/(1 + r)^{t-1} - \lambda = 0 \quad t = 1,..., n$$

$$Q - \sum_{t=1}^{n} q_t = 0$$

where Q is the total available amount of the resource supplied. Maximization can be achieved by solving the equations below:

$$\text{Max} \sum_{q_t \atop t=1}^{n} (aq_t - bq_t^2/2 - cq_t)/(1+r)^{t-1} + \lambda(Q - \sum_{t=1}^{n} q_t)$$

The solution of the above formula yields a stream of q_t which is the proposed amount of resource used up in each period t. The sum of all q_t values is equal to Q. The following example illustrates how the model solves the problem of allocating the resource efficiently. If we assume that the parameter values needed for the calculation are a = 100, b = 1, c = 20, r = 0.10, Q = 50, and the amount of resource supplied (Q = 50 units) needs to be allocated into only two separate time periods, we get the results $q_1 = 27.62$ and $q_2 = 22.38$ units. It is obvious that more of the resource must be allocated to the first period because of the existence of interest rate r. The total net benefit from the two periods is 3 228.14.

Perhaps it would be considered more equitable to allocate an average of one-half of the available resource in each period. The total net benefit of 3 221.59, however, is below that of the dynamic efficient allocation 3 228.14. Under very extraordinary circumstances, it may be decided to use up all of the resource during the first period, contrary to the principle of sustainable development. The net benefit would be only 2 750. This demonstrates that using up all of the resource to satisfy the needs of the current generation is inefficient in the long run. The net benefit of average use is a much better resource allocation than even the extraordinary use example.

Tietenberg goes on to show that the dynamic efficient allocation can be perfectly consistent with sustainability, provided the benefits are shared appropriately between generations. Some part of the extra gains should be saved for future generations. The effect of the interest rate over time means that future generations may well gain more by inheriting the savings than by the average use approach. The essential objective of the allocation is to achieve the greatest net benefit while maintaining the possibility that future generations can be at least as well off as the current one.

Tietenberg's equity formula can be modified to allocate land use for period t, when the additional population P_t can be regarded as fixed. Thus, it is clear that the marginal benefit decreases as land use q_t, or land use per capita q_t/P_{ta}, increases (Figure 4). The total net benefit of using a fixed quantity Q of the land resource over n periods can be maximized to achieve efficient land use. Thus, the allocation of land use q_t in period t is obtained by the maximization solution

$$(a - bq_t/P_{ta} - c)/(1 + r)^{t-1} - \lambda = 0 \quad\quad t = 1,..., n$$

$$Q - \sum_{t=1}^{n} q_t = 0$$

where P_{ta} is the projected additional population in period t. When the discounting rate r is zero, this becomes

$$q_1/P_{1a} = q_2/P_{2a} = q_3/P_{3a} = \cdots = q_n/P_{na..}$$

This shows that when there is no discounting (r = 0), the modified Tietenberg's equity formula provides a strict per capita equity allocation.

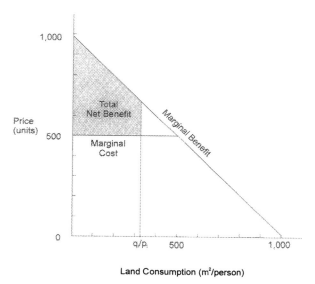

Figure 4. Marginal benefit and cost for per capita land use.

The above discussion on the equitable allocation of land resources provides some useful theoretical frameworks for sustainable land-use development planning in China. Economic development has been a major cause of agricultural land loss, with its associated negative effects of resource and environmental losses. However, the loss of agricultural land is also partly due to inadequate land-use planning and inefficient land management. The scenarios of per capita equity allocation and the modified Tietenberg's equity formula can be used as alternatives to the currently excessive land use. These two types of equity formulae can provide a basis for the rational use of land resources.

2.4. Spatial efficiency in GIS

The previous section focused on the allocation of land resources at the right time. Spatial efficiency refers to the allocation of land use in the right place — in other words, the spatial allocation of land to create the greatest net benefit. Spatial efficiency can be achieved if the following conditions are met:
1.   the best agricultural land is conserved;

2. land with the best location for development is converted to urban uses to meet necessary demand;

3. the land-use conflicts are resolved properly; and

4. land development is not carried out in a fragmented way.

Given no constraints, the spatial efficiency principle means that land with higher scores for a specific land use is selected first for that land use. Greater benefits are generated if land use is allocated to the land best suited to it. One problem is the frequent competition between different land uses for the same piece of land. The selection procedure should maintain the spatial integrity of the land to prevent fragmentation, which can result in waste of land resources because of the need for more facilities and other services.

It is very difficult, if not impossible, to realize spatial land-use efficiency with conventional methods because the amount of data is so huge. In addition, the decision itself is rather complicated because of the many criteria involved . For instance, there are obvious land-use conflicts between agricultural and urban use. Furthermore, uncertainties, which are well acknowledged in urban planning, necessitate a scenario approach to land-use planning.

GISs are becoming popular in land-use planning because of the development of modern computing technologies. A model is a substitute for an object or a system (Forrester, 1976). GIS modeling can be used to resolve complicated environmental and resource problems with the aid of powerful software and hardware functions. GIS modeling has proved to be very efficient in managing a series of complicated land-use problems. One major advantage is that it can be used easily to evaluate more than one scenario and offer various possible outcomes based on different premises. Because urban planning always involves uncertainties, there is a need to calculate and predict the possible consequences associated with those uncertainties. GIS modeling is used to provide and evaluate alternatives, and find the best one on the basis of current conditions.

GISs can be used to assist in land-use planning once the information has been collected and input. GIS can then be used to select land with higher scores for urban development and therefore suitability for land conversion. Unfortunately, another study has shown that the best location for urban development is often on the best agricultural land (Yeh and Li, 1996). Eastman *et al.* (1993) offer a heuristic solution to the problem of multiobjective land allocation under conditions of conflicting objectives. They propose handling large raster data sets through the use of procedures that have an intuitive appeal. The logic of this intuitive solution is illustrated in Figure 5.

The suitability map for each objective is created and placed as an axis in a multi-dimensional space. To find the best land for an objective, simply move the line down to a lower value until enough of the best land is located. The conflict is found in the area that is best for both objective one and objective two. A simple partitioning of the affected cells can be used to resolve the conflict (Figure 5). The decision space is divided into two regions so that cells are allocated to their closest ideal point. A 45 degree line between a pair of objectives assumes that they are given equal weight in the resolution of the conflict. The angle of the decision line can be altered by assigning different weights to competing uses.

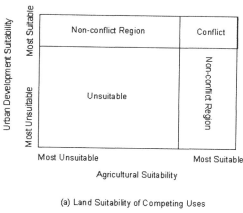

(a) Land Suitability of Competing Uses

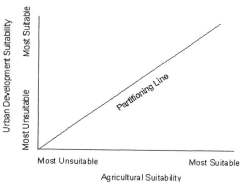

(b) Partitioning of Conflict Region

Figure 5. Competing land uses (a) and solution (b) (Eastman *et al.*, 1993).

Land allocation based simply on land suitability will lead to fragmented patterns of land use because land is heterogeneous. Fragmentation subverts the spatial efficiency principle because the cost ends up higher for either urban development or resource conservation. A neighborhood function using GISs is proposed to reduce the fragmentation in land allocation through the influence of neighborhood land uses. This is because a proposed land use will be more successful if it is compatible with neighborhood land uses.

Land suitability at location n will be affected by the land uses in neighboring localities. A suitability bonus for land use W_i will be added to a parcel's score if the use is also predominant in the neighborhood. For example, cell A would score a bonus in urban development suitability because it is surrounded by many urban land use cells (Figure 6). Cell B, on the other hand, would be penalized because its neighborhood cells are dominated by agricultural land use.

Sustainable Land Development

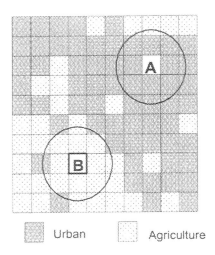

Urban   Agriculture

Figure 6. Neighborhood influences on land suitability.

The influence of neighbourhood is exerted by multiplying the original suitability $S_n(W_i)$ by a neighbourhood function $Q_n(W_i)$ for each iteration of land allocation. The new suitability set becomes

$$S_n(W_i) = S_n(W_i) \times Q_n(W_i), \tag{1}$$

where $S_n(W_i)$ is the land use suitability W_i in location (cell) n, and $Q_n(W_i)$ is the neighbourhood function. The neighbourhood function $Q_n(W_i)$ for location (cell) n is devised to indicate the contribution of neighbourhood m to the suitability of parcel n. A moving window will be used to calculate the proportion of W_j in neighbourhood m of location n.

$$P_m(W_j) = A_m(W_j)/A_m \quad m \in \Omega,$$

where $A_m(W_j)$ is the area of land use W_j in the neighbourhood m, A_m is the total area of the neighbourhood, and Ω is a rectangular neighbourhood containing $k \times k$ pixels.

The modification is made by multiplying the proportion $P_m(W_j)$ by the compatibility coefficient $p_{nm}(W_i|W_j)$. The coefficient $p_{nm}(W_i|W_j)$ is the probability that a parcel n will be in land use W_i if its neighbourhood is in land use W_j. Since all land-use types from neighbourhood m have an influence on land parcel n, the total modification will be the sum of all land-use types, expressed as

$$Q_n(W_i) = \sum_j p_{nm}(W_i|W_j)A_m(W_j)/A_m$$

The sum of the full set of $p_{nm}(W_i|W_j)$ is equal to 1:

$$\sum_j p_{nm}(W_i|W_j) = 1.$$

Equation (1) can be rewritten as an explicit iteration formula:

$$S_n^{k+1}(W_i) = S_n^k(W_i) \times Q_n(W_i). \tag{2}$$

The above formula can be used to insert the neighbourhood function in each iteration of land allocation. Fragmentation can be reduced through the definition of n iterations. Each iteration of the dynamic land allocation modifies the land suitabilities in each cell to match previous land allocation. In this way, GIS modeling encourages a compatible land-use pattern through the modification of land suitabilities.

3. Sustainable land-development modeling in Dongguan

3.1. The study area

Dongguan is located north of Hong Kong and Shenzhen and south of Guangzhou, on the eastern side of the Pearl River Delta (Figure 7). It is a new city that was upgraded from county status in 1985. In the past, it was mainly an agricultural area. Since 1985, rapid industrial expansion has far exceeded agricultural growth. The average annual industrial growth rate was 37% between 1985 and 1992, although in some of those years it has topped 45%. The annual agricultural growth rate, on the other hand, has averaged just 8.5%. Along with economic growth has come rapid urban development which has taken over much valuable agricultural land.

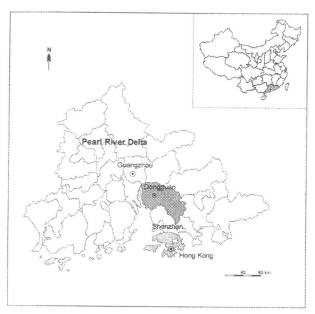

Figure 7. Location of Dongguan in the Pearl River Delta.

Land development was especially rapid after 1990 because of the sudden property boom in the Pearl River Delta. This was fueled by the Hong Kong real-estate market which led to property speculation. House prices in Dongguan were one-tenth of those of Hong Kong, so people who could not afford property speculation in Hong Kong could buy property in Dongguan for their relatives, themselves, or for investment purposes.

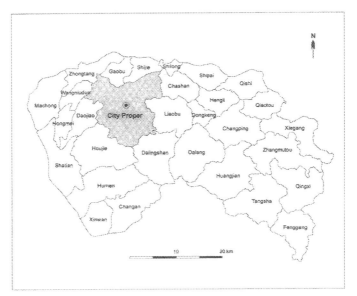

Figure 8. Towns of Dongguan.

Dongguan has a total area of 2 465 km^2. It consists of the city proper and 29 towns (Figure 8). The total permanent population in 1993 was 1.39 million, although many temporary workers from other parts of Guangdong and China live and work in the city. Land use has changed significantly throughout the whole city and much valuable agricultural land has been lost. The losses have never been estimated, however, and there is no plan to compare the benefits of future economic development against the loss of more agricultural land. The integration of remote sensing and GISs has provided planners and decision-makers with an efficient way to monitor land-use changes and formulate a sustainable development strategy to guide future urban growth (Yeh and Li, 1997).

The major land-use problem in Dongguan was how to control irrational encroachment on the best agricultural land. The integration of remote sensing and GISs was used to achieve sustainable land-use allocation. The allocation of land may be considered optimal when the aggregate social returns from its various uses are maximized (Lopez, et al., 1994). The sustainable land-development model allocates land in both time and spatial dimensions. The amount of land to be converted to urban use for each time period was based on the principle of equitable distribution between generations. Efficiency in spatial allocation was realized by reducing agricultural-suitability loss and ensuring compact development.

3.2. Amount of land supply and consumption

Remote sensing is used to carry out an inventory of the total available land resources before the land use for different planning periods is allocated. An equitable percentage of agricultural land is then chosen for land conversion to meet the demand due to population and economic growth. A high percentage of land loss would violate the principle of equitable distribution. The land available for future development is esti-mated through the use of remote sensing data. In 1988, they showed that the city had 140 524.8 ha of agricultural land. If we assume that only 30% of this total can be allo-cated for urban conversion because more land would cause too severe an environ-mental impacts, 42 157.4 ha is the optimal available area.

The city water supply can support a maximum population of between four and five million (Zeng et al., 1994). Future population growth was estimated on the basis of its historical rate. The population of the city was 1 267 605 and 1 389 232 in 1988 and 1993 respectively for an average annual population growth of 1.9% for that period. By natural increase, then, it would take the city 72 years to reach its carrying capacity.

The migrant population should also be taken into account in land-use planning. In China, planning authorities estimate the migrant worker population at a 50% discount. This means that a migrant population of 100 million is equal to 50 million of regis-tered population for in land-use planning purposes. In the worst-case scenario, the discounted additional migrant population could equal the city's registered population. The city would therefore need only about 40 years from 1988 to reach its carrying capacity.

Table 2. Optimal land use distribution with various discounting rates (r) in Dongguan for different time periods.

Years	Land use (in ha)		
	r = 0.0	r = 0.1	r = 0.2
1988 to 1993	3 741.0	6 527.1	8 400.2
1994 to 1998	4 100.0	6 535.7	8 381.5
1999 to 2003	4 493.4	6 418.2	8 101.2
2004 to 2008	4 924.5	6 136.4	7 452.1
2009 to 2013	5 397.0	5 643.0	6 291.2
2014 to 2018	5 914.8	4 879.9	3 531.2
2019 to 2023	6 482.4	3 775.4	0.0
2024 to 2028	7 104.4	2 241.7	0.0
Total	42 157.4	42 157.4	42 157.4

The modified Tietenberg's equity model was used to determine the optimal alloca-tion of the quantity of land use (42 157.4 ha) for the time period under study (40 years). It was assumed that the marginal benefit falls from 1 000 to 0 units when capital capita increases from 0 to 1 000 m^2 per capita and that the marginal cost is constant at 500 units (Figure 4). With no discounting, each generation should obtain an equal quota of land use per capita. However, actual land development cannot dis-regard the influence of discounting which reflects economic growth, inflation, and economic policy. Table 2 shows the results of the allocation of land consumption for each period, on the basis of the model with discounting rates of r = 0, 0.1, and 0.2. The results demonstrate that a higher discounting rate results in the depletion of land resources at an earlier stage. For example, when r = 0.2, the land resources will be depleted before 2018.

Table 3. Optimal land use for 1988-1993 and 1993-2005, for r = 0 in Dongguan, by town.

Towns	Additional population		Land consumption (in hectares)	
	1988 to 1993	1994 to 2005 (projected)	1988 to1993	1994 to 2005
City Proper	70 180	229 649	1 055.2	3 194.0
Zhongtang	14 198	44 093	213.5	613.2
Wangniudun	5 864	16 963	88.2	235.9
Daojiao	5 982	16 478	89.9	229.2
Hongmei	2 302	6 383	34.6	88.8
Machong	11 416	33 650	171.6	468.0
Humen	14 798	42 703	222.5	593.9
Changan	6 050	18 269	91.0	254.1
Houjie	11 360	32 048	170.8	445.7
Shatian	4 170	11 606	62.7	161.4
Liaobu	7 782	21 918	117.0	304.8
Dalingshan	9 258	30 601	139.2	425.6
Dalang	8 292	23 417	124.7	325.7
Huangjian	2 362	6 641	35.5	92.4
Zhangmutou	3 406	10 035	51.2	139.6
Qingxi	5 100	14 899	76.7	207.2
Tangsha	5 018	14 425	75.4	200.6
Fenggang	2 612	7 352	39.3	102.2
Xiegang	1 598	4 261	24.0	59.3
Changping	7 582	21 099	114.0	293.4
Qiaotou	3 308	9 037	49.7	125.7
Hengli	3 092	8 288	46.5	115.3
Dongkeng	3 050	8 308	45.9	115.6
Qishi	2 994	7 895	45.0	109.8
Shipai	4 876	13 702	73.3	190.6
Chashan	3 294	8 728	49.5	121.4
Shijie	598	1 464	9.0	20.4
Gaobu	5 656	16 790	85.0	233.5
Shilong	19 612	69 392	294.9	965.1

The optimal land-use values for the years 1988 to 1993 and 1994 to 2005 were obtained when r = 0 and r = 0.1. When r = 0 (absolute equity value), the optimal qland use values are 3 741.0 and 10 563.1 ha for the years 1988 to 1993 and 1994 to 2005. When r = 0.1 (discounted equity value), the optimal quantities of land consumption are 6 527.1 and 15 408.4 ha for the two periods. Tables 3 and 4 list the detailed optimal land use at the town level on the basis of the population distribution.

Table 4.   Optimal land use for the years 1988 to 1993 and 1993 to 2005 for r   = 0.1 in Dongguan, by town.

Towns	Additional population		Land consumption (in hectares)	
	1988 to 1993	1993 to 2005 (projected)	1988 to 1993	1994 to 2005
City Proper	70 180	229 649	1 841.0	4 659.1
Zhongtang	14 198	44 093	372.5	894.5
Wangniudun	5 864	16 963	153.8	344.1
Daojiao	5 982	16 478	156.9	334.3
Hongmei	2 302	6 383	60.4	129.5
Machong	11 416	33 650	299.5	682.7
Humen	14 798	42 703	388.2	866.4
Changan	6 050	18 269	158.7	370.6
Houjie	11 360	32 048	298.0	650.2
Shatian	4 170	11 606	109.4	235.5
Liaobu	7 782	21 918	204.1	444.7
Dalingshan	9 258	30 601	242.9	620.8
Dalang	8 292	23 417	217.5	475.1
Huangjian	2 362	6 641	62.0	134.7
Zhangmutou	3 406	10 035	89.3	203.6
Qingxi	5 100	14 899	133.8	302.3
Tangsha	5 018	14 425	131.6	292.7
Fenggang	2 612	7 352	68.5	149.2
Xiegang	1 598	4 261	41.9	86.4
Changping	7 582	21 099	198.9	428.0
Qiaotou	3 308	9 037	86.8	183.3
Hengli	3 092	8 288	81.1	168.2
Dongkeng	3 050	8 308	80.0	168.6
Qishi	2 994	7 895	78.5	160.2
Shipai	4 876	13 702	127.9	278.0
Chashan	3 294	8 728	86.4	177.1
Shijie	598	1 464	15.7	29.7
Gaobu	5 656	16 790	148.4	340.6
Shilong	19 612	69 392	514.5	1 407.8
Xinwan	3 006	9 403	78.9	190.8
Total	*248 816*	*759 496*	*6 527.1*	*15 408.4*

3.3. Land suitability

For a given parcel of land, a GIS model performed an optimal spatial land alloca-tion to determine the best land development sites. Urban and agricultural suitability maps were created before the allocation was made. Agricultural suitability for agri-culture was determined on the basis of two variables — soil type and slope. The rat-ing was based on experiments or field tests. Scores were adjustable so that the ex-perimental result could reflect the real situation. The land was rated in terms of its likely yield, with various soil and site characteristics that influence yield combined in a mathematical formula (McRae and Burnham, 1981).

Table 5. Rating scheme using the variables of soil type and slope for agricultural suitability.

Soil Type		Slope (degree)	Score	Class
Paddy soil	1) No. 23, 27, 25, 26, 21 (Yield > 9 000 kg/ha)	0-2.5	150-120	7
	2) No. 29, 17, 15, 42, 11, 22, 19, 2, 41, 4, 8, 10, 24, 20 (Yield = 7 500 – 9 000 kg/ha)	2.5-5	120-100	6
	3) No. 13, 30, 7, 18, 37, 40, 33, 28, 38, 31, 12, 9, 3, 35, 36, 14 (Yield = 6 000 – 7 500 kg/ha)	5 -7.5	100-70	5
	4) No. 32, 34, 6, 1, 39, 5, 16 (Yield = 4 500 – 6 000kg/ha)	7.5-10	70-60	4
Dry cultivated soil	1) No. 45, 46, 47, 48, 49, 52, 55, 54, 55, 56, 57, 58, 59, 60, 61, 62, 63, 64	10-15	60-25	3
Mountain soil	1) No. 44, 50, 51	15-30	25-15	2
	2)No. 43, 65	> 30	15-0	1

The procedure began with a rating of rating soil types and slope degrees on a continuous score. Table 5 provides a corresponding simplified seven-level ranking based on their suitability for agriculture: 1 - extremely unsuitable, 2 - unsuitable, 3 - less suitable, 4 - suitable, 5 - more suitable, 6 - very suitable, 7 - extremely suitable. The final score was obtained by adding the numerical continuous rating from the following formula:

$$AG\_SU = 0.5S_{soil} + 0.5S_{slope,}$$

where S_{soil} and S_{slope} are the scores for soil and slope, and AG_SU is the agricultural suitability score.

Land suitability was calculated by using GRID in ARC/INFO. The final map was created by generalizing the final score into seven categories. Thus, land which is rated 1 and 2 cannot be used for agricultural activities, land which is rated above 2 is suitable, and a raiting of 7 is most suitable for agriculture.

Urban development suitability indicated the potential suitability of land for urban uses. It was decided by two factors: distance (location) and slope. A site located close to an urban (town) centre was rated highly suitable for urban development. The cost distance rather than Euclidean distance was used to account for the effect of the availability of transportation and other costs. The cost distance, rather than the actual distance between two points, calculates the shortest accumulative cost distance of each cell to the closest target cell. The cost distance reflects the true distance better than Euclidean distance, because other factors influencing distance and cost can be considered.

In the calculation, each normal cell was assigned a cost value of one. Cells of various types of roads were assigned cost values of less than one because automobiles make the distance by road shorter. A cost value higher than one was assigned to cells of water because crossing water was more difficult than crossing land. The cost values for various types of cells are listed in Table 6.

Table 6. Cost values for various types of cells.

Cell types	Cost values
Normal cells	1.0
Water	3.0
Railway and express-way	0.1
Main road	0.2
Other roads	0.3

The COSTDISTANCE command in GRID of ARC/INFO was used to calculate the cost distance for each cell:

$$CO\_D = COSTDISTANCE(PTWN, COST)$$

where CO_D is the cost distance grid, PTWN is the grid with urban (town) centres and COST is the cost grid. The cost distance and the slope of each cell decide the urban development suitability of the cell. A cell with higher values for cost distance and slope will have a lower urban development score.

The cost distance and the slope were rated on a continuous score. Table 7 provides a corresponding simplified seven-level ranking according to their suitability for urban development.

Table 7. Rating scheme using the variables of cost distance and slope for urban development suitability.

Cost distance	Slope	Score	Suitability class
0 - 1 000	0-5	150-120	7 (extremely suitable)
1 000 - 2 000	5-10	120-100	6 (very suitable)
2 000 - 3 000	10-12	100-70	5 (more suitable)
3 000 - 4 000	12-15	70-60	4 (suitable)
4 000 - 5 000	15-17	60-25	3 (less suitable)
5 000 - 10 000	17-25	25-15	2 (unsuitable)
>10 000	>25	15-0	1 (extremely unsuitable)

The final score for urban development suitability was obtained by adding the numerical continuous rating from the following formula:

$$SU\_UR = 0.6S\_CO\_D + 0.4S\_SLOPE$$

where S_CO_D and S_SLOPE are the rating values of the cost distance and slope for urban development suitability, and SU_UR is the final rating of urban development suitability.

3.4. Modeling spatial efficiency

Given the total amount of land use, GIS modeling can be used to allocate agricultural land for urban development on the basis of the spatial efficiency rule. The five scenarios of land consumption in Dongguan during the study period were:

1988 to 1993 (past)
1) the amount of land provided is the same as that of the actual land loss (21 285.7 ha in 1988 to 1993);
2) the amount of land provided is 3 741.0 ha when $r = 0$;
3) the amount of land provided is 6 527.1 ha when $r = 0.1$;

1994 to 2005 (future)
4) the amount of land provided is 10 563.1 ha when $r = 0$; and
5) the amount of land provided is 15 408.4 ha when $r = 0.1$.

Without constraints, the most efficient allocation was to sites with higher scores of urban development suitability that could generate higher economic benefits. However, it has been argued that agricultural suitability should be involved in the model because of the need to conserve the best agricultural land. The conflict between the two objectives was resolved by means of the partitioning method discussed above (Eastman *et al.*, 1993). The partitioning line is $0.6SU\_UR - 0.4SU\_AG$. A slightly higher weight is given to urban objectives because of the greater emphasis given to urban development.

Spatial integrity needs to be implemented in the model through the use of the neighbourhood-effect function. The spatial dynamic allocation was performed with iterations, so that the neighbourhood-effect function could be integrated into the allocation model and so increase spatial efficiency. Urban development suitability was modified on the basis of the previous land-use allocation step at each iteration. For example, if a cell was surrounded by many cells which had been selected already as urban area, there was an increase in the urban development suitability of the cell because of the influence of the neighbourhood function. In the neighbourhood of a cell, three major land-use types – urban, crop, and orchard – are considered to be important influences on the urban development suitability of the cell. Other land-use types, such as water and forests, were considered incompatible with urban use. Suitability for urban development is reduced in a neighbourhood dominated by many rural types. The compatibility coefficients are listed as follows:

$p_{nm}(\text{urban}|\text{urban}) = 0.5$
$p_{nm}(\text{urban}|\text{crop}) = 0.3$
$p_{nm}(\text{urban}|\text{orchard}) = 0.2$

$p_{nm}(urban|others) = 0$

An AML program in ARC/INFO GRID was developed for the detailed implementation of sustainable land allocation. The program enables the sustainable land-
development model to improve spatial efficiency by taking land suitability and spatial
integrity into consideration.

3.5. Results and discussion

The modified Tietenberg's equity formula will achieve more efficiency in using land
resources through the maximization of the total net benefit. Per capita equity guarantees that the same amount of per capita consumption is reserved for all generations.
More resources are allocated to future generations for per capita equity distribution
because of population growth. The modified Tietenberg's equity formula is rather
weak because more resources are consumed by the existing generation due to discounting. However, the modified Tietenberg's equity becomes the per capita equity-
distribution when the discounting factor is not used ($r = 0$). A higher discounting rate
results in the depletion of land resources at an earlier stage. It is important to choose
an appropriate discounting factor in the allocation of land resources to avoid the depletion of the resource. The emphasis on a land supply based on equity between generations is useful for a better allocation of land resource distribution. The model can
be used also to examine past land consumption to see how far the actual land development deviates from the rational land development generated by the model.

Urban expansion and rural urbanization are proceeding at an astonishing rate,
with a significant amount of agricultural land lost each year in Dongguan (Yeh and Li,
1997). The urban sprawl areas are clearly visible in satellite images. If there is no
controlling function to guide land-use patterns in the direction of sustainable development, agricultural land in Dongguan will be depleted very soon and serious environmental problems will follow.

The sustainable land development model was used to generate the optimal land
allocation for the years 1988 to 1993 and 1994 to 2005. Figure 9 shows the actual
land loss and the optimal allocation of the land conversion generated by the model
with the same land use total of 21 285.7 ha. Figures 9 and 10 show the results of the
optimal allocation of land development for the years 1994 to 2005 to achieve the
greatest efficiency in spatial and time dimensions for the modified Tietenberg's equity
formula when $r = 0$ (no discounting) and $r = 0.1$. The two scenarios allow 10 563.1
and 15 408.4 ha of land loss for land development in this period (1994 to 2005).

The comparison between the actual land-loss pattern and the optimal land-loss
pattern (based on the modified Tietenberg's equity formula) was used to examine the
land-use problems in Dongguan between 1988 and 1993. It was found that the actual
land loss in Dongguan at that time exceeded the expected proportion, which was calculated on the basis of the inter-generational equity principle. The actual land loss was
as high as 21 285.7 ha, while the expected loss was only 9 687.4 ha and 2 488.2 ha,
when based on the modified Tietenberg's equity formula and per capita equity distribution, respectively. In both cases, the actual agricultural land use was also much
higher than expected — by a factor of between 2 and 9. The city should therefore

keep land development under strict control and so reduce the amount of land loss for the sake of future generations.

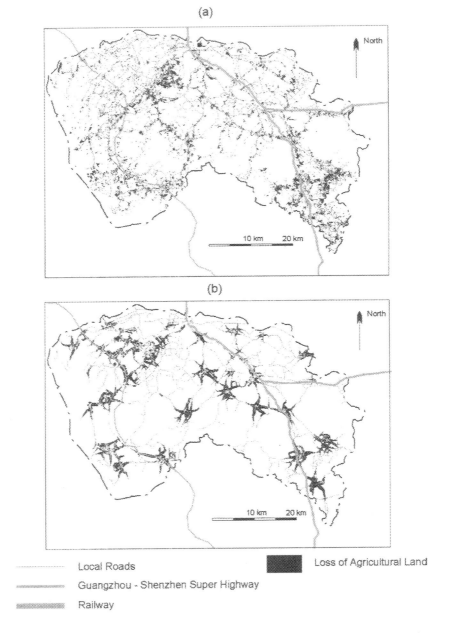

(a)

(b)

Local Roads

Guangzhou - Shenzhen Super Highway

Railway

Loss of Agricultural Land

Figure 9. Comparison between actual (a) and optimal (b) land loss with the same amount of total land use between 1988 and 1993.

(a) discounting rate r = 0

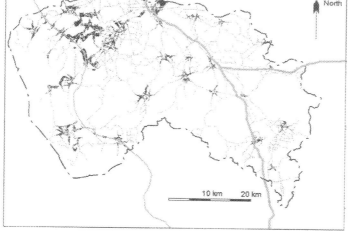

(b) discounting rate r = 0.1

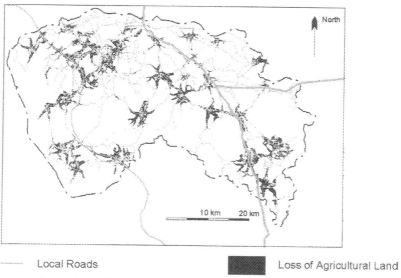

Figure 10. Optimal land development according to the sustainable land development
model between 1994 and 2005.

Even with the same actual land loss, better efficiency could be achieved if land development was based on spatial efficiency principles. It was found that the actual development pattern lacks proper land-use planning so that the cost was greater than it should have been. The comparison between the actual land loss pattern and the optimal development pattern highlights these land-use problems. Two simple indices, compactness (Ebdon, 1985) and agricultural suitability loss (S_{loss}), can be used to evaluate the efficiency between the two patterns. Given the same amount of land consumption, efficient land allocation should use a pattern of more compact land uses and less agricultural suitability loss. The compactness index CI = Area / Perimeter. CI was calculated through the use of ARC/INFO GRID commands. The larger the value of CI, the more compact the development was.

Loss of good-quality agricultural land is another aspect of land loss. The impact of land loss varies according to its agricultural suitability. If a town loses a large proportion of fertile land, the cost to agriculture is greater than for a town that loses less fertile land. So it is more appropriate to use agricultural-suitability loss instead of area loss in the evaluation of land loss. The total loss of the agricultural suitability is

$$S_{loss} = \sum_i \sum_j S(i, j)$$

where S_{loss} was the total suitability loss, $S(i,j)$ the suitability for agricultural type j at location i where the land loss took place. S_{loss} was calculated through the use of ARC/IFRO GRID commands: the higher the value of S_{loss}, the more valuable the lost agricultural land was.

Table 8 shows that the actual land loss is highly fragmented because its compactness index is less than half of that of the optimal model. There is also a decrease in suitability loss if the optimal land-development model is used.

Table 8. Comparison of the efficiency between actual and optimal land loss with the same amount of land use between 1988 and 1993.

	Compactness index (CI)	Suitability loss (S_{loss})
Actual development	9.4	9.26E+06
Optimal development	23.5	8.90E+06

A further comparison between the concurrence of the actual land conversion and the optimal land conversion was carried out through the use of the Summary function of GIS, with analysis in ERDAS IMAGINE. Table 9 indicates that only about one-fifth of the actual land conversion occurs at the exact locations expected by the optimal model. This means that a large proportion (80.1%) of the actual land conversion falls outside the ideal optimal locations. That cost is unnecessarily high for the development pattern because it is not spatially efficient.

The sustainable land-allocation scenarios can provide some guidelines for Dongguan's future land development. If the model is applied, the city can avoid chaotic and wasteful allocations of land resource. Land allocation cannot be determined by demand alone; otherwise excessive land loss will occur. This was witnessed in the land development that occurred in Dongguan between 1988 and 1993. Future land development planning should take the constraints of available land resources carefully into account to achieve sustainable land development for the city.

Table 9. Concurrence matrix of actual and optimal agricultural land loss with the same amount of land drain between 1988 and 1993 (in ha).

Optimal development	Actual development		
	Converted	Not converted	Total
Should be converted	4 098.7 (19.3%)	17 187.0 (9.4%)	21 285.7
Should not be converted	17 187.0 (80.7%)	204 882.5 (90.6%)	222 069.5
Total	21 285.7 (100.0%)	222 069.5 (100.0%)	

By testing different development scenarios and land-consumption parameters, planners and government officials in China can use the sustainable land-development model as a decision-support system. Planners in other areas of the world that are also under the pressure of rapid urban growth also will find the model useful. It is intended to suggest areas where future urban development should take place, and should therefore be used to develop strategic plans to meet the objectives of sustainable development for their cities.

References

ACCA (1993). *Introduction to China's Agenda 21*, Administrative Centre for China's Agenda 21, Beijing: China Environmental Science Press.

Chang, S.D. and Y.W. Kwok (1990). "The Urbanization of Rural China", in R.Y.W. Kwok Y. W. Parish, A.G.O. Yeh, and X.Q. Xu, (eds.) *Chinese Urban Reform: What Model Now?*, London: M. E. Sharpe, 140-157.

Chisholm, M. (1964). *Rural Settlement and Land Use*. New York: John Wiley.

Common, M. (1988). *Environmental and Resource Economics: An Introduction*. London: Longman,.

Dasmann, R.F. (1964). *Wildlife Management*. New York: John Wiley.

Eastman, J.R., P.A.K. Kyem, J. Toledano, and W. Jin, (1993). *GIS and Decision Making*. Worcester, M.A.: Clark Labs for Cartographic Technology and Geographic Analysis.

Ebdon, D. (1985). *Statistics in Geography*. Oxford: Basil Blackwell.

Edwards, R.Y. and C.D. Fowle (1955). *Transactions of the 20th North American Wildlife Conference*. Washington, D.C.: Wildlife Management Institute.

Forrester, J.W. (1976). *Principles of Systems*. Cambridge, Massachusetts: Wright-Allen Press.

International Union for the Conservation of Nature (1980). *World Conservation Strategy*. Gland, Switzerland: International Union for the Conservation of Nature.

Lopez, R.A., F.A. Shan, and M.A. Altobello (1994). "Amenity benefits and the optimal allocation of land", *Land Economics* 70(1), 53-62.

McRae S. G. and C. P. Burnham (1981). *Land Evaluation*. Oxford: Clarendon Press.

Markandya, A. and J. Richardson (1992). "The Economics of the Environment: An Introduction", in A.Markandya and J. Richardson, (eds.) *Environmental Economics*. London: Earthscan, 7-25.

Meadows, D.H., D.L. Meadows, J. Randers and W.W. Behrens (1972). *The Limits to Growth*. New York: Universe Books.

Platt, R.H. (1972). "The Open Space Decision Process: Spatial Allocation of Costs and Benefits". The University of Chicago, Chicago, *Research Paper*, 142.

Tietenberg, T. (1992). *Environmental and Natural Resource Economics*. New York: Harper Collins.

U.S. Department of Housing and Urban Development (1978). *Carrying Capacity Action Research: A Case Study in Selective Growth Management*. Oahu, Hawaii, Final Report.

World Commission on Environment and Development (1987). *Our Common Future*. Oxford: Oxford University Press.

Xu, X.Q. (1990). "Urban Development Issues in the Pearl River Delta", in R.Y.W. Kwok, W. Parish, A.G.O. Yeh, and X.Q. Xu, (eds.) (1990) *Chinese Urban Reform: What Model Now?* London: M. E. Sharpe, 183-196.

Yeh, A.G.O. and X. Li (1994). "Land Use Analysis and Evaluation Using GIS and Remote Sensing in the Formulation and Implementation of Sustainable Regional Development Strategy", in *Proceeding of Symposium on Space Technology and Application for Sustainable Development*. United Nations Economic and Social Commission for Asia and Pacific (UNESCAP) and State Science and Technology Commission of China, Beijing. 2-4-1 to 2-4-14.

Yeh, A.G.O. and X. Li (1996). "Urban Growth Management in the Pearl River Delta - an Integrated Remote Sensing and GIS Approach", *The ITC Journal* 1, 77-86.

Yeh, A.G.O. and X. Li (1997). "An Integrated Remote Sensing and GIS Approach in the Monitoring and Evaluation of Rapid Urban Growth for Sustainable Development in the Pearl Rive Delta, China", *International Planning Studies* 2(2), 193-210.

Zeng, T.C., Z.B. Liu, L. Li, and Y.J. Li (1994), "Strategies for Urbanization in Dongguan", in Economic Department of Dongguan Government, *Strategies toward the 21st Century for Dongguan*, Guangzhou: Huachen Press, 28-47 (in Chinese).

5 WATER RESOURCE MANAGEMENT
A Case Study for EcoKnowMICS
Eduardo D. Gamboa

1. Introduction

Behind every decision support system (DSS) is a data system representing its problem domain, and behind every good decision is a shared understanding of this underlying data system. Understanding means being able to establish unambiguously a correspondence between the data and the domain that the data represent. Economic data systems must be understandable not only to the economists who build them and to the policymakers who use them – but also to the vast majority of ordinary people whose daily lives are affected by the policies that they engender. A shared understanding among builders and users provides the checks and balances that ensure the accuracy, integrity, and internal consistency of economic data systems.

Modern data systems, however, have become incomprehensibly complex, especially to nontechnical people. Although they use formal tools (account tables, matrices, equations) to organize and structure data into precise and easily digestible pieces, they are unable to bridge the gap between the highly abstract world of economic concepts and data, and the real world of people, households, firms, goods, and services. Consequently, economic data are seldom questioned or examined during the decision-making process because of the difficulty in understanding their complexity and their source (see Duncan and Gross, 1995, for an evaluation of the problems and opportunities faced by modern statistical data systems).

Data systems naturally grow larger and more complex to support the growing diversity of user perspectives and purposes (Dunn, 1974). This is beginning to put a strain on the representational capabilities of statistical data systems. Consequently, amidst the volume and sophistication of data produced by modern, computerized economic data systems, there are embarrassing pockets of poverty in our understanding of the data.

Concerns about information overload can be traced to a mismatch between the computer-enhanced capability of data systems to compile and present large amounts of data, and the naturally limited capacity of human users to absorb and comprehend the resulting deluge. This chapter attempts to show how information technology (IT) can improve our ability to navigate successfully through oceans of data (Partridge, 1996). IT includes object-oriented technology for building the software components of economic data systems, and World Wide Web (Web) technology for distributing and sharing these software components.

Specifically, this paper proposes the use of standardized, shareable, and reusable software components for building a Web-based virtual economy to complement existing economic data systems. This virtual economy is a software representation of the economic domain and serves as a vehicle for promoting a shared understanding of economic data. The software components are implemented as Java programming language objects in Web applets, which are computer programs executed inside a Java virtual machine installed in almost all Web browsers. Thus, any user with a Web browser can collaborate on the virtual economy with other users connected to the Web.

As with any new technology, the most effective application areas for IT are those where it can exert maximum leverage. Economic planning, for instance, is an inherently social activity designed to reconcile conflicting vested interests. For example, a plan to increase the minimum wage may conflict with another plan to reduce the unemployment rate. The technological challenge is to build decision-support tools that can help to clarify, reconcile, communicate, and eventually share the meaning of economic data under different contexts and for different purposes. The case study, of a project in the Philippines, illustrates the use of a virtual economy by local governments preparing investment plans for the water supply sector.

2. Uses of a shared understanding of data

2.1. Linking producers and consumers in the data value chain

Most economic data come from a long chain of data-transformation processes performed by different data producers at different times and locations. The first step in obtaining household data at the national level, for instance, is usually the design of a household survey. Different survey designs yield varying degrees of accuracy and reliability of data. Within a particular survey design, different levels of interviewer expertise will have similar implications. Using sample data from interviews, the statistical estimation procedure – selected from possibly several alternative procedures – determines to a large extent the reliability of the final population estimates.

If the purpose of the household survey is to feed data to an articulated system of national accounts, the statistical estimates will have to undergo a reconciliation process to ensure the balancing of the accounts. Consumers, who are often unfamiliar with this multilayered process and may be beguiled by the neatness and elegance of the accounting presentation, tend to be uncritical of the economic data. This dangerous tendency can be minimized if we make it easier for producers and consumers alike to understand the origins of the data by using visual navigation paths through all the data transformations (for a detailed discussion of these issues, refer to Duncan and Gross, 1995).

2.2. Linking micro and macro data

The economic system is studied at two levels: microeconomy and macroeconomy. Microeconomy is concerned with the behavior of individual households and firms as decision-making units; macroeconomy, on the other hand, is concerned with the behavior of artificially created sectors as decision-making units. Microeconomic data systems coexist with macroeconomic data systems in the same way that microeconomic theories coexist with macroeconomic theories. The basic sources of macroeconomic data, however, are necessarily the individual households and firms that constitute the microeconomy. Consequently, the integrity of macroeconomic data depends on the validity of the original microeconomic data, and the conceptual and computational bridges that link them (Postner, 1988; Bloem, 1990).

The system of national accounts, for instance, can provide comprehensive and consistent data for macroeconomic simulations because it views economic agents as homogeneous firms and households that can be grouped conveniently into abstract statistical sectors. Although this view fits neatly the aggregation requirements of the accounting rules, it presents a highly sanitized, stylized, and possibly misleading picture of the real world of heterogeneous firms and households. An alternative to the macroeconomic simulation using aggregated accounts data is the microeconomic simulation using data from individually differentiated micro-units. The aggregated outputs from microeconomic simulations become inputs to macroeconomic analyses (Orcutt, 1961).

Consistency of data at the macroeconomic level does not necessarily mirror an underlying consistency at the microeconomic level. Participants in a transaction may have different views of how the transaction should be recorded. One may record the transaction as an earned income payment, the other as a transfer payment. Aggregating data on such transactions can lead to statistical inconsistencies at the macroeconomic level. Forcible reconciliation of inconsistencies to satisfy accounting requirements is a necessary evil that can be made more palatable to users by making the reconciliation procedures easier to understand.

2.3. Linking observation, description, and simulation data

Observation, description, and simulation data are three basic but fundamentally different types of data commonly found in data systems. Observation data capture the actual states and events of economic agents as perceived and recorded by the agents themselves, e.g. the actual price and quantity of commodity_X purchased, recorded, and reported by household_A. Description data are abstracted from observation data to support a particular analytical view, e.g. the average price and quantity of commodity_X purchased by household_A during the past year, as derived by estimation procedures based on a particular statistical theory. Simulation data are generated from a mathematical model with an underlying explanatory theory, e.g. the expected price and quantity of commodity_X to be purchased by household_A. Comparing the hypothetical simulation data with actual observation data tests the underlying theory.

How to handle missing or incompatible data always has been a problem for compilers of interlocking data systems such as the system of national accounts. These accounts are designed to present a comprehensive view of the whole economy while enforcing consistency, compatibility, and comparability of data from disparate sources. It has become standard practice to fill gaps in the accounts by substituting description and simulation data for the missing observation data. This, however, raises questions about the validity of testing theories using data generated by other theories (Richter, 1994 and Holub and Tappeiner, 1997). Users need to understand the separate meanings of the data clearly before they can understand their combined meaning properly.

2.4. Linking theory and data

Because theoretical concepts are seldom directly observable, their corresponding statistical data are difficult to find. Proxy variables, which are directly observable and also measurable, are used to bridge this gap. Choosing from among alternative proxy variables, however, can be somewhat arbitrary without some measure of conceptual "closeness of fit". A thorough understanding of the concepts underlying both theoretical and proxy variables is necessary to build and use such a measure (Eisner, 1989). Computational matching of theoretical concepts and empirical observations are discussed extensively in Thagard (1988).

3. Shared understanding of a domain representation

3.1. What is a domain representation?

A domain representation is a description of "what a domain is", for the purpose of understanding its data in terms of the objects that generated them. Because it is not an explanation of "why a domain is", it does not make use of theoretical explanatory objects, and therefore it maps unambiguously onto the domain. In other words, each object in the representation refers to a corresponding object in the domain. An example of a domain representation in economics is the United Nations System of National Accounts (SNA). It aims to provide answers to the fundamental economic questions listed in Table 1 (United Nations, 1993).

Table 1. Domain questions and answers.

Questions about the domain	Answers from the domain representation
Who does what?	Transactor_A engaged in activity_X
By what means?	Type of resource stock of transactor_A
For what purpose?	Type of activity_X
With whom?	Transactor_B engaged in activity_X
In exchange for what?	Type of resource stock of transactor_B
With what changes in stocks?	Resource flows between transactor_A and transactor_B through activity_X

In principle, the SNA captures, organizes, and presents data on all economic transactions of all economic agents residing within the geographical boundaries of an economic domain (typically a country) during a particular period of time (typically a year). The fundamental questions that guided the construction of the SNA identified the following generic objects as basic to a representation of the economic domain: transactors, activities, stocks of resources, and flows of resources. References to time and space objects aid users in mapping from an economic domain onto its representation, and vice versa. Economic data are organized around these objects.

3.2. Uses of an economic domain representation.

The four generic uses of an economic domain representation form a closed loop that links the domain, its representation, and the user (see Table 2). The loop involves three mappings. The first mapping, Visualization, is a two-way mapping between the graphical notation of a representation and the user's own internal visual representation. This mapping enables the user to understand visually the economic domain and at the same time manipulate graphically its representation. The second mapping, Observation, is used to classify and measure entities in the economic domain and to record these measurements in the representation. The third mapping, Implementation, guides planners and policymakers who manage the economic domain by using the results of simulations performed on the representation.

Table 2. Uses of an economic domain representation.

Visualization	To communicate a shared understanding of the economic domain by using a standard graphical notation that helps map the domain representation onto the user's own internal visual representation.
Observation	To classify and measure the actual state and actual changes in the state of the economic domain by using a mapping from the domain onto the domain representation.
Simulation	To reflect hypothetical changes in the state of the economic domain by using computed changes in the domain representation.
Implementation	To effect desired changes in the state of the economic domain by using a mapping from the domain representation onto the domain.

3.3. From SNA to virtual economy – a paradigm shift?

In economic data systems, the most basic element is an ordered pair (a tuple with two placeholders). Let this ordered pair be called a property consisting of <label, value>, where a label is a string of characters and a value is a number implicitly understood as the magnitude of the statistical measurement described in the label. A property represents an atomic or indivisible fact about the economic domain.

Examples: gross value added (gva) of establishments
   property1: <gva of establishment_1 in million US dollars in 1997, 100>
   property2: <gva of establishment_2 in million US dollars in 1997, 200>
   property3: <gva of enterprise_B in million US dollars in 1997, 300>
Each property in the examples above can be understood even in isolation, because each provides information on the following dimensions of meaning:

name: gross value added;

value: (million US dollars): 100, 200, 300;

owner: establishment_1, establishment_2, enterprise_B;

time reference (year): 1997; and

space reference (address): establishment_1, establishment_2, enterprise_B.

Clearly, to capture all the facts about the economic domain will require an inordinate number of properties containing repetitive information in their labels. For efficient organization and storage of properties, formal mechanisms such as the account table, the accounting matrix, and the algebraic equation have been adopted for use in economic data systems. One can view these mechanisms as structured containers of properties with strict membership rules and formal, built-in operators for manipulating the property values.

An account table, for instance, has rules for classifying a property as either a credit or a debit, and mathematical operators to compute the sum of credits and the sum of debits, and to ensure that the total credits equal the total debits. An accounting matrix has rules for assigning a property to a cell to represent a relationship between the entities identified in the corresponding row and column. As a mathematical construct, the matrix has operators for manipulating the contents of the cells. An algebraic equation, on the other hand, represents a mathematical relationship that has inherent rules for determining which properties can substitute for the terms of the equation, and built-in operators for computing the equation.

Although a property by itself is already useful, knowing the relationships between properties can be even more useful. In the example above, the fact that *property3* can be computed as the sum of *property1* and *property2* follows from the fact that establishment_1 and establishment_2 are parts of enterprise_B. This relationship, however, is not explicitly represented in any of the properties. Consequently, account tables, matrices, and algebraic equations, although efficient structures for containing and manipulating properties, provide poor representations of structural relationships in the problem domain.

Figure 1 shows how a domain object model can overcome the representational deficiencies of account tables, matrices, and algebraic equations. The static object model has four constructs: class, object, object.property, and relationship. The object is a representation of an actual, identifiable entity in the economic domain (e.g. establishment_1 represents an actual business establishment). The class represents the type of an object (e.g. an enterprise is a type of economic agent and non-financial enterprise is a subtype of enterprise). The object.property is a representation of an attribute of an object (e.g. value added is an attribute of an establishment). The relationship can represent four types of relationships in the economic domain: nonfinancial enterprise *is-a* enterprise; enterprise_B *is-instance-of* nonfinancial enterprise; establishment_1 *is-part-of* enterprise_B; and *is-member-of* the manufacturing sector.

Much like an account table or a matrix, the object model is also a structured container of properties, wherein an object.property can contain one or more properties, representing, for instance, a single value or successive values of the same attribute over time or across space. An advantage of the object model is its ability to represent much more detailed and complex structural relationships than those represented by an account table, matrix, or equation. Thus, the object model provides a clear "snapshot" of an economic domain at a particular point in time.

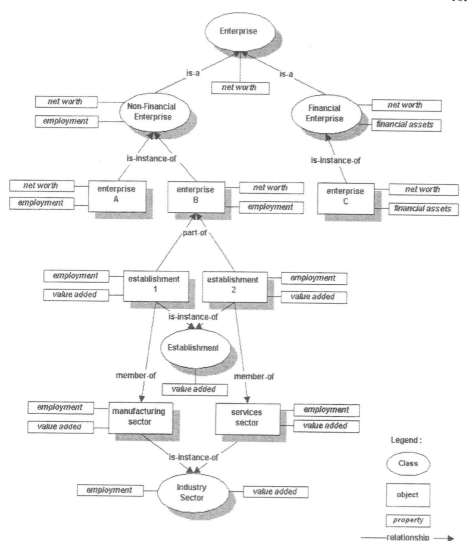

Figure 1. Domain object model.

Another advantage of the object model is that it can incorporate methods for changing its state. The state of an object is represented by the set of object.property values that describe the object at a given point in time. The methods of an object are computational procedures for deriving new object.property values for the object. The methods execute in response to events (e.g. the end of a time period) and set the new state of the object. The execution of methods can be repeated for successive time periods to generate time-series data. The addition of methods converts the static object model, which focuses on structural relationships, into a dynamic object model, which represents the changes in the state of objects over time (object models of enterprises are discussed in Gale and Eldred, 1996).

The most important advantage of object models is that they can be implemented directly into software using an object-oriented programing language such as Java. The objects in the object model map directly onto the objects in the software. The methods of the objects can map directly onto declarative business rules that can be processed by a rules engine using logical inference. A declarative rule is an English-like statement with the basic structure: when <object.property value> is needed and if <condition> is true, then perform <actions>. The <actions> may include the computational algorithm for deriving the value of the object.property to which the method is attached (Ross, 1997).

If the object model represents entities in the economic domain, and the methods embody knowledge about the dynamic behavior of these entities, then the object model becomes a software model of the economic domain – in other words, a virtual economy. Because the interactions between software objects can mimic the interactions between real entities in the economic domain, the virtual economy can simulate accurately the operations of an actual economy.

A comparison of alternative representation tools and how they support user needs is shown in Table 3.

Table 3. Comparison of alternative representation tools.

	Account table	Algebraic equation	Virtual economy
Visualization	No	No	Yes
Observation	Yes	Yes	Yes
Simulation	No	Yes	Yes
Implementation	No	No	Yes

An account table maps the economic domain onto credit/debit entries and beginning/ending balances of the table. It is an effective tool for implementing identification, classification, and measurement procedures. It lacks, however, a graphical notation for visualizing the economic domain. Moreover, because an account table is designed primarily to handle additive composition and decomposition of data values based on definitional relationships, it is inadequate for deriving changes in data values based on complex, nonadditive behavioural relationships.

An algebraic equation maps the economic domain onto variables of the equation. The emphasis is on the formal and precise computational procedures for deriving data values and changes in these values over time. The algebraic equation surpasses the account table by providing computational structures for complex mathematical relationships. The formal, precise, and standard (but often esoteric) mathematical notation, however, is incapable of providing direct and unambiguous mappings onto the economic domain. Although it can give users a glimpse of what would be, it fails to help users to visualize what is and to effect what should be in the economic domain. Boland (1991) discusses the pitfalls in interpreting mathematical functions as explanations.

A virtual economy maps the economic domain onto a framework of software objects that can be distributed over the Web. The emphasis is on a shared framework for visualizing, observing, simulating, and implementing a virtual economy using standard Web browsers. The virtual economy is the only representation tool that provides a graphical notation for visualizing a one-to-one mapping from real domain objects onto the software objects representing them. This same notation helps users to navigate from the abstract world of the virtual economy back to the concrete world of the real economy.

4. The EcoKnowMICS virtual economy

The previous section introduced the concept of a virtual economy and showed how it makes a shared understanding of economic data possible by focusing on the domain objects that generate the data. This section discusses the architecture of the virtual economy of EcoKnowMICS (Economic Knowledge Management, Integration, and Communication System). Building — and using — this virtual economy requires the expertise of different types of users, each of whom may have different concerns and speak different "languages" to express these concerns. Without a common communication tool to unify builders and users, cooperation can be difficult. The primary motivation behind EcoKnowMICS is to support collaborative work on the Web aimed at building, using, and maintaining a virtual economy.

4.1. The domain views: ROADMAP

Economic data can play different roles in different contexts. The contexts vary with the perspective or view of the user. Different users have different views of the economic domain. EcoKnowMICS supports seven types of domain views: Registration, Observation, Abstraction, Description, Manipulation, Application, and Prescription (ROADMAP). Figure 2 shows the types of users and the corresponding views supported by the layered software architecture of the virtual economy.

A data role is a software object that contains domain data that it presents within the context of a particular domain view. A domain view is a software object that contains the data roles that represents a particular user perspective. A data role has business rules (script) that specify its performance within the domain view. This involves deriving the data for the role by applying its business rules to the data of other roles within one — or across different — domain views.

The SNA is an example of a domain view in the Description layer (United Nations, 1993). All data in the SNA are interpreted within the context of this domain view. The consumption expenditure account item, representing the outflow of funds from the household sector in exchange for the inflow of consumption goods, is an example of a data role within the SNA domain view. Using standard accounting procedures (business rules), this data role derives the dollar amount of consumption expenditure (for a given location and time period) and presents it to users as data.

The valid use of a data role within a domain view layer is specified by its function. The valid source of a data role is specified by its form, which can be mapped onto the function of another data role in the next view layer down (Figure 3). This interlayer mapping between the form and function of data roles provides the computational linkage between view layers, and the visual mechanism for tracing the sources and uses of data across view layers.

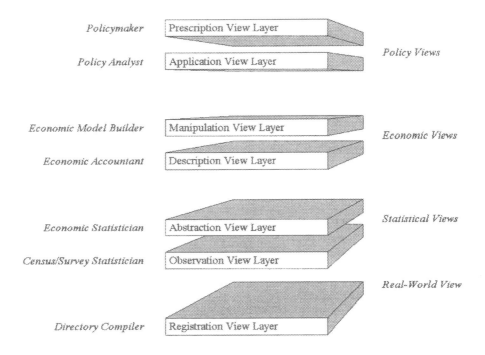

Figure 2. Users and view layers of the virtual economy.

The layered architecture provides maximum flexibility to the virtual economy. Each domain view layer is completely "pluggable", replaceable, and reusable. Builders and users can build customized libraries of domain view layers to suit particular needs, as shown in Figure 4.

4.1.1. Registration view layer. This layer supports the domain view of directory and registry compilers concerned with raw legal and administrative data from real-world entities embedded in the physical, social, cultural, and legal fabric of the economy. Compilers identify, locate, and register households, firms, and government institutions. These economic agents are represented in the virtual economy by data roles called Registered Units, which are created and destroyed in accordance with the births and deaths of their real-world counterparts. The Registered Unit data roles are listed in a directory, and are used primarily to prepare population listings and to archive data from censuses, surveys, and case studies. Fergie (1986) discusses the need for central register systems to support economic accounting.

4.1.2. Observation view layer. This layer supports the domain view of census and survey statisticians concerned with statistically designed data derived from raw legal and administrative data. Every economic agent in a census, survey, or case study is represented by an Observed Unit data role, which derives its data from a corresponding Registered Unit data role.

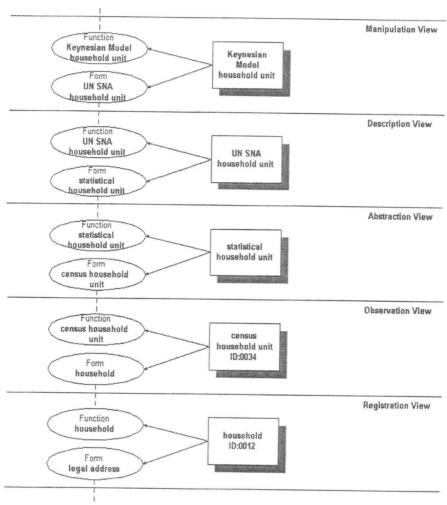

Figure 3. Data flow across view layers.

4.1.3. Abstraction view layer.

This layer supports the domain view of economic statisticians concerned with summary data distilled from individual sample data and other observation data, and used to describe statistical artefacts. Conceptually, these artefacts are abstract homogeneous entities created to represent the three postulated types of economic agents: ownership-control unit, enterprise, and establishment. A Statistical Unit data role is created for each artefact. Ownership-control units are concerned with equity and are involved in power relationships. Enterprises are concerned with liquidity and are involved in financing relationships. Establishments are concerned with productivity and are involved in production relationships. An ownership-control unit may have one or more component enterprises. An enterprise may belong to only one ownership-control unit, but may have one or more component establishments; however, an establishment may belong to only one enterprise (see Bloem, 1990, for a detailed discussion of statistical units).

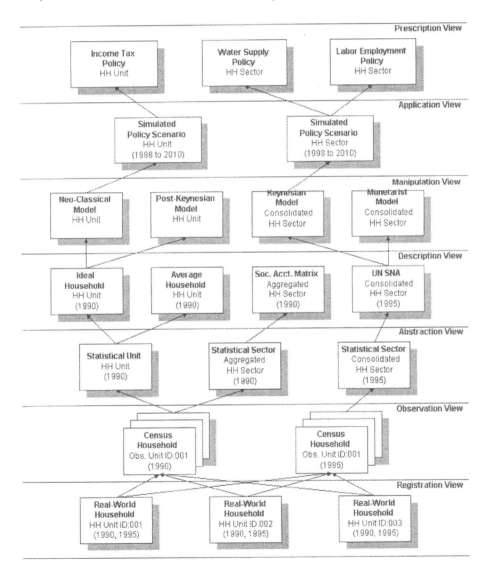

Figure 4. Libraries of reusable domain view layers.

4.1.4. Description view layer. Explanations derived from economic laws and theories are applicable only to descriptions that are generalized from specific observations. The generalizations to be explained are the changing states of economic agents (micro-units and macrosectors) engaged in activities that generate resource flows from stocks. This layer supports the domain view of economic accountants concerned with the consistent and integrated recording of economic stocks and flows of agents (e.g. the national income accounts, balance of payments accounts, input-output tables, and others). An Accounting Micro-Unit data role is created to represent an economic agent. Multiples of these can be created to serve as a representative sample of all agents belonging either to a specific macroeconomic sector or to the whole economy. Ruggles and Ruggles (1975) originally proposed a similar approach. Each Accounting Micro-Unit data role maps onto its corresponding Statistical Unit data role to provide a conceptually consistent Description view of its behavior across time (time-series). The grouping of Accounting Micro-Units into Accounting Macro-Sectors corresponds to the grouping of Statistical Units into Statistical Sectors during particular time periods. The grouping is valid only for the specified time periods.

4.1.5. Manipulation view layer. This layer supports the domain view of economic model builders concerned with the cause and effect linkages that explain and predict the interactions between economic agents. It also provides the mechanisms for generating alternative scenarios from these interactions under different policy assumptions. A Model Micro-Unit is created for every economic agent whose behavior is to be simulated. Similarly, a Model Macro-Sector is created to simulate the behavior of an economic sector. The behavioural parameters of the simulation model are estimated from the time-series data (past behaviour) of the corresponding Description data roles (e.g. Accounting Micro-Unit and Accounting Macro-Sector).

4.1.6. Application view layer. This layer supports the domain view of policy analysts concerned with plausible outcomes (scenarios), the costs likely to be incurred under each scenario, and the distribution of benefits to recipients and burdens to providers before and after policy implementations. Analysts use these scenarios to evaluate policy implications. A Simulated Micro-Unit data role is created to store the various scenarios generated by its corresponding Model Micro-Unit data role under different initial and run-time conditions. Similarly, a Simulated Macro-Sector data role is maintained to store the scenarios generated by its corresponding Model Macro-Sector data role.

4.1.7. Prescription view layer. This layer supports the domain view of policymakers concerned with policy choices, priorities and trade-offs. Activities supported by this layer include the identification of policy problems, evaluation of policy outcomes, and search for the optimum policy solution. A Policy Micro-Unit data role is created to represent each economic agent that is relevant to a Policy Model domain view. To compare and evaluate alternative policy scenarios, a Policy Micro-Unit data role is mapped onto a Simulated Micro-Unit data role, which provides the scenarios generated by its corresponding Model Micro-Unit data role. Similarly, a Policy Macro-Sector data role, representing an economic sector, is created and mapped onto a Simulated Macro-Sector data role. Moss (1980) discusses an information strategy for presenting and analyzing welfare outcomes using accounts and behavioural models.

4.2. Global views: representation and computation

The classification categories (classes) of data roles specified in the local domain views (ROADMAP) represent specific user perspectives. The data roles are classified also into categories specified in two global domain views: representation and computation.

4.2.1. Representation roles. A representation role indicates the role of data in a domain model. The agents in the economic domain are represented in the virtual economy by domain models that involve transactors holding stocks of resources. These transactors interact with each other through activities that generate flows to and from stocks at specific times and places. EcoKnowMICS uses three types of domain models to represent three types of domain data: FAST (Flow, Activity, Stock, Transactor) — for operations data, TIME (Time Initiated Model of Events) — for time-reference data, and SPACE (Spatial and Physical-Arrangement Classification of Entities) — for space-reference data.

FAST data roles contain domain operations data on flows, activities, stocks, and transactors. A flow data role contains data on resource movement between a transactor and an activity during a period of time (e.g. consumption expenditure of the household sector in the goods market for one year). An activity data role contains data on the mechanism that regulates a resource movement (e.g., a goods market). A stock data role contains data on resource holdings of a transactor at a point in time (for example, the financial assets of the household sector at the end of last year). A transactor data role contains data on agents that control resource holdings and generate resource movements (for example, household sector). *StockFlowResource* data roles contain descriptions of the types of resource holdings (stock) and resource movements (flow).

TIME data roles contain domain time-reference data that enable users to organize domain operations data according to time-initiated events. A *TimeUnit* data role contains data identifying a time-period of standard duration (e.g. year 1996). A *TimePath* data role contains data identifying a time period represented by a set of successive *TimeUnit* data roles of standard duration (e.g. months Jan1980 to Dec1996, years 1980 to 1996). A *TimeLine* data role is a built-in *TimePath* data role, containing data that identifies all calendar days within a system-specified reference time period (e.g. days 01Jan1901 to 31Dec2000). A *TimeFrame* data role contains data that identifies a time period represented by one (or more) *TimePath* data roles that in turn identify one or more different time periods.

SPACE data roles contain domain space-reference data that enable users to organize domain operations data according to the type of spatial location and the type of spatial relationships between locations (e.g. contains, part of, overlaps). A *PointSite* data role contains data identifying a point location. A *HouseholdSite* data role contains data identifying the location of a household. An *EstablishmentSite* data role contains data identifying the location of an establishment. A *FacilityPointSite* data role contains data identifying the location of a facility, such as an oil well. A *LineSite* data role contains data identifying a linear location. A *RoadwaySite* data role contains data identifying the location of a roadway. A *WaterwaySite* data role contains data identifying the location of a waterway. A *FacilityLineSite* data role contains data identifying the location of a facility line, such as an oil pipeline. An *AreaSite* data role contains data identifying an area location. A *GeoPoliticalSite* data role contains data identifying the location of the territory of a geopolitical unit. A *CountrySite* data role contains data identifying the location of the territory of a country. A *CitySite* data role contains data identifying the location of the territory of a city. A *ProvinceSite* data role contains data identifying the location of the territory of a province. A *Municipal-itySite* data role contains data identifying the location of the territory of a municipal-ity. A *BarangaySite* data role contains data identifying the location of the territory of a barangay (village). An *OtherAreaSite* data role contains data identifying the location of the territory of a nongeopolitical unit. A *RuralUrbanSite* data role contains data identifying the location of a territory classified as either urban or rural. A *FacilityAr-eaSite* data role contains data identifying the location of the territory occupied by a facility, such as an oil field. A *PointSite* may be part of a *LineSite*, which in turn may be part of an *AreaSite*. An *AreaSite* may *contain PointSites, LineSites,* and other *Ar-eaSites*. An *AreaSite* may also overlap *LineSites* or other *AreaSites*.

4.2.2. Computation roles. A computation role indicates the role of data in a compu-tation model that enables users to distinguish between raw data and computed data, and trace how data are computed within the virtual economy. An initialized data role contains data obtained directly from either a user or another data role. A postulated data role contains exogenous data obtained directly from a user (e.g. user-supplied assumptions in a simulation). A mapped data role contains exogenous data obtained directly from a data role in another view (e.g. national account production data copied from data estimated from a survey-of-establishments sample data).

An derived data role contains data derived from either analytical or empirical proce-dures. An analytical data role contains data derived from a purpose-specific analytical relationship. A smoothed data role contains data derived from the elimination of un-wanted variations in a time series. A spliced data role contains data derived from joining two or more time series. An interpolated data role contains data derived from outlying data in a time series. An extrapolated data role contains data derived from preceding data in a time series. A converted data role contains data derived from the application of a conversion factor (e.g. cost derived from quantity using unit price as conversion factor). An estimated data role contains data derived from a representative sample data (e.g. total household sector income estimated from sample household incomes). A composed data role contains data derived from combining component values. An aggregated data role contains data derived from the gross value of a com-bination (e.g. household sector payments, including inter-household payments). A consolidated data role contains data derived from the net value of a combination (e.g. household sector payments) to other sectors (excluding inter-household payments). A Decomposed data role contains data derived as a residual from splitting a combined value into its component values. A disaggregated data role contains data derived as a residual component of an aggregated value. A deconsolidated data role contains data derived as a residual component of a consolidated value.

An empirical data role contains data from an empirically derived behavioural rela-tionship (e.g. employment as a function of wages and interest rate). A revalued data role contains data that are observable in the domain, but are adjusted for accuracy of coverage (e.g. revalued household sector financial assets). A reconciled data role contains data that are observable in the domain, but are adjusted for consistency of coverage (e.g. adjusted household sector income). An imputed data role contains data that are not observable in the domain, but are estimated for completeness of coverage (e.g. services of owner occupied housing imputed as additional income of the house-hold). An attributed data role contains data that are observable in the domain, but are redefined and reused for another purpose (e.g. employer's pension fund contribution as a component of employee income). A proxy data role contains data that are esti-mated from observable data in the domain and then defined to correspond to non-ob-servable theoretical concepts (e.g. gross domestic product).

Together, the data roles, domain views, and business rules constitute a virtual economy, with the business rules to provide the codified economic knowledge un-derlying its behaviour, consistent with specified constraints, policies, and procedures.

4.3. Graphical notation

EcoKnowMICS provides a visual modeling tool using a graphical notation imple-mented in standard Web browsers. It can present data roles as standardized graphical objects that users can visually combine and manipulate as building blocks of a shared virtual economy. This enables geographically dispersed users to interact simultane-ously through standard and easily accessible Web facilities.

Graphical images and diagrams play an important role in explaining complex rela-tionships (Bradford, 1997). The use of diagrams for explaining relationships, as well as for rigourous deductive reasoning, is explored in Hammer (1995). The following diagrams illustrate the basics of EcoKnowMICS graphical notation. Actual use of these diagrams is illustrated in the case study.

In the simplest case, the cotransactor is not relevant, and the diagram shows the re-source flow exchange activity as both sink and source (Figure 5). The diagram does not have to show the destination of the primary flow and the origin of the reciprocal flow. In the example, the transactor is a bakery, which exchanges its bread stocks for money stocks through an exchange activity (i.e. the final consumption market).

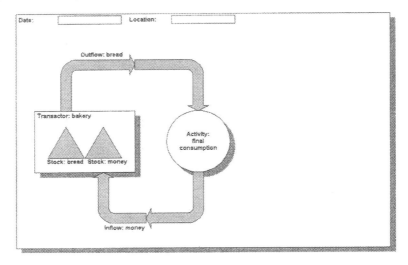

Figure 5. Exchange activity as sink and source.

The next diagram is used if both transactor and cotransactor are relevant but only the primary flow is required, or to show nonreciprocal transfer of resources (Figure 6). Figure 7, on the other hand, shows the primary and reciprocal flows, and their origins and destinations (i.e. the transactor and cotransactor). The operations of the economic domain can be represented completely by this basic set of diagrams.

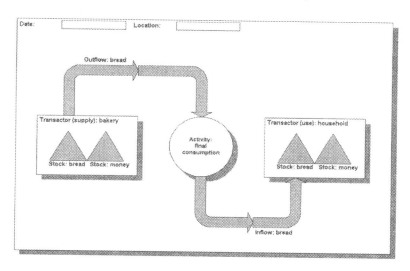

Figure 6. Nonreciprocal activity.

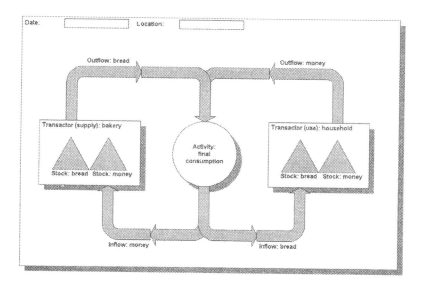

Figure 7. Exchange activity with origins and destinations of flows.

5. Case study: support system for water supply sector investment planning

5.1. Background: multiple decision-making contexts in decentralized planning

In the private business sector, some large enterprises own establishment units distributed across wide geographical areas, with each unit operating as an independent profit center. Although decision-makers at the various establishments are autonomous, they all share a common perspective embodied in an enterprise-wide business model, and they all follow the same enterprise-wide business rules governing the generation and use of corporate data. In this context, enterprise-wide DSSs are indeed feasible, and are actually widely used.

The geopolitical reality, however, is far from this ideal corporate reality. The raw data used in planning are seldom generated for specific decision-making purposes, but are by-products of the day-to-day administrative operations of government agencies. Countrywide, intersector, and interagency models and rules are necessary to specify and unify the interpretation and use of raw administrative data; however, this has yet to be recognized and appreciated by most government agencies.

Although the reason for being of most planning data rarely corresponds to the reason for using the data, planners rely on them because it makes more economic sense to reuse existing data than to undertake customized data gathering and processing for every decision-making activity. Every bit of data generated has a cost, and every time data are reused, their value is enhanced in relation to the fixed cost of generating them.

Reusing data without understanding it properly, however, can be more dangerous than having no data at all. Without data, perhaps no decisions can be made, but with the wrong data, harmful decisions can be made. Any support system should deal with the existing data first and provide users with the tools to get the most out of them. Perhaps the most urgent need is for a tool to help users understand the data, why they were generated (purpose) and how (process).

Decentralization, along with the consequent devolution of administrative functions, has resulted in a gradual and selective transfer of responsibilities from the national government to local government units. Some of these responsibilities include sector plan preparation, implementation, and monitoring. In a decentralized setting, planners at both the national and local levels are faced with problems unique to distributed decision-making, data gathering, and data processing. The preparation and implementation of a water supply sector plan, for instance, involves the decisions of   different entities at different times and locations. This case study focuses on the system support requirements of decentralized planning as perceived by local geopolitical units, i.e. the provinces, the municipalities within provinces, and the barangays (villages) within municipalities.

The provincial government is responsible for preparing a sector plan for each conventionally defined developmental sector (e.g. water supply and sanitation, health, housing, education, etc.). At the initial planning stage, the provincial governments decide how to prioritize the needs of the different sectors, and the needs of the constituent municipalities and barangays within each sector. During budget deliberations, the national government decides how to allocate investment funds to the provinces based on its own criteria. The Department of Public Works and Highways (DPWH), as a line agency, decides on the disbursement schedule of allocated funds and on the type, location, and installation schedule of the water supply facilities to be provided to the provinces.

Often, the preparation and implementation of a sector plan is packaged as a project to be partly financed by an external funding agency (such as the World Bank [WB] or the Asian Development Bank [AsDB], etc.), either through a loan or grant agreement with the national government. Although this type of project is nominally managed by a line agency, the major decisions – from the initial planning to the implementation – are influenced by recommendations made by outside consultants hired by the agency responsible for the loan or grant. Naturally, these operational decisions must comply with guidelines set by the lending agencies, which generally encourage the full use of the loaned funds in the shortest time possible. These guidelines, however, may conflict with the desire of provincial governments to align their rate of withdrawal with their limited capacity to absorb and make effective use of the borrowed funds.

Even though all these decisions are about the same sector plan and the same plan domain, they are made from vastly different perspectives. From the above discussion, we can identify four: provincial government, national government, line agency and funding agency. Each perspective embodies a specific purpose that determines what data are needed and in what form. Building an effective DSS for sector planning is feasible only if the distribution of data gathering and processing are congruent with the distribution of the data needs arising from these various decision-making perspectives.

To determine the demand for water, for instance, the official number of persons residing in a geopolitical unit can be obtained from the regular population censuses conducted by the National Statistics Office (NSO) for mandated time intervals (every five or ten years). The population counts for intervening years, however, have to be estimated. To determine the supply of water, an inventory of water supply facilities that were actually installed in the province can be obtained from the DPWH engineering districts that installed them. The number of facilities that are actually operational, however, has to be estimated.

For both supply and demand, the purpose for which the data are being compiled heavily influences how they are estimated. To improve its share of investment funds, a province tends to be liberal in its population growth estimates, and conservative in its estimates of the number of available water supply facilities. On the other hand, the local DPWH engineering district projects a favourable image if it has installed a large number of facilities and maintained them in good operating condition. It is not uncommon to find that values for basic planning data, like population or the number of operational deep wells in a province can differ, depending on the source.

Similarly, to prioritize the need for water among competing areas within a province, provincial planners have to estimate the relative prevalence of water-borne diseases in these areas, using statistical indicators generated from data gathered and collated by the Department of Health. The incidence of diarrhea is an example of health-related data that may be used in generating planning indicators. The operational definition of what constitutes a case of diarrhea may differ, however, between the agency that recorded the original raw data, the agency that defined and computed the indicator, and the provincial planner, the eventual user of the indicator. In the realm of planning data, what you see may not be exactly what you will get.

5.2. The planning support problem: managing data, information, and knowledge

A common concern among sector planners has been the scarcity of timely and reliable data. To address this concern requires viewing data as resources for input to the decision-making process, and recognizing the need to manage them.

Planning involves the following data-intensive activities:

- *describing* the actual situation in the sector (historical statistical data);

- *predicting* alternative future situations in the sector (model simulation data); and

- *prescribing* a desired situation in the sector (plan implementation data).

The first activity deals with statistical measurements describing the actual past and present conditions in the sector. The second activity involves using a model of the sector to generate probable sector scenarios in the future. The third activity involves analyzing and interpreting descriptive statistical and predictive simulation data before prescribing a plan of action to attain a desired outcome for the sector. In all these activities, the timeliness, accuracy, and reliability of raw and derived data determine the effectiveness of the decisions made.

Even in an ideal world of perfectly accurate statistical measurements, the value of statistical data to decision-makers diminishes when data coming from different sources are inconsistent because of differences in measurement procedures, or incompatible because of differences in the concepts underlying their definitions. It diminishes also when data for a critical variable are missing for some of the years in the plan time frame, or incomplete for the set of geopolitical units in a plan domain. An overabundance of data, on the other hand, can be a curse, rather than a blessing, if decision-makers lack the filtering and focusing mechanisms to cull only relevant data. Finally, as conditions change in the sector, failure to keep data current eventually results in obsolescence. Creating and enhancing the value of data is at the core of the data management problem facing decision-makers.

Processing data and having access to it are two different things. Information comes from the interpretation of data; useful information, however, comes only from the right interpretation of the right data by the right person at the right time. Statistical data recorded on paper and stored by different institutions in different filing facilities are difficult to locate, retrieve, and combine for analysis and interpretation; when data is organized and stored systematically in easily accessible electronic media, however, these tasks are simplified. When a decision-maker is conducting an interprovincial comparison of investment plans, for instance, data access and presentation are of crucial importance. Clearly, how the data are organized, stored, and presented for interpretation will determine to a large extent the efficiency of retrieval that is possible and, consequently, the quality of interpretation that is attainable. The commonly perceived inadequacy in the quantity and quality of decision-support information is actually a consequence of a festering information management problem.

Closely related to data and information management problems is the inadequacy of development planning expertise at the local levels of government. The result is that high-quality information remains unrecognized, unappreciated, and unused. Understandably, the level of expertise at the national level is higher than that at the local levels. Replicating this scarce planning expertise at the national level and distributing it to the local levels, where it is needed most, is part of an emerging knowledge management problem.

The severity of these problems was highlighted during the preparation of provincial master plans for the water supply and sanitation sector. The WB requires the preparation of provincial master plans as a prerequisite to the implementation of water supply and sanitation facilities under the first water supply, sewerage, and sanitation sector project (WSSS). To assist the provinces receiving WB assistance, the Danish International Development Agency (DANIDA), the United Nations Development Programme (UNDP), the Japanese International Cooperation Agency (JICA), and the DPWH provided funds to hire outside consultants to help the provincial planning and development officers to prepare the master plans.

This lack of local technical expertise meant that consultants often had to be responsible for more than one of the provincial plans at a time. Each consultant, however, usually had his or her own view of how a plan should be prepared and presented. Consequently, the resulting plans differed in form and content.

These differences would not have mattered much if the plans had been prepared, used only once, and then discarded. Planners at both the local and national levels, however, need to update and compare plans across time (time-series analysis) and across provinces (cross-section analysis). Time-series analyses are used for implementation activities at the provincial level, while cross-section analyses help set priorities for provincial needs at the national level. The generally accepted plan-preparation procedures, with their multiple, consultant-based planning methodologies and data presentation formats, made it difficult for planners to analyze, compare, and implement provincial master plans.

Moreover, it soon became apparent that without proper technology transfer, local governments were becoming increasingly dependent on outside consultants and national agencies such as the Department of the Interior and Local Government (DILG). This dependency inhibited initiatives to develop local expertise in planning and data management.

To break the stranglehold of dependency, the WB funded a technical assistance project with the DILG aimed at developing a database approach to the preparation of the master plans in January 1995. The objective was to build a provincial planning database with standard input data requirements, planning rules, and output report formats. This database, along with appropriate computer hardware, software, and training, would be distributed to the provincial governments for use in generating standard, and therefore comparable master plans. The project emphasized the need to institutionalize the operations of the database to ensure its sustainability. Three pilot provinces (Benguet, Pangasinan, and Palawan) participated in the project.

Earlier failures with the database approach indicated that the database would be institutionally unsustainable if its support function was narrowly focused on plan preparation and comparison. The urgency and frequency of plan preparation and comparison (at most once every four or five years) was insufficient to justify the need for regular, and necessarily local, "care and feeding" of the database. A more sustainable function was to provide generic support for institutional day-to-day decision-making activities, of which plan preparation and comparison were only two among many. In other words, the database had to be deeply embedded in the geo-political institutional fabric, and not be viewed simply as a tool for *ad hoc* uses.

One implication of this shift in functional focus was a widening of the range of potential users of the database. Initially, the primary target users were national and provincial planners and policymakers, who, with the help of the database and standard planning rules, would be able to generate standard plans for any local government level and allocate investment funds. With the shift in focus, the range of users expanded to include almost everybody in sector development from the national government to the barangay level.

Another implication was greater flexibility in the design and use of the database. Earlier assumptions about the need for adequate levels of computer literacy among planners and policymakers gave way to the more pragmatic need to design for the lowest level of expertise expected among potential users. Flexibility and accessibility of design thus became more important design criteria than optimum performance.

Consequently, the traditional monolithic design of planning databases, wherein data storage, retrieval, manipulation, and presentation were tightly integrated in a single system for maximum efficiency, was deemed to be inappropriate. Rather than providing one complex tool, the database developers decided to produce a "toolkit" of simple, stand-alone, but complementary tools that would give users the flexibility to choose a specific tool, or combination of tools, that fit their particular purposes and expertise. The tools included database, spreadsheet, and geographic information system (GIS) programs for data retrieval, data analysis, and thematic mapping. They were collectively referred to as a planning support system.

A third implication was the need for transparency of the support system. Although complex and rigid, one advantage of a monolithic support system is that users can perform a fixed number of well-defined functions quite well, without needing to understand the internal mechanisms of the system. With a toolkit, users have more flexibility, but they can perform poorly if they have an inadequate understanding of the individual tools and their interactions. In the latter case, it is the responsibility of the user to select the appropriate tools and to coordinate their use.

The system design choice was between concentrating the intelligence to perform a predetermined complex task in a monolithic system and distributing the intelligence to many simple tasks among many users supported by many simple, stand-alone tools, and using a framework to combine the simple outputs to achieve a complex outcome.

A sector master plan is a highly complex document generated from many simple tasks, such as counting the actual number of households and water supply facilities, projecting these counts into the future, determining the shortfall in service facilities, preparing reports, analyzing reports, and others. Basic spreadsheets or databases can support many of these tasks. These simple tasks, however, would have to be integrated within a well-defined and well-understood framework that combines the outputs from each task to produce a final, more complex product.

Concern about system maintenance was a major factor influencing the decision to adopt the framework and toolkit approach. Because of the ongoing decentralization, devolution of government functions, and standardization of planning methodologies, it was considered premature and inappropriate to build a complex monolithic support system. Rapidly evolving user requirements could have proven fatal to complex systems that required a high level of technical expertise to maintain and modify. With a graphical and easily understandable framework linking the simple tasks and showing how they combine into complex tasks, users could enjoy long-term benefits from a simple, flexible, and maintainable toolkit-based support system.

5.3. The EcoKnowMICS Solution: "Visual Planning"

Recognizing the planning-support problems of local governments, the International Development Research Centre (IDRC) funded a project to adapt the EcoKnowMICS framework to the requirements of provincial planners.

This framework evolved from an earlier IDRC project in 1988, which involved a domain analysis of the economic system using the then newly-emerging object-oriented and expert-system technologies. The primary motivation behind this project was the need for an alternative graphical representation of the economic system to provide users with visual traces of data sources and data derivation procedures. The main project activities were to:

1.   identify and catalogue generic economic objects and their interrelationships, and

2.   implement them as domain object models and rules in an expert system.

Both the object models and the rules relied heavily on graphics for visual presentations to users.

In adapting the EcoKnowMICS framework to sector planning, the object models served not only as communication tools, to help users reconcile and unify the various decision-making perspectives in the planning process, but also as cognitive tools, to help users understand the underlying planning methodology. By using an explicitly graphical notation to represent the structure and processes of the economic system, the models proved to be more intuitive and understandable to nontechnical users than the conventional economic accounting models with their implied accounting rules. This case study shows how the IDRC-funded adaptation of the EcoKnowMICS framework complemented the work done in the WB technical assistance project with the DILG.

The WSSS plans generated over the past years differed not only superficially in the presentation format, but also substantially in the planning procedure or methodology. From the national perspective, which is concerned mainly with interprovincial comparisons, the present WSSS plan preparation process had to be rationalized to ensure valid comparisons. To rationalize (or re-engineer) the process required a level of technical expertise which was not readily available, either at the national or provincial levels. Without this expertise, the benefits from rationalization would be difficult to obtain and sustain. The EcoKnowMICS framework provided the initial guidelines to rationalize the planning methodology.

The first stage of the process was to arrive at a consensus on what data the plans should present to allow for quick and accurate comparisons of provinces. The purpose of these comparisons was to determine relative needs, which could guide the allocation of investments among the provinces. Because comparisons implied shared interpretations of data, the building of a data dictionary – incorporating standard definitions of data items and classification categories – was a necessary first step.

The data dictionary linked the plan data (raw and derived) to the planning procedures (methodology). It listed the operational meaning for each item of data, defining how it was derived from other data items, or how it could be observed directly. Moreover, it showed which data items had the same names or conceptual definitions, but different operational meanings, and which had different names and conceptual definitions, but the same operational meanings. Examples of entries in the data dictionary are shown in Table 4.

Table 4. Water supply planning data dictionary.

Type of water supply facility	Conceptual definition	Operational definition
Well	Level-1 water supply facility, drilled or dug into the ground, with lining material, and used to retrieve groundwater.	A drilled or dug hole that provides water from beneath the surface of the ground, with or without the use of a pump.
Deep well	Drilled well (with pump), to a depth of 20 m or more.	Well with Magsayasay or Malawi pump.
Shallow well	Drilled well (with pump) to a depth of 20 m or less.	Well with pitcher pump.
Data item		
DwStk.count	Inventory of stock of deep wells.	The number of deep wells in barangay San Jose as of end of 1994 as counted by the barangay head.
SwStk.count	Inventory of stock of shallow wells.	The number of shallow wells in barangay San Jose as of end of 1994 as counted by the barangay head.
WellStk.count	Inventory of stock of all wells.	=dwStk.count + swStk.count.

During the next stage of the rationalization process, a standard planning methodology was culled from the best features of existing methodologies. It was subjected to validation by users and tested for internal consistency. To facilitate validation and testing, the methodology was codified as formal rules stored in a rule base and implemented in software. Expressing the methodology formally as rules made it easy to understand and revise; packaging it in software made it easy to replicate and disseminate to different users at various locations.

This rule base approach eliminated the tedious and error-prone translation from human logic to computer logic by expressing the planning rules of a methodology in written, English-like statements, intelligible to both people and computers. The rules were more coherent than the macrostatements in the Lotus and Excel spreadsheet programs typically used by consultants in training provincial planners (see Ross, 1997, for a description of business rules). They helped users understand the methodology by making its logical structure and (usually) hidden and implied assumptions explicit. Improved understanding eased the burden on users to evaluate and reconcile the various methodologies currently in use. Moreover, the object models (representing the plan domain) and the rules (representing a planning methodology) together facilitated the training of provincial planners by applying a generic formal planning methodology to the particular conditions of their province.

To minimize the loss of information due to gross averaging of data at the provincial level, planners prepared two types of plans: an individual plan and a consolidated plan. The underserved population (shortfall) was determined for each individual plan domain. An individual plan domain could be a barangay or a sub-municipality area, either all urban barangays or all rural barangays. These individual plans were consolidated for the whole province.

The purpose of this case study is simply to illustrate the potential benefits of using the EcoKnowMICS framework; therefore only procedures considered crucial to a sector plan are presented. Although in practice the same plan domain model can support all three components of a sector plan (i.e. investment, implementation, and monitoring), this case study covers only the preparation of the investment plan. Moreover, it focuses only on the rules for determining the underserved population for a level-1 type of water supply facility.

The main transactors in the economic domain are shown in the whole-part hierarchy (Figure 8).

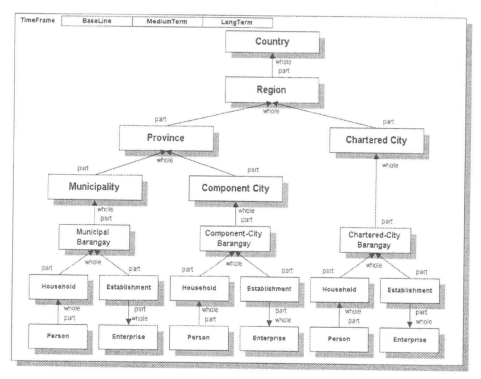

Figure 8. FAST object model: geopolitical, business, and household transactors.

A province is made up of municipalities and component cities, which are in turn made up of barangays. Provinces and chartered cities comprise a region, and all regions together constitute the whole country. WSSS investment plans are prepared for each barangay, which is the lowest-level geopolitical unit. This means that the underserved population is determined for individual barangays. The barangay plans are then consolidated at the provincial level. The investment funds from the national government are then allocated based on a prioritization of provincial needs.

The stocks that play a part in determining the amount of investment in a barangay are shown in Figure 8. The demand for investment will depend on the relative growth of these stocks. If the population growth (person stock) exceeds the growth in the water supply (water stock), the underserved population will increase. This will result in a demand for additional investment in water supply facilities (facility stock) subject to the constraint imposed by the availability of financing (funds stock).

The size of the stocks will vary over time as a result of incoming and outgoing flows generated by different activities. Each activity shown in Figure 9 is both a sink and a source. If outflows exceed inflows, then the stock will decrease; if inflows exceed outflows, then the stock will increase. The basic planning problem is how to forecast accurately the size of each stock.

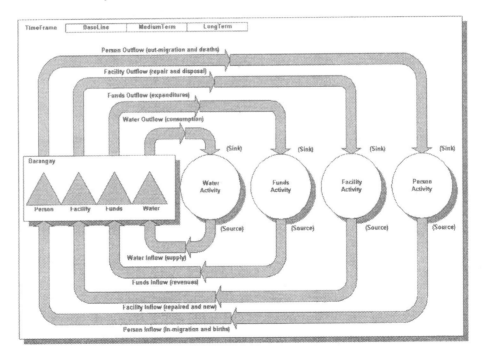

Figure 9. FAST object model: WSSS plan domain flows, activities, stocks, and transactor.

The water demand is determined by the population to be served. Vital and migration activities affect population growth. Water consumption activity determines the water demand. A fixed population with increasing per capita consumption will increase the demand for water. A growing population with a fixed per capita consumption also will increase water demand.

The water supply is determined by the stock of facilities. The types and subtypes of activities generate the inflows and outflows of facilities in a barangay. The facility maintenance activity generates the inflow and outflow of facilities that determine the facility stock at any point in time. The facility stock represents facility-years of service life. This stock decreases with normal use over time or when facilities are pulled out of service for repair or disposal. The stock increases when repaired or new facilities are installed for service. The water supply activity generates the flow of water available for consumption.

The facility stock represents the total facility-years available for service at a point in time. The facility disposal and repair activities are sinks for outflows of facility-years in the form of physical facility units. The facility water-supply activity is a sink for outflows of facility-years in the form of wear and tear resulting from use in operations. The facility installation activity is a source of inflows of facility-years in the form of repaired and new facilities installed for service. The main constraint to maintaining a sufficient stock of facilities is the availability of funds. Funds are depleted through expenditure and other funds-outflow activities. Funds are augmented through revenue and other funds-inflow activities. The facility installation activity, a subtype of expenditure activity, affects also the facility stock.

The funds stock represents the funds available to the geopolitical unit for disbursements. The expenditure and other funds-outflow activities are sinks for outflows of funds. The revenue and other funds-inflow activities are sources of inflows of funds.

The central concern of the planning activity is the availability of water. The facility/water supply activity is concerned with the provision of water. The person/water consumption activity is concerned with the use of water for personal purposes. The water stock represents the amount of water available to the geo-political unit at a point in time. The person/water consumption activity is a sink for outflows of water.

The types and subtypes of activities shown in Figure 10 illustrate how the activities are integrated. The facility installation activity is a subtype of both expenditure activity and facility maintenance activity. The water supply activity is both a subtype of facility activity and water activity. Finally, the person/water consumption activity is a subtype of both water activity and person activity. This graph shows how the various activities together determine the water supply and demand in a geopolitical unit.

The person/water consumption activity converts the target-served population into the underserved and the fully served populations (Figure 11). If the stock of water supply facilities is sufficient, the underserved population will be zero; if insufficient, the underserved population will be greater than zero. The aim of investment planning is to minimize the underserved population.

The facility installation activity converts funds into water supply facilities. Minimizing the underserved population means maintaining a sufficient stock of water supply facilities, and a sufficient stock of funds to finance the repair of non-working facilities and the purchase of new facilities.

The whole-part hierarchies of the person and facility stocks show how these stocks are broken down into types that are meaningful to water supply planners. The person stock represents the population living within the target geopolitical unit. The target-served population is computed based on a plan-specified target percentage of the population. This population is then divided into the underserved and the fully served, based on the existing facility stocks. The water supply facility stock is one among several facility stocks in the geopolitical unit (e.g. sanitation).

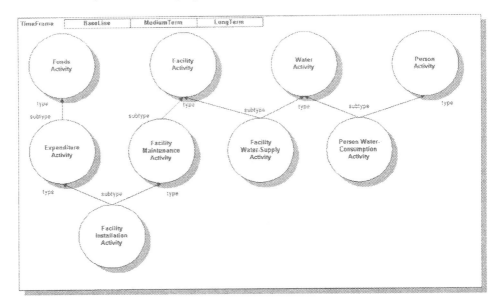

Figure 10. FAST object model: integrated plan domain activities.

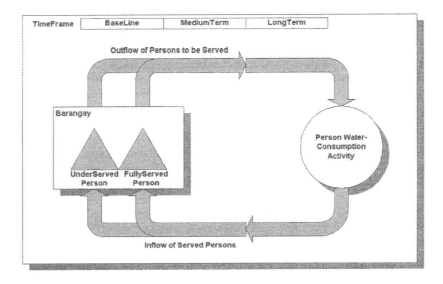

Figure 11. FAST object model: service level of water supply facilities.

The WSSS plan (prescription layer) is applicable to a generic geopolitical unit. In this case study, the geopolitical unit is the barangay (registration layer).   The ROADMAP view layers shown in Figure 12 illustrate the organization and flow of data required to generate simulated population data for a particular barangay. Census data (observation layer) from the NSO and DILG provided the time-series data for the years 1980, 1990, and 1992 (abstraction and description layers). These data were used to estimate the parameters of a population projection model (description and manipulation layers) of the barangay for the years 1993 to 2010 (application layer).

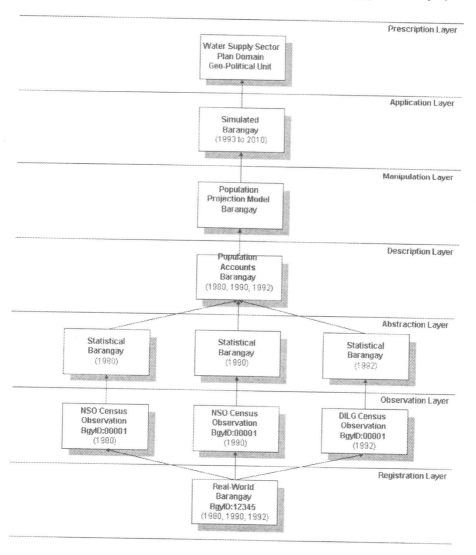

Figure 12. EcoKnowMICS ROADMAP view layers: WSSS plan domain.

5.4. Web-based plan domain object model and rule base

When the EcoKnowMICS framework was first adapted in 1995, the database was implemented in Access, Microsoft's desktop database management system. The domain object model and planning rules were implemented in Neuron Data's Nexpert Object, an object-oriented expert system shell. Access is a relational, not an object-oriented software; it contained raw planning data that could not be mapped directly to the objects in Nexpert Object, which managed the planning rules. The technology of the time allowed only a partial implementation of the tight integration offered by the framework.

There was also a serious logistical problem to be solved. Everyone involved in preparing the sector plan needed copies of the runtime versions of both the Access domain database and the Nexpert Object planning rule base. This meant that changes – whether modifications or upgrades – in any of these necessary items would have had to be replicated in all the copies that had already been distributed, to ensure the consistency and compatibility of data and rules across all planners, in all locations. The financial and administrative burden was formidable.

During the latter part of the 1990s, however, the Internet – and in particular the Web – became widely used. Java, an object-oriented programing language, became a standard programing language for the Web. Neuron Data developed a Java version of Nexpert Object called Elements Advisor (Neuron Data, 1998). The main advantage of Elements Advisor is that it can be installed on a Web server and retrieved as needed by clients anywhere on the Web. It is now fairly reasonable to assume that planners have easy access to the Web, so Elements Advisor neatly solves the problem described above.

The domain object models and planning rules can be maintained in the central Web server location so that their latest versions are available to any Web client, at any location, and at any time. Moreover, Elements Advisor's ability to connect directly to both relational and object-oriented databases using Java (IBM, 1998) enables tight integration between rules and data in a single-software solution. At this stage, the Java objects and rules are only experimental. Major funding will be needed to develop and deploy fully functional Java applets and applications on the Web. The working assumptions and definitions of the prototype are listed below.

1. The WSSS plan service area within a geopolitical domain includes the plan target-served population.

2. The plan target-served population is defined as a percentage (plan target percentage) of the total domain population.

3. Although all persons in the plan target-served population are served by existing water supply facilities, some are fully served and some are under-served (shortfall), according to official nationwide standards of water consumption.

4. The underserved population is defined in the plan as that portion (residual) of the target-served population that is not fully served by existing facilities.

5. The fully served population is estimated using a facility-based approach, in which each type of water supply facility is rated according to the maximum number of households that can be fully served by one facility of that type. For example, if

the rating for a deep well is a maximum of 15 households, and that well is serving 20 households, then it is estimated that 15 households are fully served and five households are underserved.

6. The plan has three time periods: base-line (1993), medium-term (1994-1998), and long-term (1999-2010). The plan target percentages are applied to the domain population estimated for 1998 and 2010 to determine the target-served population in those years.

7. To compute the underserved population in 1998, it is assumed that the 1993 (base line) fully served population remains constant until 1998. To compute the under-served population in 2010, it is assumed that the medium-term plan has been implemented, and that the fully served population in 1998 (which is assumed to be equal to the plan target-served population in 1998 after implementing the plan) remains constant until 2010.

8. The plan covers the three types of water-supply facilities. A level-1 facility is a point-source facility such as a deep well, a shallow well, or a developed spring. A level-2 facility is a communal system, which consists of a rudimentary distribution system connecting a water source to a public water faucet in a common area. A level-3 facility consists of a centralized storage and distribution system connected to private, individual faucets in households.

6. Concluding remarks

This paper shows how the convergence of computer and communication technologies can be harnessed to build Web applications that enhance the understanding and sharing of economic data. It requires, however, a major paradigm shift; builders and users of economic data systems must not only model the data (e.g. economic account tables), but also the domain that generates the data (e.g. virtual economy).

EcoKnowMICS provides object models, rule bases, and a graphical notation as building blocks of a Web-based virtual economy. The object models represent the structure and state of the economic domain. The rule bases codify the knowledge underlying the behavior of domain objects. As the case study demonstrates, the graphical notation helps to demystify the economic system by presenting a stylized picture of its components and their interrelationships. Finally, the Web packages and distributes the objects, rules, and data of the virtual economy. All these elements together form EcoKnowMICS, which helps decision-makers to visualize and observe what is, simulate what if, and implement what ought to be.

References

Bloem, A. M. (1990). "Units in National Accounts and the Basic System of Economic Statistics," *The Review of Income and Wealth* 36(3), 275-288.

Boland, L.A. (1991). *The Methodology of Economic Model Building*, Routledge, New York, NY, USA, 121-125.

Bradford, P. (1997). *Information Architects*, Graphis, Inc., New York, NY, USA.

Duncan, J. W., A. C. Gross (1995). *Statistics for the 21ˢᵗ Century, Proposals for Improving Statistics for Better Decision Making*, Irwin Professional Publishing, Burr Ridge, IL, USA.

Dunn, E. S. Jr. (1974). *Social Information Processing and Statistical Systems - Change and Reform*, John Wiley & Sons, Inc., New York, NY, USA.

Eisner, R. (1989). "Divergences of Measurement and Theory and Some Implications for Economic Policy," *The American Economic Review*, March 1989.

Fergie, R. (1986). "Statistical Units Standards and Central Register Systems: Keys to the Development of Economic Accounting," *The Review of Income and Wealth* 32(1),49-68.

Gale, T., J. Eldred (1996). *Getting Results With the Object-Oriented Enterprise Model*, SIGS Books, New York, NY, USA.

Hammer, E. M. (1995). *Logic and Visual Information*, Center for the Study of Language and Information, Stanford, CA, USA.

Holub, H. W., G. Tappeiner (1997). "Modeling on the Basis of Models," *The Review of Income and Wealth* 43(4), 505-510.

IBM (1998). *Visual Age for Java Enterprise 2.0 Documentation*, North York, Ontario, Canada.

Moss, M. (1980). "Social Challenges to Economic Accounting and Economic Challenges to Social Accounting," *The Review of Income and Wealth* 26(1), 1-17.

Neuron Data (1998). *Elements Advisor 2.0 Documentation*, Mountain View, CA, USA.

Orcutt, G. H. (1961). *Microanalysis of Socioeconomic Systems: A Simulation Study*, Harper & Brothers, New York, NY, USA.

Partridge, C. (1996). *Business Objects, Re-Engineering for Re-Use*, Butterworth-Heinemann, Oxford, Great Britain.

Postner, H. (1988). "Linkages Between Macro and Micro Business Accounts: Implications for Economic Measurement," *The Review of Income and Wealth* 34(3), 313-335.

Richter, J. (1994). "Use and Misuse of National Accounts From a Modeling Perspective", *The Review of Income and Wealth* 40(1), 99-110.

Ross, R. G. (1997). *The Business Rule Book, Classifying, Defining and Modeling Rules*, Database Research Group, Inc., Boston, MA, USA.

Ruggles, R., N. Ruggles (1975). "The Role of Microdata in the National Economic and Social Accounts," *The Review of Income and Wealth* 21(2), 203-216.

Thagard, P. (1988). *Computational Philosophy of Science*, MIT Press, Cambridge, MA, USA.

United Nations (1993). *System of National Accounts*, New York, NY, USA.

6 DECISION SUPPORT FOR INCENTIVE STRATEGIES
A Rural Development Application in Central Africa
Benoit Gailly and Michel Installé

1. Introduction

This chapter describes an application of a new decision support tool to select incentive strategies that use multicriteria hierarchical models. Based on an algorithm developed by the authors (Gailly, 1994), it was used, in combination with a standard interactive-multicriteria-optimization algorithm, to solve a regional development problem in Burundi.

This tool allows the analyst to deal easily with two hierarchical levels of decisions, here called "upper-level" and "lower-level", each with different preference structures and each influencing the other. The objective of this decision support system (DSS) is to help upper-level decision-makers (public authorities, "the leaders") choose the best incentive strategies, while taking into account both their own objectives and the reactions of lower-level decision-makers (farmers, "the followers") to the incentives.

This tool was implemented through a software designed to investigate the impact of various incentives strategies that could be implemented by regional public authorities to improve the rural development of a northern province in Burundi. This was done by modeling the preference structures of the private farmers involved and evaluating the various criteria that were considered by the public authorities.

2. Presentation of the decision-support tool

In this section, the mathematical model corresponding to the hierarchical multiactor and multicriteria problems considered in this paper will be presented first. Then, the proposed resolution algorithm will be described briefly, along with the required restrictive hypotheses. Finally, some information about the software used to implement the developed algorithm will be provided. For more detailed technical information on the resolution algorithm, see Gailly (1994) and Gailly *et al.* (1997).

2.1. The hierarchical multicriteria model

The model examined here is based on the following bilevel programing model P, which takes account of one upper-level decision-maker and n lower-level decision-makers:

(P) $\text{Max}_y F(x_1^*, ..., x_n^*, y)$ (1)

   s.t.: $G(x_1^*, ..., x_n^*, y) \leq 0$ (2)

          $y \in Y$ (3)

and for i = 1,..., n,

(P$_i$) $f_i(x_i^*, y) = \text{Max}_x f_i(x_i, y)$ (4)

   s.t.: $g_i(x_1, ..., x_n, y) \leq 0$ (5)

where"

   P$_i$ denotes the ith lower-level problem;

   $y \in R^m$ is the decision variable for the upper-level decision-maker (the incentive);

   $x_i \in R^{m'}$ is the decision variable for the ith lower-level decision-maker;

   $x_i^* \in R^{m'}$ is the optimal reaction for the ith lower-level decision-maker;

   $F : R^{m+nm'} \to R^p$ and $f_i : R^{m'+m} \to R^{p'}$ are upper- and lower-level objectives, respectively;

   p and p' are the number of objectives of the upper- and the lower-level decision-makers, respectively;

   $G : R^{m+nm'} \to R^q$ and $g_i : R^{m+nm'} \to R^{q'}$ are upper- and lower-level constraints, respectively;

   $Y \subseteq R^m$ is the set of feasible upper-level decision variables; and all functions are assumed to be continuous.

Hence, in this problem, the lower-level decision-makers choose the value of their decision variables (the reactions x_i^*) according to their objective and constraints (f and g), in a way that is influenced by the upper-level decision variables (the incentive y). Upper-level decision-makers want to find the best way to influence them in terms of their own objectives and constraints (F and G). Note that this problem is implicit (equations 4 and 5) and therefore cannot be solved through the use of standard optimization techniques.

Several authors already have considered hierarchical problems, which mainly have to do with the so-called "Stackelberg theory game" (Nijkamp and Rietveld, 1981; Sherali *et al.*, 1983; Bialas and Karwan, 1984; Bard and Moore, 1990). The bilevel model we introduce here differs from those previously presented (Simaan, 1977; Cruz, 1978; Aiyoshi and Shimizu, 1981; Nijkamp and Rietveld, 1981; Bard, 1983) by allowing

1. the user to consider decision models with multiple, nonaggregated upper-level objectives;

2. a new set of solutions to be generated on-line at each iteration according to successive updated versions of the problem; and

3. distinctive upper- and lower-level constraints to be considered simultaneously.

2.2. General resolution of the problem: hypotheses and basic ideas

The basic idea of the method is to decompose the resolution process into two phases. During the first, the initial implicit problem is converted into an equivalent explicit one, where the optimal reactions x_i^* are defined by explicit relations instead of implicit ones. During the second, given these explicit relations between the incentives and the reactions, any standard multicriteria-optimization procedure may be performed on the explicit model.

In order to be able to perform these two phases (explicitation and optimization), the following restrictive hypotheses are needed:

(H1) the objectives f_i of each follower may be reduced to a single objective through, for example, a weighted summation method (Vincke, 1989), and this objective is supposed convex in xi and linear in y;

(H2) the constraints g_i of each follower depend only on the decision variable x_i and on the incentive y and are supposed linear; and

(H3) the upper multiobjective problem (equations 1 to 3) is convex; that is, F and G are convex functions and Y is a convex set.

In the following, only the first phase will be made explicit because the second phase is implemented by using standard optimization procedures, as stated above.

We first note that, as the lower-level decision-makers are independent of each other (hypothesis H2), they can be considered as the components of one single "super follower" whose optimal decision vector is expressed as $x^* = [x_1^*, \ldots, x_n^*]$. Then, given hypotheses H1-H3, it can be shown (Gailly, 1994; Gailly et al., 1997) that there exists a finite number J of disconnected subsets Y_j of Y and continuous functions $h_j(.)$, such that

$$Y_1 \cup Y_2 \cup \ldots \cup Y_J = Y$$

for $j = 1, \ldots, J$, if $y \in Y_j$, then $x^* = h_j(y)$.

In other words, the reactions x^* of the followers to any feasible incentive y can be explicitly computed as the value taken by a "piecewise" function $h(.) = \{h_1(.), \ldots, h_J(.)\}$, defined over the incentive feasible set Y. The determination of the subsets Y_j and the functions $h_j(.)$ is explained in Gailly et al. (1997). It is implemented through an iterative procedure using optimality conditions, such as dual and primal feasibility conditions, or first-order Kuhn-Tucker optimality conditions (Fiacco, 1976). Once the subsets Y_j and the functions h_j have been identified, the implicit bilevel "piecewise" multiobjective optimization model P can be replaced by an equivalent multiobjective model P*:

$$(P^*) \qquad \text{Max}_y\ F(x^*, y) \qquad\qquad\qquad (6)$$
$$\text{s.t.} : G(x^*, y) \leq 0 \qquad\qquad\qquad (7)$$
$$\text{for } j = 1, ..., J, \text{ if } y \in Y_j, \text{then } x^* = h_j(y). \qquad\qquad (8)$$

2.3. Implementation of a DSS for the convex linear hierarchical case

The general methodology presented above was implemented in the form of a DSS for the particular case of a convex linear hierarchical multicriteria optimization model formulated as follows:

$$(PCL) \qquad \text{Max}_y\ F(x_1^*, ..., x_n^*, y)$$
$$\text{s.t.} : \quad G(x_1^*, ..., x_n^*, y) \geq 0$$
$$y \in Y,$$

and for $i = 1,..., n$:

$$(PL_i) \qquad c_i(y)x_i^* = \text{Max}_x\ c_i(y)x_i$$
$$\text{s.t.} : \quad A_i x_i \geq b_i(y)$$

where: n is the number of followers ;
$F(.)$, $G(.)$, y, Y, and x are defined as in Section 2.1;
$c_i(y)$, $i = 1, ..., n$, are the followers' objectives;
A_i, $i = 1, ..., n$, are the matrices of the followers' constraints;
$b_i(y)$, $i = 1, ..., n$, are the right-hand-side terms of the followers' constraints; and
$c_i(.)$ and $b_i(.)$, $i = 1, ..., n$, are linear functions of y defined over Y.

For this case, it may be shown (Gailly et al., 1997) that the explicitation phase is quite easy to implement: the functions $h_j(.)$, which are linear in y, and the corresponding subsets Y_j, whose boundaries are given by linear constraints, may be determined by using the primal and dual feasibility conditions corresponding to the formalism of the standard Simplex algorithm applied to the lower-level problems PL_j and $j = 1,...n$.

The optimization phase uses standard interactive multicriteria optimization software (Cao et al., 1990) that allows first for the computation of the payoff matrices inside each subset Y_j. Then, within the nondominated subset Y_j, efficient solutions are iteratively computed and proposed by the analyst who assists the upper-level decision-makers in making an interactive choice of various aspiration and veto levels for the upper-level criteria values.

2.4. The corresponding software

Both the explicitation phase and the optimization phase described above have been implemented in a software package called IMPACT that runs in a Windows 95 environment.

This software was designed in such a way that it can be used without any knowledge about computers: it is organized as a highly interactive hierarchical tree of windows that specialize in the various data from the problem to be processed. Hence, one window may present data for the upper-level decision-makers, another for lower-level decision-makers, and so on. The results of the interactive and iterative optimization procedure may be displayed and/or printed in both analytical and graphic forms.

3. Application

The hierarchical decision-support tool described above was used to investigate the impact of various incentive strategies that could be implemented by regional public authorities to improve rural development of a region in the Republic of Burundi. The region consists of three municipalities (Muruta, Kayanza, and Matongo) of the northern province of Kayanza, with a total area of 26 500 hectares.

On the basis of the current conditions in the region and its potential for demographic and economic improvement, the public authorities selected two objectives for rural development: 1. to increase the land areas devoted to tea production, the latter being a substantial means for the country to earn hard currency, and 2. to intensify trade in six basic foods (green peas, beans, wheat, sweet potatoes, potatoes, and corn) within the region, as well as between the region and other parts of the country. This strategy appears to have the potential to increase the productivity of the region and minimize erosion through better correlation between the types of crop and the climate/geography of the region.

First, in order to investigate the impact of various incentives that could be implemented by the public authorities, a multicriteria hierarchical model was built to provide both a tractable and realistic mathematical simulation of the corresponding real-life problem. To do this, it was necessary

1. to identify a lower-level decision model that could be reduced to a linear single-criterion model simulating the preference structure of the local farmers; and

2. to identify an upper-level decision model that could quantify the upper-level objectives of the public authorities.

3.1. The lower-level decision model

The setup of this linear single-criterion model was done in two steps. During the first step, a linear multicriteria model was built that was based on various field inquiries, as well as available quantitative and qualitative data about soil quality and the demographic and physical characteristics of the studied region (Gailly, 1992; Ndimubandi, 1993).

Seven major crops were included in the model: green peas, beans, wheat, sweet potatoes, potatoes, corn, and tea. They were distributed over three growing seasons (rainy, dry, and short dry). The farms were divided into four typological groups on the basis of the soil suitability established for each of the seven crops. These four groups represent the lower-level decision-makers' units. The decision variables at the lower level were the fields allocated to each crop within the four groups and the three growing seasons; the proportion of these fields reserved for local consumption; the proportion of fields left fallow; and the manpower transfers between the four groups.

After discussions with local farmers and observation of their land-use strategies, the following criteria were selected as being potentially important:

1. the suitability of crops in terms of market demand and export opportunities –
    these criteria are identical to the criteria (TRADE and DEV) used to quantify the
    objectives of public authorities; for an evaluation, see section 3.2;

2. sensitivity to climatic and economic risks – the sensitivity of each crop to these
    risks was expressed as a sensitivity index on a scale ranging from one (not sensitive) to 28 (totally sensitive); the contribution of each crop to this risk criterion is
    equal to its field area times its sensitivity index; and

3. the soil erosion rate – erosion rate coefficients were estimated for each crop
    within the four farm groups; that is, for each class of soil suitabilities. For more
    information, see Ndimubandi (1993).

The constraints were characterized by the limitation of available financial resources, the minimal energy and protein nutritional supplies needed for subsistence, the minimal level of local consumption[1] for each crop, and the limitations of available manpower and land.

The quantitative technical coefficients of the model (such as productivities, nutritional values, and prices) were evaluated on the basis of the available data and balanced with respect to the characteristics of each typological farm group. The qualitative technical coefficients (such as the sensitivity to climatic and economic risks) were estimated by comparing opinions after inquiries and interviews with local experts. Finally, the right-hand-side terms of the linear inequalities of the decision model (such as available manpower and financial resources, and the minimum protein content) were estimated on the basis of the available average data for the region and balanced with regard to the geographic and demographic characteristics for each typological group. The resulting linear multicriteria model contained approximately 200 variables and 100 constraints.

During the second step, an interactive multicriteria analysis of the model was performed with software developed by the authors (Cao and Installé, 1990). Several representative solutions were generated and compared with the current strategies of the farmers. When all the contributions of local and foreign experts (agricultural scientists, economists, local public authorities, and so on) on the generated solutions had been taken into account, the initial multicriteria model was iteratively refined, tuned, and ultimately reduced to a single criterion model with the following features :

1. the local consumption rate was estimated at about 80%;

2. the sensitivity to climatic risk may be estimated by applying the maximum risk for one single crop to all crops;

3. the criteria of soil erosion rate and sensitivity to climatic risk can be replaced in the model by constraints setting maximum acceptable levels; and

4. the export opportunities and the suitability of crop production to market demand are not taken into account by the farmers.

Considering these features, the model was reduced to a single criterion. It describes sensitivity to economic risk: that is, the probability of a food shortage as a result of a lack of financial means

The constraints include the maximum allowable soil erosion rates; maximum allowable climatic risk for a single crop; available financial resources; minimum protein and energy nutritional content; minimal level of local consumption for each crop; available manpower and manpower transfer between typological groups; and the available arable land. This reduced model contains approximately 80 variables and 40 constraints.

3.2. The upper-level decision model

The definition of such a model requires that
1. the rural development objectives of the upper-level decision-makers be identified and quantified in the form of criteria;

2. the potential incentive strategies that can be implemented by the public authorities be identified and their impact on the criteria of the lower-level decision-makers be determined; and

3. the constraints on those strategies be evaluated.

The upper-level objectives have been listed already. The objective of increased tea production was quantified as the current monetary value of the tea produced in the region. It was denoted by the symbol DEV and expressed in US dollars.

The objective to increase trade in the six basic food crops was denoted by the symbol TRADE. Implementation took place in two steps. First, on the basis of inquiries at local markets and field interviews, a "market demand intensity" index was given to each of the six food crops, ranging from one (no market demand) to seven (very high market demand). Second, the current value of TRADE was expressed as the sum of the contributions of each food crop to market demand, those contributions being equal to the corresponding field area under cultivation by the corresponding market demand intensity index.

Various incentive strategies were identified that considered the sociocultural, political, and legal contexts, as well as the limited availability of technical and financial public resources. The following two were selected because of their potential efficiency: technical assistance programs to improve the farmers' productivity (S.1), and creation of nonagricultural jobs to provide farmers with extra financial resources (S.2).

3.3. Analysis and results

In this section we present an analysis of the impact of the incentive strategies S.1 and S.2 on the criteria DEV and TRADE.

3.3.1. Strategy S.1: technical assistance. This incentive consists of supplying farmers with technical assistance which improves their productivity. Hence, the incentive may be quantified in terms of an equivalent supply of man-days to the farmers. Its impact on the farmers' decision model may be modelled in terms of an increase of the right-hand-side term of the constraints limiting the available manpower. It was decided to study the impact of such technical assistance within each of the four farms groups, the incentive limited to an equivalent of 200 000 man-days, which would amount to a 5% increase of the manpower productivity of the whole region. Some of the results that were generated will be presented first, and then they will be interpreted in terms of incentive strategies and their influence on the private farmer's behaviour.

As a result of the explicitation phase of the resolution method (see Section 2.2), the feasible set Y of the upper-level decision variables (the amount of technical assistance given to each of the four farm groups) was divided into nine subsets. Of these, four were identified as producing lower-level reactions that were potentially interesting with regard to the upper-level objectives. The characteristics of these four subsets are summarized in Table 1. During the optimization phase of the resolution method, this useful information allows for the interactive choice of the most suitable incentive strategy in terms of the upper-level criteria.

Table 1. Subset characteristics.

Subset	DEV_{max} (USD)	$TRADE_{max}$ [%]	$y_{1,min,max}$ [man-days]	$y_{2,min,max}$ [man-days]	$y_{3,min,max}$ [man-days]	$Y_{4,min,max}$ [man-days]
I	194 050	6.77	0 to 153 000	0	0	0 to 137 100
II	194 050	7.87	57 500 to 153 000	0	0	0 to 98 900
III	181 350	8.77	0 to 180 700	0	0 to 35 000	0
IV	170 350	8.88	0	0 to 200 000	0 to 65 900	0

$y_{i,min,max}$ is the range of variation of the technical assistance for the given subset, provided to the ith farm group.

DEV_{max} stands for the maximum increase of tea production and is given by $DEV_{max} = DEV_{opt} - DEV_0$ [USD/year] where DEV_{opt} is the optimal value of DEV on the subset; DEV_0 is the value of DEV without technical assistance. $TRADE_{max}$ stands for the maximum relative increase in trade [%]:

$$TRADE_{max} = (TRADE_{opt} - TRADE_0) / (TRADE_\infty - TRADE_0),$$

where:

$TRADE_{opt}$ is the optimal value of TRADE on the subset;

$TRADE_0$ is the value of TRADE without technical assistance; and

$TRADE_\infty$ is the value of TRADE with unlimited technical assistance.

Interpreting some of the data in Table 1, we noticed that, instead of giving absolute values for the upper-level criteria DEV and TRADE, it is more informative to present the relative increase of those criteria. This provides a better representation of the impact of the incentive on the upper-level objectives.

The interpretation of the data in Table 1 is not trivial because only incomplete information about the subset characteristics is provided, and only ranges for values are given, although they are "piecewise" linear and were computed during the explicitation phase of the methodology. For example, Table 1 shows that in subset I, technical assistance is provided only to the first and last group of farms, with maximum values of 153 000 and 137 100 man-days, respectively. In that subset, the maximum increases of the upper-level criteria would be, respectively: DEV_{max} = USD194 050/year and $TRADE_{max}$ = 6.77%. However, the results provide a good illustration of the degree of freedom left to the upper-level decision-maker in choosing, during the optimization phase, the incentive strategies that would result in an acceptable trade-off between the two criteria and the rate of use of the incentives.

As an example of the use of the optimization phase, the public authorities decided to identify a satisfying compromise in the subset III of the incentive set Y. A standard interactive multicriteria optimization software identified an acceptable compromise, with the following incentive values for each of the four groups of farmers: 71 100, 0, 20 100, and 0 man-days, respectively. This result corresponds to the optimal technical-assistance strategy for the farmers of the four typological groups. This strategy gave rise to the following increase in the criteria values:

$$DEV_{max} = USD\ 181\ 150/year$$

and

$$TRADE_{max} = 6.93\%.$$

The investigations and analyses carried out on the basis of these results have identified some key features of the incentive strategy, which are discussed below.

Table 1 shows that there is a substantial trade-off between the two upper-level objectives. The incentive strategies belonging to subsets I and II are more favourable to an increase in tea exports (DEV criterion), while those belonging to the subsets III and IV are more favourable to the increased trade (TRADE criterion). More precisely, an analysis of the variations of the criteria over the different subsets (not given here) shows that the optimization of tea production assumes that all technical assistance is provided to the first farm group.

Unsurprisingly, this group corresponds to a part of the region that has particularly favourable agricultural conditions for the cultivation of tea. On the other hand, an increase of trade requires technical assistance be divided among farm groups II and III, which face globally favourable agricultural conditions for the seven considered crops. Note that in both cases, group IV, which corresponds to a part of the region with quite unfavourable agricultural conditions, does not receive any technical assistance when the upper-level objectives are optimized.

There is a saturation effect for technical assistance to groups I, III, and IV because the maximum value of such technical assistance over the considered subsets never reaches its ceiling of 200 000 man-days. For example, any increase in technical assistance to the farmers of group III over and above a level estimated (for their group) at approximately 65 900 man-days, has no effect whatsoever on their behaviour.

The poor value obtained for the increase of the TRADE criterion illustrates the difficulties involved in trying to improve the current trade situation of the region.

3.3.2. Strategy S.2: creation of nonagricultural jobs. The second case study looked at the impact of incentive strategy S.2, which provides farmers with the opportunity to find a paying job in a nonagricultural sector (in a packaging firm associated with tea exports). A direct consequence of this incentive strategy for the farmers is the possibility of added income from a daily salary. This would allow them to buy subsistence food on the market and the chance to reorient their farm production toward crops like tea, which are more suitable to the soil on their land and less economically risky. The creation of nonagricultural job opportunities, therefore, might favour both objectives of upper-level decision-makers.

To integrate this incentive into the hierarchical decision model designed for S.1, the following modifications were introduced:

1. creation of new decision variables at the lower level, noted $w_1, \ldots,$ and w_4, where w_i is the amount of nonagricultural paid man-days allocated to the i-th group of farmers;

2. introduction of these variables to the available manpower constraints and farmers' financial resource constraint equations, to account for the financial income they generate; and

3. creation of an upper-level decision variable y equal to the total number of paid nonagricultural man-days offered by the public authorities with the following condition:

$$w_1 + w_2 + w_3 + w_4 \leq y,$$

y = upper level that was fixed at 100 000 man-days.

In order to evaluate the impact on the results of the wage levels offered for the nonagricultural jobs, an analysis was carried out, first, for two different wages, 120 Fbu/day and 240 Fbu/day.[2]

The implemented software used to solve the explicitation phase of the resolution method (see section 2.2), divided the feasible (one-dimensional) set Y of the upper-level decision variable (the amount of paid nonagricultural man-days offered by the public authorities) into 11 subsets. Within each subset, the relationship between TRADE and DEV and the number of offered nonagricultural man-days was investigated. In Figures 1 and 2, the black and white squares on the graphs represent wages of 120 and 240 Fbu/man-day, respectively; the graph plots the absolute increase for these criteria (the difference between their actual value and the value obtained without any availability of nonagricultural jobs). Note that some of the subsets were too small to be shown on the graphs.

In Figures 1 and 2, each interval between two consecutive dots corresponds to a subset of Y in which the farmers' reactions and, as a consequence, the upper-level objectives (DEV and TRADE) are linear.

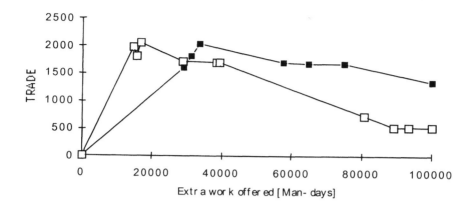

Figure 1. TRADE variations with regard to the offer of nonagricultural jobs.

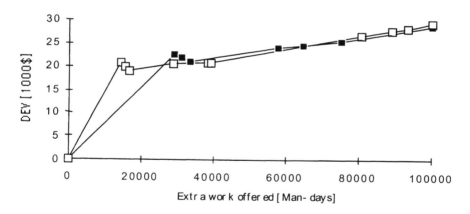

Figure 2. DEV variations with regard to the offer of nonagricultural jobs.

Next, we investigated the simultaneous availability of external jobs at both wages (some at 120 Fbu/man-day, denoted by y_1, and some at 240 Fbu/man-day, denoted by y_2). The influence of the combined incentives on the upper-level objectives was also studied. The investigation resulted in a partition of the incentives space (both amounts were limited to 100 000 man-days) into 12 different subsets. The result is shown in Figure 3, where only 11 subsets could be represented because the last one was too small. This last result might be used to identify the farmers' reactions to any combination of both external activities or to an external activity at a wage of between 120 and 240 Fbu/day (which amounts to the same thing). That point is discussed below.

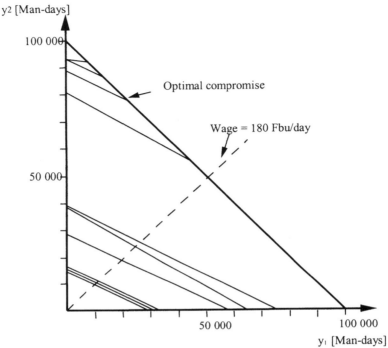

Figure 3 : Partitioning of the incentives space [y_1 , y_2].

The potential impact of the two different incentive strategies (providing external work at 120 and 240 Fbu/day) on the two upper-level objectives was analyzed through the information given in Figures 1, 2, and 3.

It was observed first that, with an offer of external jobs that ranged from zero man days to 20 000 man days, there was an important increase in both DEV and TRADE criteria. This was due to a shift in the variety of crops produced by some groups of farmers: the supplementary income earned from their nonagricultural jobs allowed them to devote more of their land to tea production, a low-risk crop, because they were able to buy more food for personal consumption in the markets.

Second, at around 20 000 man-days of available nonagricultural jobs, a saturation effect was observed because the farmers did not want to devote a large part of their land to a single crop. Their decision-making took into account the maximum climatic risk associated with each crop, which was limited by a maximum value.

Finally, if farmers earn even more from nonagricultural jobs, they use the extra income to substitute green pea with sweet potato crops. They do this because sweet potatoes are less subject to local economic risks. However, the substitution does not meet the public authorities' objective of answering market demand, which causes a drop in the TRADE criterion. Note that the upper-level objective of increasing tea production still grew slightly because the model assumed that the nonagricultural jobs resulted in an overall increase in value added of the crops devoted to export.

The results generated from the simultaneous availability of jobs at both wages may be interpreted in the following manner. The effect of the extra work offered at a wage different from 120 and 240 Fbu/day is shown by the dotted line plotting the individual farmer's' reaction to a wage p of 180 Fbu/day in Figure 3. The line is represented by the equation:

$$120 \, y_1 + 240 \, y_2 = p \, (y_1 + y_2),$$

where p is the offered wage ($120 < p < 240$).

Evaluating the farmers' reactions to the incentives along that line provides a full description of the farmers' response to external work at a wage equal to p. Given the farmers' reactions to both incentives (y_1 and y_2), their value can be derived easily from the generated information.

On the basis of the information generated by the investigation of the combined impacts of both incentives, an interactive multicriteria analysis was performed for both the upper-level variables (y_1 and y_2) and the upper-level objectives (DEV and TRADE). The optimal incentive values were found to be $(y_1; y_2) = (17\ 400; 82\ 600)$ man-days, corresponding to an optimal compromise between both objectives.

The result implies that the corresponding optimal salary p is equal to $(120 \, y_1 + 240 \, y_1) / (y_1 + y_2) = 219$ Fbu/day. In other words, the offer of two simultaneous extra work opportunities of 17 400 man-days at 120 Fbu/day and 82 600 man-days at 240 Fbu/day produces the same effect as the provision of a single extra work opportunity of 100 000 man-days at 219.12 Fbu/day.

This second application shows that variations in the terms on the right-hand-side of the equation (y_1 and y_2 values may range from zero to 100 000 man-days) can be used to study the impact of incentive strategies that correspond to varying technical coefficients in the decision model.

4. Summary

This chapter has presented and discussed the application of a new decision-support tool using multicriteria hierarchical models to select incentive strategies in a region experiencing rural development problems.

We were able to consider two classes of decision-makers, the public authorities and local farmers, with different policy objectives that affected both of them. We examined the impact of two types of incentive strategies that could be implemented by public authorities to select the one that most closely met their objectives.

The approach used in these two case studies could be extended easily to other cases with different types of incentive strategies.

Acknowledgments

The research described in this chapter was carried out by the Belgian Programme on Inter-university Poles of Attraction, initiated by the government of Belgium, Prime Minister's Office for Science, Technology, and Culture. Responsibility for these findings rests solely with the authors. This work has also been supported by a research contract from the European Community under the program "Science and Technology for Development" (Res. Contr. No. STD2 - CT90 - 0321).

Endnotes

1.   The rate of local consumption is the minimum proportion of food the farmers wish to produce
         themselves instead of buying it.
2.   1 US$ = ± 180 Fbu (1994).

References

Aiyoshi E. and K. Shimizu (1981). "Hierarchical decentralized system and its new solution" by a barrier method, *IEEE Trans. SMC*, SMC-11, N° 6.

Bard J.F. (1983). "An efficient point algorithm for a linear two-stage optimization problem", *Operational Research* 31, 670-684.

Bard J.F. and J.T. Moore (1990). "A Branch and Bound algorithm for the bilevel programming problem", *S.I.A.M.J. Sci. Stat. Comp.* 11, 281-292.

Bialas W.F. and M.H. Karwan. (1984). "Two level linear programming", *Management Science* 30, 1004-1020.

Cao D. and M. Installé (1990). "An interactive multiple criteria decision making algorithm for the planification of agro-socio-economic systems", *Proceedings of the 11th World Congress of IFAC*, Tallinn, USSR 12, 139-144.

Cruz J.B., Jr. (1978). "Leader-follower strategies for multilevel systems", *IEEE Trans. AC* 23(2).

Fiacco A.V. (1976). "Sensitivity analysis for nonlinear programming using penalty methods", *Mathematical Programming* 10, 287-311.

Gailly B. (1992). "Modèle multicritère de simulation d'exploitations agricoles. Présentation et description", *Rapport technique, CESAME,* Louvain la Neuve, 26 pages.

Gailly B. (1994). *Decision Support Tools for Incentives Strategies Using Hierarchical Multi-criteria Optimization*, PhD thesis, CESAME, Université Catholique de Louvain, Belgium.

Gailly B., F. Boulier, M. Installé (1997). " A new decision support tool for the choice of incentive strategies using hierarchical multicriteria models", submitted for publication to the *Journal of Multi-criteria Decision Analysis*.

Ndimubandi J. (1993). "Modèle d'optimisation multicritère; application à la région de Kayanza", *Technical report*, ECRU, Université Catholique de Louvain, Belgium.

Nijkamp P. and N. Rietveld. (1981). "Multi-objective multi-level policy model: an application to regional and environmental planning", *European Economic Review* 15, 63-89.

Sherali H.D. *et al.* (1983). "Stackelberg-Nash-Cournot equilibria: characterization and computation", *Operations Research* 19, 217-235.

Simaan M. (1977). Stackelberg optimization of two levels systems, *IEEE Trans. SMC* 7(4).

Vincke Ph. (1989). *L'aide multicritère à la Décision*, éds. de l'Université Libre de Bruxelles, Bruxelles, Belgium.

7 EFFICIENT STRATEGIES
An Application in Water Quality Planning

Alexander V. Lotov, Lioubov V. Bourmistrova
and Vladimir A. Bushenkov

1. Introduction

New information technologies like multimedia, virtual reality, and geographic information systems (GISs) within the framework of decision support systems (DSSs) present planners with new and exciting opportunities. These include enhanced DSS efficiency through the display of graphic information, which allows rapid, integrated assessment of one or more decision alternatives. This chapter describes a graphics-based decision support technique that provides information on the outcomes of a very large (or infinite) variety of possible decision strategies, and helps the decision-maker select the best option. This technique directly supports the two steps of the decision choice phase, as shown in Figure 1. By means of decision screening, this technique selects a small number of strategies and subjects them to further detailed exploration and what-if analysis. It has both static and dynamic graphic components.

Figure 1. Decision choice steps.

Section 2 introduces the applications of this technique in support of water quality management. Section 3 describes some key environmental problems and discusses the usefulness of simplified integrated models for decision screening. Section 4 introduces the feasible goal method (FGM) and interactive decision map (IDM) techniques. Section 5 outlines the methodology of the search for effective environmental strategies. In Section 6, the DSS developed for Russia's Resurrection of the Volga

River program is used as a case study to illustrate both the methodology and the FGM/IDM technique. Section 7 describes Web-based applications of the technique and the Appendix provides a mathematical description.

2. Environmental decisions

The applications of this decision support technique are based on a methodology developed for an integrated assessment of environmental problems. The methodology described is applied in a DSS for water quality planning in river basins, and is used to help provide efficient and effective solutions for water quality problems.

Many computer-based DSSs provide what-if analysis for only a few decision alternatives. Unfortunately, environmental problems often have a very large (or infinite) number of alternative solutions. This means that the decision-maker has the complicated task of selecting only a few alternatives for the what-if analysis. Experts usually are asked to develop several decision alternatives so that decision-makers need not face the perhaps thousands of options to arrive at their decision. The strategies experts suggest will reflect their experience, perceptions, and goals, which may well differ from those of the decision-makers. This can result in deadlock during the implementation phase, especially if decision-makers are forced to choose from among strategies which do not reflect their opinions or interests.

The importance of decision screening in water management problems has been stressed by Dorfman (1965), who articulated the need for simplified models. Often, single-criterion optimization is considered adequate to help to screen the decision alternatives. For environmental problems, however, there is rarely a single criterion situation: decision screening must be based on multiple-criterion methods. Examples of their use in water management are described in Cohon and Marks (1975), Cohon (1978), Loucks et al. (1981), Moiseev (1982), and Louie et al. (1984).

Certain aspects of environmental problems complicate the development of efficient strategies to resolve them. They are summarized below.

- Environmental problems often involve competing interests of different social groups.

- Environmental decisions are made by politicians who tend to view problems on a case-by-case basis, rather than considering the whole picture. In addition, their personal priorities may sometimes obscure their official objectives.

- Politicians have little time to study a large volume of information on any given topic; they need information presented in a concise and comprehensive way.

- The public must be kept informed about environmental decisions made on their behalf: environmental problems affect everyone.

- Environmental modeling requires the integration of knowledge from a number of other disciplines, including ecology, economics, geography, and the social sciences. Also, the environmental model should include descriptions of its subsystems, including treatment of pollutants, transport of pollutants, and the effect of pollutants on the ecology, as well as the economic impacts, environmental measures, and so forth.

The methodology presented in this chapter satisfies all of these requirements. Introduced in 1994 (Lotov, 1994; Lotov *et al.*, 1997b; Lotov, 1998), it is based on two main principles:

1. strategy options are screened through the application of integrated mathematical models that combine simplified mathematical descriptions, expert judgements, and empirical data; and

2. strategy options are screened by a user by means of the FGM and IDM techniques that support the exploration of efficient frontiers among aggregated economic and environmental choice criteria, and the identification of a feasible goal.

3. Integrated environmental problem models

The application of simplified integrated mathematical models to screen various strategy options is an important part of the methodology described in this paper. An integrated mathematical model combines simplified mathematical descriptions of the subsystems with expert judgments and empirical data. An integrated model is needed because the original mathematical model that describes the subsystems of a particular environmental problem (for example, discharges of pollutants, treatment of pollutants, transport of pollutants, impacts on the ecosystem, health impacts) may be diverse, and usually has been constructed by specialists from different fields. Since every scientific field has its own mathematical language, combining, or even coordinating these models is problematic. Hence the need to simplify and integrate the models, and sometimes supplement them with expert judgments and empirical data.

3.1. Constructing simplified integrated models

The most universal way to build an integrated model is to establish an approximation for the input-output dependencies, or parameterization, of the original model. Consider a stationary linear model describing the regional transport of a single pollutant from several pollution sources. Assume that a linear model of partial derivatives can be used to evaluate the pollutant concentration at any point in the region and that the pollution discharge per unit time is constant. If formulated correctly, the problem of determining the concentration of the single pollutant is solved.

If pollutant discharge data are not known, the point source method can be used; source functions describing the concentrations that result from the individual source and the unit rate of discharge are constructed. Since the pollutant transport model is linear, the concentration at any given point (for example, at a monitoring point) may be calculated as the sum of the products of the source function values for real discharges. In other words, the concentration of the pollutant at this point is a linear function of pollution discharges, denoted by the scalar product of the influence coefficient vector and the discharge vector.

Given the finite number of monitoring points where the pollutant concentrations are of interest, concentration values can be calculated by multiplying the influence matrix (its rows are provided by influence coefficient values) by the discharge vector. In the linear case, influence matrices can be calculated precisely by using the values of the source functions. In the nonlinear case, influence matrices can only approximate the dependencies of the pollution concentrations on the discharges; how they are constructed depends upon the particular scientific field. The universal approach is based on the application of regression analysis of data on input-output dependencies, obtained by the simulation of nonlinear models. Along with the approximation of an influence matrix, the simulation provides applicability ranges for its elements.

It is often impossible to construct a single influence matrix whose applicability range covers all possible inputs. In this case, several influence matrices may be developed. Each matrix is related to a certain applicability range; all of the matrices cover all possible inputs. Note that within the framework of environmental issues, an applicability range can describe a variety of input values that do not change the environmental system qualitatively. If the limits of this kind of applicability range are violated, it may not be possible for the system to exist in the same qualitative form.

If there is no adequate mathematical model for a subsystem, an influence matrix can be constructed through regression analysis of experimental or historical data. Sometimes, experts can provide both an influence matrix and its applicability range. This chapter does not describe other forms of simplified models as they are not universal and are often based on specific features of the particular system.

A combination of influence matrices and other simplified descriptions, balance equations, and restrictions imposed on variables contributes to an integrated model that describes an environmental system. Simplified integrated models are typically less precise then the original models; however, precision is not of paramount importance, as integrated models are used only to screen feasible strategies. Consequently, the rough description offered by a simplified integrated model may be sufficient to identify a feasible goal. Moreover, it is assumed that any strategy predicated on a simplified model will be verified and improved later, through simulations with adequate subsystem models.

3.2. An example

Simplified integrated models were produced with the DSS used by Russian environmental engineers who had developed reasonable strategies for water quality improvement in small river basins. A detailed description of this system is provided by Lotov and colleagues (1997a).

A stretch of river was divided into a number of segments, with a monitoring station at the downstream end of each segment to monitor the concentrations of several dozen industrial pollutants. Polluting industrial enterprises were grouped by production technology and pollutant output; municipal water services were grouped in the same fashion. About 20 polluting industries and municipal water services were included in the integrated model. The industrial enterprises and municipal services were grouped by river segment, in line with the experts' recommendation. This grouping of the polluting enterprises was an important simplification.

Discharge treatment processes by an industry or a water service also were described in a simplified form. The discharge treatment model describing the decrement in the cost of wastewater treatment for an industry or service depends on the specific wastewater purification technology. The model was based on a database of the discharge treatment technologies, also developed by the experts.

Pollutant transport was described approximately by an influence matrix that made it possible to compute the concentrations of pollutants at the monitoring stations by multiplying the influence matrix by the discharge vector. In general, the coefficients of the influence matrix can be developed through simulation of detailed pollutant transport models; in this case, however, they were estimated by the experts.

4. The FGM/IDM technique: an informal introduction

This section offers an informal introduction of the FGM/IDM technique on the basis of the simplified integrated model outlined in the Introduction. One decision variable of the model describes the volume of discharge from a particular industry or service that is treated with a particular wastewater purification technology in a particular river segment. A collection of such variables constitutes a water quality improvement strategy. Several criteria options can be used in the search for a preferred strategy. This illustration of the FGM/IDM technique incorporates three criteria (see Figure 2): fish stock depletion ("Fishing"), general pollution ("General"), and the cost of the project.

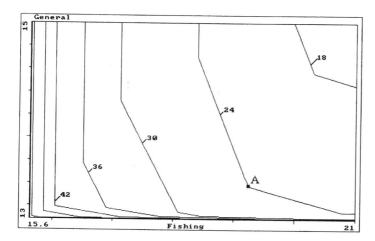

Figure 2. A decision map.

4.1. Interactive decision maps

The FGM/IDM technique is displayed as a decision map which shows the efficiency relationship among the criteria. In this case, the values for the pollution criteria are represented as "Fishing", or fish stock depletion along the horizontal axis and "General", or general pollution along the vertical axis (Figure 2). The decision map consists of several intervals, each related to a certain cost (indicated near each interval). The cost is given in millions of USSR 1988 rubles.

The interval marked with point A, for example, is related to the cost of 24 million rubles. Any point on the graph (denoting the combination of both criteria) that is located to the right of the interval is feasible or, in other words, indicates a strategy that costs no more than 24 million rubles. Conversely, any point to the left of the interval is not feasible at that cost.

The interval represents the limit of feasible combinations for the two pollution criteria values at the given cost. This limit is of particular interest to decision-makers. It is called the efficient frontier, because its shape shows how much fish stock depletion is related to the increase in general pollution, if the cost is limited to 24 million rubles. Above point A on the interval, relatively small increments of fish stock depletion are needed to decrease the value of general pollution. The slope of the interval changes drastically at A. The efficient frontier thus illustrates the efficient trade-offs between the two pollution criteria for the given cost. It is worth mentioning that the notion of efficiency trade-off clearly differs from the notion of value trade-off used by Keeney and Raiffa (1976); the latter implies subjective compensation of losses in one criterion by gains in the other.

Comparing one efficient frontier to another, one can see how the cost increment affects the number of feasible combinations. For example, if the cost is restricted to 18 million rubles, the value of general pollution cannot be less than about 14.6 and the value of fish stock depletion cannot be less than about 20. Adding six million rubles provides more opportunities to decrease both pollution criteria. For example, at 24 million rubles, the general pollution value of 13.4 becomes feasible. On the other hand, if the cost is increased to 30 million, fish stock depletion can be decreased to a value of about 18.5. Thus, the decision map provides a rough guide on the efficient trade-offs among all three criteria.

A decision map thus plots the efficient frontiers between two criteria, for several values of a third. The FGM/IDM technique applies the concept of a modified decision map (Haimes et al., 1990). In the standard decision map technique, several cross-sections of the efficient frontier of the feasible set in criterion space (FSCS) are shown at once. Though the modified decision maps are similar to standard decision maps, they have several advantages – most important, their computing method.

Decision maps are not prepared in advance. First it is necessary to determine an approximation of the Edgeworth-Pareto hull (EPH) of the FSCS. The EPH is a set in the criterion space that, in addition to the feasible criterion values, also contains all dominated criterion values. The algorithm used to obtain the EPH is provided in the Appendix. For linear models, the EPH is a convex set in the criterion space. After the EPH is determined, decision maps are obtained by computing and displaying efficient frontiers (cross-sections) of the EPH. Since the EPH in the criterion space is approximated in advance, several decision maps can be depicted or animated in real-time. Later, references to decision maps will imply the modified version without any additional indication.

Decision maps have several features in common with topographical maps. Since the efficient frontiers of decision maps do not intersect (although they may sometimes coincide), they resemble the height contours of a topographical map. The value of the third criterion in a decision map is similar to height in a topographical map. Knowledge of topographical maps will aid in the analysis of decision maps. It is easy to view the variety of feasible combinations of the first and second criterion in relation to the value of the third ("places higher than..."). It is also easy to understand which values of the third criterion are feasible for a given combination of the first and second criterion ("height of this point is between..."). The proximity of efficient frontiers in the vicinity of a point indicates a slope, as in a topographical map. A slight shift in the efficient frontier results in a substantial change in the value of the third criterion.

4.2. Feasible goals method

Once the exploration of a decision map is complete, the decision-maker may wish to identify a preferred feasible combination of the criterion values (a feasible goal). This is the main feature of the FGM. If point A (Figure 2) were identified as a feasible goal, the wastewater treatment plan that resulted in the identified values of cost and pollution criteria can be computed automatically (see the Appendix for details). If the IDM technique is applied, a preferred feasible goal can be identified on a decision map with a simple mouse click. Decision-makers do not need to perform complicated interactive procedures to extract their preferences. Methods that are not related to the decision-makers' preferences, but are aimed at the display of the efficient set in criterion space, are named multiple-criterion generation methods (Cohon, 1978). Within this framework, decision-makers are given full freedom of choice with regard to the efficient combinations of the criteria.

The application of multiple-criterion generation methods (including the FGM/IDM technique) may be premised on the position of modern psychology that people base their decisions on their mental models of reality, which – although fairly crude – give an integrated picture of reality (Lomov, 1984).

Not only are mental models often crude, but they are also untrue in many respects. One of the purposes of decision support, therefore, is to enhance a user's capacity to formulate mental models. Knowledge about available options in a clear graphic form can help decision-makers understand it clearly at both the conscious and the subconscious levels. In the context of multiple-criterion generation methods, the FGM/IDM technique depicts knowledge about efficient frontiers for selected criteria in precisely this kind of unambiguous, graphics-based format. The technique does not force users to answer multiple questions about their preferences; rather, it helps decision-makers improve their decisions through a clearer understanding of the situation.

5. Methodology

The main steps of this methodology to find efficient environmental strategies are listed below.
1. Experts conduct an initial qualitative analysis of the problem. They develop a list of subsystems to be considered within the framework of the study, as well as a list of possible criteria (performance indicators).

2.   Information on previous studies of the subsystems (including their mathematical modeling) and their interaction is collected. Researchers develop a list of variables that describe the interaction among the subsystems.

3.   The ranges of possible variations of the variables are identified.

4.   Mathematical models of the subsystems are developed (or adapted) and calibrated. If such models do not exist and cannot be developed in time, experimental and historical data are collected for the statistical analysis of the input-output relations.

5.   Influence matrices and other simplified descriptions of the subsystems are developed ,and the applicability ranges are evaluated. If the applicability range of a single influence matrix of a subsystem's model does not contain the ranges of its input variables, the ranges are squeezed or several influence matrices are developed whose applicability ranges cover the ranges of the input variables. The integrated model is constructed by the unification of the influence matrices, and by the imposition of restrictions on the variables.

6.   The constraints on possible strategies are specified, allowing the extraction of a set of feasible strategies.

7.   Decision-makers select criteria from the list and impose limits on all or some of the performance indicators. The EPH of the FSCS is approximated.

8.   Decision-makers explore various decision maps using the IDM software, based on the FGM/IDM technique. From this they gain an understanding of the efficient trade-offs among the criteria. At this point the decision-makers can identify one or several feasible goals; the related strategies then are computed automatically.

9.   The most effective and efficient strategies are displayed to the decision-makers. In the case of spatial strategies, GISs may play an important role.

10. Decision-makers select one or more strategies that will provide a starting point for the detailed development of environmental projects. In contrast to the screening stage, the development of environmental projects is based on the simulation of detailed adequate models.

This is a simplified illustration; in reality, various snags can prevent the process from running smoothly. For example, if a developed strategy violates constraints that had not been included in the model in advance, these constraints may be introduced later, but the imposition will alter the range of feasible strategies. The process would start again at step 6. If the exploration of a strategy shows that a simplified (or even an original) model of a subsystem is imprecise, it would have to return to step 2 and continue the research by developing the subsystem's model.

This methodology was applied to the development of environmental strategies for ground water management (Kamenev et al., 1986; Lotov et al., 1997b), atmosphere pollution abatement (Bushenkov et al., 1994; Lotov et al., 1997b), and global climate change (Lotov, 1994; Lotov et al., 1997b).

6. A DSS for water quality planning in large rivers

As noted above, the FGM/IDM technique was implemented in the water quality planning support system developed for the Resurrection of the Volga River program. Pollutant transport data (the influence matrices) were obtained through the use of IRAS simulation software developed by Loucks and colleagues (1985). The technique helped decision-makers identify feasible goals for water quality and devise strategies for investment allocation among multiple regions in the river basin.

6.1. Criteria

The model has several thousand variables. Only 100, however, were considered as possible criteria (performance indicators). These variables were displayed at the very beginning of the study (Figure 3). The approximation technique used in the DSS allows the user to select between two and seven criteria from the hundred. Users also may specify limits on the values of all selected indicators.

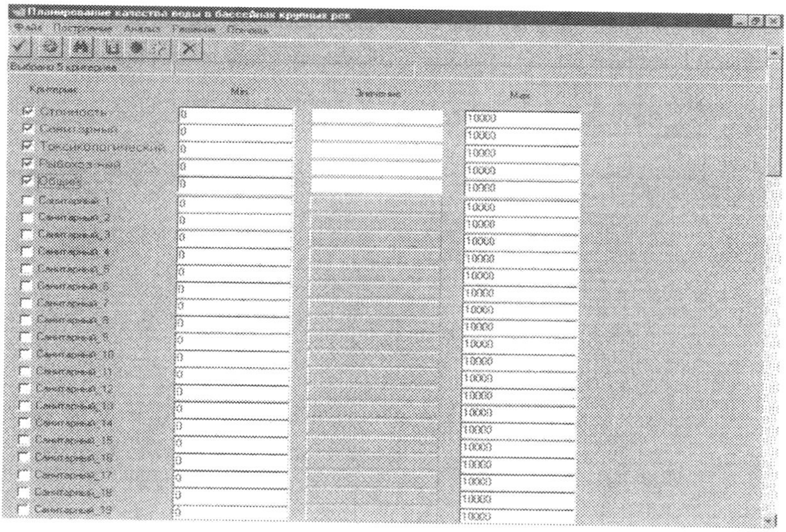

Figure 3. Selection of the first five performance indicators.

In Figure 3, the first five indicators are chosen as criteria and no constraints are imposed on their values. These criteria are the cost of the project and four aggregated water quality criteria. The cost is calculated as the sum of all costs related to the application of particular discharge treatment technologies. The aggregated water quality criteria describe pollution levels at the monitoring stations. As more than twenty pollutants are being tested for, they are grouped into aggregated environmental indicators. The values of these indicators are measured in relative units. The desired value of an aggregated environmental indicator is equal to one. Among monitoring stations, the maximal value of an indicator is used as a pollution criterion. Since four pollution groups were being studied, four criteria were developed: (1) fish stock depletion indicator (FDI), (2) general indicator (GI), (3) toxicological indicator (TI), and (4) sanitation indicator (SI).

The system verifies the compatibility of the restrictions defined by bounds on performance indicator values. If no feasible solution can be found, the user must change the bound values. If they are compatible, the subsystem is ready to approximate the EPH. When the approximation is complete, the subsystem for the visual exploration of decision maps can be invoked (methods to solve problems with thousands of variables are described in the Appendix).

6.2. Initial display

At the very beginning of the visual analysis, the subsystem displays a picture containing one slice of the EPH. Two axis criteria are chosen automatically. Figure 4 depicts one slice depicting cost and sanitation indicators. Since this case study includes five criteria, the initial values of the other three criteria are fixed; their values are indicated on three scroll-bars under the picture of the slice. The efficiency frontier of the slice consists of two linear segments separated by a small kink. A click on any point of the picture reveals that point's coordinates. In Figure 4, a point on the efficient frontier was selected with coordinates 273.036 and 47.308.

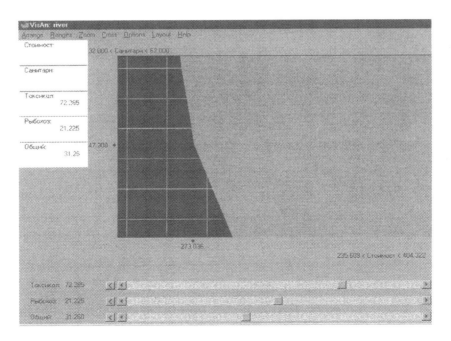

Figure 4. Efficient frontier for cost (horizontal axis) and sanitation (vertical axis) indicators.

It is possible to change the values of the fixed pollution criteria by moving the "thumbs" of the scroll bars. This will change the efficient frontier. Animation of the picture (automatic movement of the "thumbs") is also possible.

6.3. Decision map

The decision map in Figure 5 illustrates the efficient trade-offs between the GIs, SIs, and the cost. GI values are shown on the horizontal axis; SI values are shown on the vertical axis. Cost values are shown by means of different colours. The colour/cost correspondence is indicated in the palette beneath the decision map (in millions of 1988 rubles). The efficient frontiers for the GIs and SIs are the edges of the differently shaded areas.

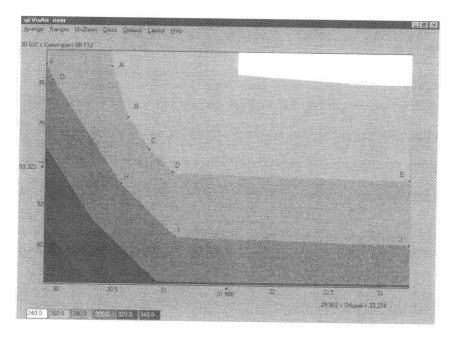

Figure 5. Efficient frontiers for general (horizontal axis) and sanitation (vertical axis) indicators.

The lightest shading corresponds to all the combinations of GI and SI values that are feasible at a minimum cost of 240 million rubles. With 240 million rubles spent on pollution abatement, the GI value cannot be less than about 31.6, and the SI value cannot be less than about 80.5. These values, as well as the coordinates of any other point given in the decision map, can be shown with a mouse click at the desired point.

The cost increase to 260 million rubles yields a larger set of feasible GI and SI values. The minimal feasible GI value decreases to 30.5 (point A). The SI value at point A cannot be less than 84.5. Alternatively, at 260 million rubles the SI value decreases to 57 (point E). SI = 57 raises the GI value to 33.3. The shape of the frontier shows that with 260 million rubles available it is not possible to decrease the GI and SI to their minimal values simultaneously.

Points A, B, C, D, and E divide the efficient frontier for 260 million rubles into four segments. Trade-offs between the GI and SI change at points B, C, and D, so the unit decrement of the SI results in different increments of the GI at different points along the frontier. Depending on the preferences of the user, therefore, different points are likely to be chosen.

It is apparent that the cost increment results in an expansion of the area of efficient solutions. Every extra 20 million rubles expands the range of feasible GI and SI values. A GI value of 29.9 is not feasible at a cost of 260 million; a minimum of 280 million rubles is needed to reduce general pollution to that level. In this way, the decision map makes it easier to understand the relation between cost and water-quality improvements.

6.4. Matrix of decision maps

A user wishing to explore five criteria at once must use a matrix of decision maps (Figure 6). First, the user must arrange the criteria, that is, select two criteria whose values will correspond to the rows and columns of the matrix. In Figure 6, TI values correspond to the columns,and FDI values correspond to the rows. Any column is subject to a certain constraint imposed on the TI value (TI is not greater than…); these constraints are indicated above the columns. Any row is subject to a certain constraint imposed on the FDI value; these constraints are provided to the right of the rows.

A decision map located in a particular column and row is subject to constraints imposed on the TI and FDI values. There are only four rows and four columns in Figure 6, but it is possible to use as many rows and columns as the screen resolution permits. The constraints imposed on the TI and FDI values can be changed easily. The user can choose any single map in the matrix with a mouse click for more detailed exploration. If the user clicks on the decision map in Figure 6 which indicates TI = 100 and FDI = 26, the picture presented in Figure 5 will appear.

The top decision map in the third column represents a TI value no greater than 90 and a FDI value no greater than 26. Since the area that corresponds to the cost of 240 million rubles is absent from the decision map, these TI and FDI levels cannot be achieved at that cost. In this decision map, the cost areas begin at 260 million rubles; therefore, these levels of TI and FDI can be achieved only at that minimum cost.

In contrast, the bottom map in the third column represents an FDI value no greater than 16. The minimal cost in this case is 300 million rubles. In the decision map above that (FDI values no greater than 19.3), the minimal cost decreases to 280 million rubles. In the decision map that represents an FDI value no greater than 22.6, the minimal cost decreases to 260 million rubles and the efficiency frontiers that correspond to the greater cost increments move to the left. Therefore, a broader set of combinations of GI and SI values becomes feasible.

The efficient trade-offs among general and sanitation indicators also change from one row to another. This is additional information that may be important to a decision-maker.

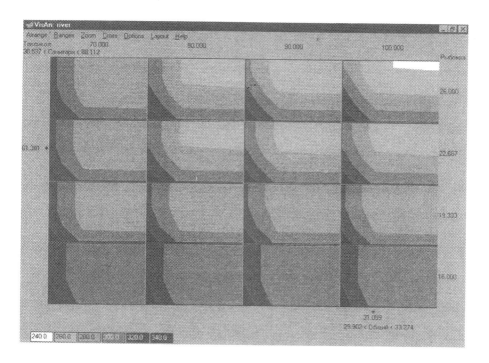

Figure 6. Matrix of decision maps.

The decision maps in the upper row of the matrix share a common FDI value of no greater than 26. The TI values change to no greater than 70, 80, 90, and 100 respectively. The minimal costs in these decision maps are 280 million rubles for TI no greater than 70; 260 million for TI no greater than 80; 260 million for TI no greater than 90; and 240 million for TI no greater than 100.

A more detailed analysis may entail a comparison of feasible values of axis criteria and the efficient frontier between them for different decision maps in the same row. Once again, these values can be found as coordinates of clicked points on a decision map. For example, let us compare the minimal GI values that are feasible at the minimal cost of 280 million rubles on the decision maps across the upper row. These values are:

GI = 30.7 for TI no greater than 70 (the related value of SI = 60.9);
GI = 30.4 for TI no greater than 80 (SI = 70.3);
GI = 30.0 for TI no greater than 90 (SI = 79.1); and
GI = 29.9 for TI no greater than 100 (SI = 84.5).

Since the approximation of the EPH is conducted in advance, decision maps are displayed relatively quickly. An updated matrix of decision maps is displayed a few seconds after the values of the fourth and the fifth criteria have been changed. It is also possible to change quickly the list of the three criteria for which trade-offs are to be displayed.

When comparing decision maps in the matrix for properly chosen fixed values of the fourth and fifth criteria, the user can understand the influence of these two criteria on efficient trade-offs among the first three. Hence, the efficient trade-offs among all five criteria displayed by decision map matrices, are shown in an interactive context.

In a case where six or seven criteria are necessary, more scroll-bars can be provided under the decision map matrix. The user can change the values of these criteria by moving the "thumbs". The matrix of decision maps may be animated. This time-consuming animation is possible because of the prior approximation of the EPH (The FGM/IDM software can be obtained at http://www.ccas.ru/mmes/mmeda/soft/).

6.5. Feasible goal and strategies

Users have full freedom to choose the point in the criteria space with regard to the efficient frontier. By choosing one of the decision maps in a matrix, the user identifies the values of the row and the column criteria. After that, the procedure identifies the feasible values for the other three criteria. The particular values are displayed for any decision selected on the map. The subsystem for decision computing then bases a strategy for water quality improvement on the identified feasible goal. The strategy can then be displayed in maps generated by a GIS module of the system (Figure 7). Monitoring stations are marked on the map with numbered icons.

To receive information about investment allocation along a certain segment of the river and pollution concentrations in its related monitoring station, the user must click on the appropriate icon.

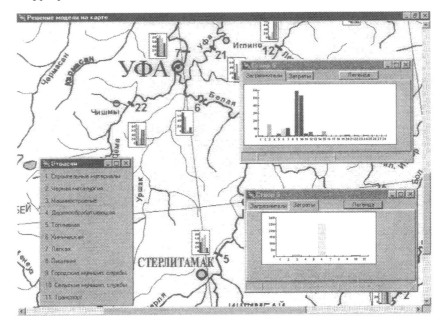

Figure 7. A segment of the river basin map generated by a GIS module.

In this case, the icon for segment number 5 was selected, which generated a display of additional windows. The upper window shows the concentrations of 28 pollutants in the monitoring station (after the discharge treatment facilities are completed). The lower window displays information on investment allocation; in this case, it illustrates the funds invested in 11 industries. The window on the left lists the industries themselves (1. construction materials production; 2. ferrous metallurgy; 3. machinery construction; 4. woodworking industry; 5. fuel industry; 6. chemical industry, and others).

The DSS also gives the user an opportunity to study the strategy from another point of view. Users are provided with aggregated information on all the values at all of the monitoring stations. Diagrams are used to describe how pollution indicators change from one monitoring station to another.

6.6. Applications

The next step is the installation of the DSS in the regional centers located in the Volga River basin. The system will help regional authorities to develop strategies that are profitable for them and acceptable to the Ministry of Natural Resources, as well as to the other regions. In this form, the system will support environmental negotiations.

In Thiessen and Loucks (1992), a negotiation support system can be categorized according to its function as a negotiation preparation system or as a negotiation process support system. The former supports prenegotiation strategic planning, while the latter facilitates the negotiation processes. The above application of the DSS incorporates its function as a negotiation preparation system. In this form, it cannot be used to support a negotiation process.

7. The FGM/IDM technique on the Web

Within the framework of the FGM/IDM technique, interaction between the user and the computer is quite simple. For this reason, it can be implemented on computer networks such as the Internet. Now, new and exciting opportunities that did not exist earlier can be provided for network users. In particular, FGM/IDM-based Internet resources can provide direct access to information on a myriad of possible decision alternatives. This access is important in various fields; however, from the societal point of view, its most important asset is the ability of the FGM/IDM technique to support an Internet-based search for preferred strategies that may solve important public problems. The technique also can help to obtain unbiased information on these solution options; in other words, information that is not screened by experts and other mediators (Williams and Pavlik, 1994). Moreover, the FGM/IDM technique supports an independent search for a preferred strategy to help ordinary people be actively involved in public discussions on matters that affect them directly.

A prototype Internet resource (http://www.ccas.ru/mmes/mmeda/resource) demonstrates this idea on the basis of a problem considered in detail in Lotov et al., (1997). This prototype allows the user to study the problem of water quality planning in a region with developed agricultural production. If the agricultural production increases, it may spoil the water quality in a lake. Therefore, reasonable strategies for agricultural production and water management must be developed.

The efficient trade-offs among several criteria are displayed in three premade decision maps. The user can choose an appropriate decision map and identify a preferable feasible goal; all that is required for the user to develop a preferred feasible strategy is Internet access and a Web browser.

The Internet resource described here is only a prototype of the future Internet resources that will use the FGM/IDM technique to its full capacity. The application of resources of this kind will help to develop strategies regarding various public problems, such as regulating a national economy or the resolution of local, regional, national, or even global environmental problems. It will help people to understand the complications that stakeholders face in the process of trying to solve public problems and it may prompt them to try to influence the decision-making process.

8. Appendix: The FGM/IDM technique

In the framework of the FGM/IDM technique, the generation of efficient frontiers and the screening of feasible decisions are based on the algorithms of a universal mathematical approach called the generalized reachable sets (GRS) method (Lotov, 1973). The GRS method is used to approximate and display a variety of attainable output vectors for a large (or infinite) variety of possible input values. In decision problems, the computational algorithms of the GRS method transform feasible decisions into feasible combinations of criterion values — the FSCS or its EPH.

8.1. The EPH of the FSCS

Let the decision variable x be an element of the decision space W (e.g., of finite-dimensional space R^n; in this case, decision vectors x are considered). Let the set of feasible decisions X from the space W be given. Let the criterion vector y be an element of linear finite-dimensional space R^m. If it is assumed that criterion vectors y are related to decisions by a given mapping $f: W \rightarrow R^m$, then the FSCS is defined as the variety of criterion vectors that are attainable if all the feasible decisions are used:

$$Y = \left\{ y \in R^m : y = f(x), x \in X \right\}.$$

Let us suppose that a user is interested in the increment of the criterion values y. In this case, a criterion point y' dominates (is better than) another criterion point y, if $y' \geq y$ (i.e. $y_i' \geq y_i$ for $i=1,2,...,m$) and $y' \neq y$. Then, the nondominated frontier of the FSCS is defined as the variety of nondominated points of $y \in Y$, or points for which the dominating points do not exist in Y (the set of dominating points is empty):

$$P(Y) = \left\{ y \in Y : \left\{ y' \in Y : y' \geq y, y' \neq y \right\} = \varnothing \right\}.$$

The nondominated frontier is often identified as the Edgeworth-Pareto set.

Another set in criterion space is used to determine a nondominated frontier of the FSCS that has a simpler structure than the FSCS, but has the same nondominated frontier. This set is called the EPH of the FSCS. EPH is defined as

$$Y^* = Y + (- R_+^m),$$

where R_+^m is the nonnegative cone of R^m. Both Y and Y^* have the same nondominated frontier:

$$P(Y^*) = P(Y).$$

Note that the dominated frontiers disappear in Y^*.

Decision maps are calculated as collections of two-dimensional (two-criterion) slices (cross-sections) of the EPH. Let u denote the values of two selected criteria and z^* denote the fixed values of the remaining criteria. Then, a two-dimensional slice of the set Y^* related to z^* is defined as

$$G(Y^*, z^*) = \{u : (u, z^*) \in Y^*\}.$$

It is important to note that a slice of the EPH contains all combinations of the values of the two criteria that are feasible if the values of the remaining criteria are not worse than z^*.

In the IDM/FGM technique, decision maps help to identify a preferable feasible goal. Once identified, y' is regarded as the "reference point" (Wierzbicki, 1981), that is, an efficient decision is obtained by solving the following optimization problem:

$$\min_{1 \le j \le m}(y_j - y_j') + \sum_{j=1}^{m} \{\varepsilon_j (y_j - y_j')\} \Rightarrow \max,$$

for $y = f(x), x \in X,$

where $\varepsilon_1, \ldots, \varepsilon_m$ are small positive parameters. Since the goal is close to the EPH, the efficient decision results in criterion values that are close to the goal.

8.2. Approximation of the EPH

Decision maps are computed and depicted quickly with the FGM/IDM technique, since the EPH is approximated in advance. If the EPH is a convex set, its approximation of the EPH is based on iterative methods combining convolution methods (Fourier, 1826) and optimization techniques. The EPH is approximated by the sum of the cone $(- R_+^m)$ and a polytope approximating the FSCS. A short description of the convolution-based methods is provided in Lotov (1996). If the EPH is a nonconvex set, an approximation of the FSCS and its EPH is more complicated, but is nonetheless more possible (Kamenev and Kondratiev, 1992; Bushenkov et al., 1995).

Linear finite-dimensional models are usually applied in the integrated assessment of environmental problems. The space W is a linear space R^n, the set X is a polytope of R^n, and the mapping $f: R^n \rightarrow R^m$ is linear. Usually, the set X is specified by a linear constraint system:

$$X = \{x \in R^n : H\,x \le h\},$$

where H and h are a given matrix and vector, respectively. The linear mapping

$$f : R^n \to R^m$$

is specified by a matrix F, such that $f(x) = F\,x$.

The FSCS

$$Y = f(X) = \{\, y \in R^m : y = F\,x,\ H\,x \le h \,\}$$

can be approximated by a polytope Q, constructed in the form

$$Q = \{\, y \in R^m : D\,y \le d \},$$

where D is a matrix and d is a vector to be computed.

8.3 Approximation of the FSCS

As stated above, the EPH is approximated by the sum of the cone $(-R_+^m)$ and a polytope approximating the FSCS. Methods for the approximation of the FSCS and its EPH are closely related (details are provided in Chernykh, 1995).

Let us consider the general problem of the approximation of a convex compact body C from R^m. It is assumed that the values of the support function of the body C can be computed, which is defined by the formula on the unit sphere of directions

$$g_C\,(u) = \max\,\{<u, y>: y \in C\},$$

where $<.,.>$ is the scalar product. In the case of the FSCS for a linear problem, the value of the support function can be determined by solving a linear programing problem. Here, two methods are described for the approximation of convex compact bodies. Both of them are iterative; they allow a sequence of convex polytopes P^0, P^1, ..., P^k, ... with an increasing number of vertices. Every polytope in the sequence is based on the previous one, using evaluations of the support function for the approximated body C. These methods construct a sequence of convex polytopes $P^k \in R^m$ such that

$$\lim_{k \to \infty} \delta(P^k, C) = 0,$$

where $\delta(.,.)$ is the Hausdorff distance between any two compact sets from R^m:

$$\delta(C_1, C_2) = \max\,\{\, \sup\,\{d(x, C_2): x \in C_1\,\},\ \sup\,\{d(x, C_1): x \in C_2\,\}\,\},$$

and $d(.,.)$ is the distance in R^m.

Note that these are methods to approximate the body C to the greatest degree of accuracy; that is an important difference when these methods are compared with those that use a simplex or ellipsoid to estimate a body. Obtaining this degree of precision, however, comes at a high price: the complexity of an approximating polytope increases rapidly if the accuracy of approximation and the dimension m increase. Therefore, methods are needed that have an optimal rate of convergence and an optimal complexity of approximating polytopes.

To introduce the concepts of optimal rate of convergence and of optimal complexity of approximating polytopes, it is necessary to consider a "standard" polytope that provides the best approximation of the convex compact body C. Let $P(C)$ denote the class of convex polytopes whose vertices belong to the boundary of C. Let $P_N(C)$ denote the subclass of polytopes with the number of vertices no greater than N. A classical tenet in the theory of convex sets dictates that there always exists a polytope $P_N \in P_N(C)$, such that

$$\delta(C, P_N) = \delta(C, P_N(C)) = \inf\{\delta(C, P) : P \in P_N(C)\}.$$

This polytope P_N is called the best approximation polytope; thus

$$\lim_{k \to \infty} \delta(C, P_N) = 0.$$

For any convex compact body C with twice continuously differentiable boundary and positive principle curvatures, there exist positive constants k_C, and K^C such that

$$k_C N^{\frac{2}{1-m}} \leq \delta(C, P_N) \leq K^C N^{\frac{2}{1-m}}.$$

Incidentally, this formula shows that a body with a smooth boundary may be approximated by polytopes with reasonable accuracy only if $m < 8$. It is interesting to compare the sequence of the best approximation polytopes generated by a particular approximation method to the sequence of polytopes of the best approximation itself.

Two iterative methods for the approximation of convex bodies by polytopes are described here. The first one, called the Estimate Refinement (ER) method, is a typical example of an iterative method.

8.4. The Estimate Refinement method

Let $U(P)$ be the finite set of outer unit normals to the facets of a polytope P. The set $U(P)$ is defined if the polytope P is specified as the solution set of a system of linear inequalities.

Following is a description of the $(k+1)$-th iteration of the ER method. Prior to the iteration,

$$P^k \in P(C)$$

should be constructed as the solution set of a linear inequalities system; then the $(k+1)$-th iteration will consist of two steps.

Step 1. Find the element $u^* \in U(P^k)$ that provides the maximal difference between the values of the support function of the body C and the support function of the polytope P^k:

$$g_C(u^*) - g_{p^k}(u^*) = \max\{ (g_C(u) - g_{p^k}(u)): u \in U(P^k) \}.$$

Find a point p^* from the boundary of the body C, such that $<u^*, p^*> = g_C(u^*)$.
Step 2. Construct the convex hull of the point p^* and the polytope P^k:

$$P^{k+1} = \mathrm{conv}\{p^*, P^k\}$$

in the form of the solution set of a linear inequalities system.

It is assumed that the initial polytope $P^0 \in P(C)$ has been specified prior to the application of the method. The construction of the initial polytope is discussed in Chernykh (1992).

It was demonstrated in Kamenev (1994) that for bodies with a smooth boundary, the ER-generated sequence of polytopes P^k has the same rate of convergence as the best approximating polytopes P_N, $N = m+1$, $m+2$, $m+3$,.... Therefore, the ER method is optimal with respect to the number of vertices. Other important properties of the ER method are provided in Kamenev (1994).

In real-life environmental problems (such as those studied in this paper), the set X belongs to a space R^n of a high dimension, where the value of n is several thousand. The evaluation of the support function is related to the solution of a linear programing problem; with thousands of variables, this can be a lengthy process. Methods that have been developed until recently (including the ER method) require multiple evaluations of the support function during an iteration. Thus, it became important to develop a new method that would be optimal with respect to the number of evaluations of the support function. Such a method was proposed by Kamenev (1996), called the Converging Polytopes (CP) method.

8.5. The Converging Polytopes method

In the framework of the CP method, a body C is approximated using two polytope sequences. One consists of inscribed polytopes with the tops belonging to the body boundary, and the other consists of circumscribed polytopes. These polytope sequences come closer and closer together during the process of approximation.

Let $Q(C)$ denote a set of convex polytopes whose facets touch the boundary of the set. Once $P^k \in P(C)$ and $Q^k \in Q(C)$ are constructed, the *(k+1)*-th iteration will involve the following two steps:
Step 1. Find the element $u^* \in U(P^k)$ that provides the maximal difference between the support functions of the inscribed and the circumscribed polytopes:

$$g_{Q^k}(u^*) - g_{p^k}(u^*) = \max\{(g_{Q^k}(u) - g_{p^k}(u)):u \in U(P^k)\}.$$

Find a point p^* from the boundary of the body such that $<u^*, p^*> = g_C(u^*)$.
Step 2. Construct the convex hull of the point p^* and the polytope P^k:

$$P^{k+1} = \mathrm{conv}\{p^*, P^k\}$$

in the form of the solution set of a linear inequalities system. Put

$$Q^{k+1} = Q^k \cap \{y \in R^m : <u^*, y> \leq g_C(u^*) \}.$$

It is clear that only one evaluation of the support function is required for each iteration. Therefore, the CP method requires a smaller number of evaluations of the support function than vertex selection methods do.

An experimental comparison of the CP and ER methods was performed, related to the approximation of various ellipsoids of different dimensions. The study showed that the CP method generates polytopes that have approximately the same number of vertices for the same precision of approximation as the polytopes generated by the ER method. This is a significant result, since the CP method requires a substantially smaller number of evaluations of the support function in order to obtain a polytope with the same number of vertices.

The application of the CP method makes it possible to construct the EPH in a reasonable amount of time, in spite of a lengthy evaluation of the support function. In turn, this facilitates the construction of a DSS for water quality planning in large rivers.

Acknowledgment

This paper was supported in part by the Russian Foundation for Basic Research under grant No 98-01-00323.

References

Bushenkov, V. A., O. L. Chernykh, G. K. Kamenev, and A. Lotov (1995). "Multi-dimensional Images Given by Mappings: Construction and Visualization", *Pattern Recognition and Image Analysis* 5(1), 35-56.

Bushenkov, V., V. Kaitala, A. Lotov, and M. Pohjola (1994). "Decision and Negotiation Support for Transboundary Air Pollution Control between Finland", Russia and Estonia, *Finnish Economic Papers* 7(1), 69-80.

Chernykh, O.L. (1992). "Construction of a Convex Hull for a Set of Points as a System of Linear Inequalities", *Zh. Vychisl. Mat. Mat. Fiz.* 32(8), 1213 - 1228 (in Russian).

Chernykh, O.L. (1995). "Approximation of Pareto Hull of Convex Set by Polyhedra." *Zh. Vychisl. Mat. Mat. Fiz.* 35(8), 1285-1294 (in Russian).

Cohon, J. (1978). *Multiobjective Programming and Planning.* New York: Academic Press.

Cohon, J. L. and D. M. Marks (1975). A Review and Evaluation of Multiobjective Programming Techniques, *Water Resource Research* 11(2).

Dorfman, R. (1965). Formal Models in the Design of Water Resource Systems, *Water Resources Research* 1(3), 329-336.

Fourier, J. B. (1826). "Solution d'une Question Particulière du Calcul des Inegalités." *Oeuvres II*, 317-328.

Haimes, Y.V., K. Tarvainen, T. Shima, and J. Thadathil (1990). *Hierarchical Multiobjective Analysis of Large-Scale Systems.* New York: Hemisphere Publishing.

Kamenev, G. K. (1994). "Study of an Algorithm for the Approximation of Convex Bodies." *Zh. Vychis Mat. Mat. Fiz.* 34(4), 608-616 (in Russian).

Kamenev, G. K. (1996). "Algorithm of Converging Polytopes." *Zh. VychisL Mat. Mat. Fiz.* 36(4), 134-147 (in Russian).

Kamenev, G. and D. Kondratiev (1992). "A Method for the Analysis of Non-closed Non-linear Models", *Mathematical Modeling* 4(3), 105-118 (in Russian).

Kamenev, G. K., A. V. Lotov, and P.E.V. van Walsum (1986). *Application of the GRS Method to Water Resources Problems in the Southern Peel Region of the Netherlands*, CP-86-19. IIASA, Laxenburg, Austria.

Keeney, R. L. and H. Raiffa (1976). *Decisions with Multiple Objectives: Preferences and Value Tradeoffs*. New York: Wiley.

Lomov, B. F. (1984). *Methodological and Theoretical Problems of Psychology*. Nauka Publishing House, Moscow, Russia (in Russian).

Lotov, A. V. (1973). An Approach to Perspective Planning in the Case of Absence of Unique Objective. *Proceedings of Conference on Systems Approach and Perspective Planning*. Computing Center of the USSR Academy of Sciences, Moscow, Russia.

Lotov, A. V. (1994). *Integrated Assessment of Environmental Problems*. Computing Center of Russian Academy of Sciences, Moscow, (in Russian).

Lotov, A. V. (1996). Comment to the Paper by D. J. White "A Characterization of the Feasible Set of Objective Function Vectors in Linear Multiple Objective Problems", *European Journal of Operational Research* 89(1), 215-220.

Lotov, A. V., V. A. Bushenkov, and A.V. Chernov (1997). Feasible Goals Method in Web: Experimental INTERNET Resource for Development of Independent Strategies, http://www.ccas.ru/mmes/mmeda/resource/.

Lotov, A. V., V. A. Bushenkov, and O. L. Chernykh (1997a). "Multi-criteria DSS for River Water Quality Planning", *Microcomputers in Civil Engineering* 12(1), 57-67.

Lotov, A. V., V. A. Bushenkov, G. K. Kamenev, and O. L. Chernykh (1997b). *Computers and the Search for Balanced Trade-off: Feasible Goals Method*. Moscow: Nauka Publishers, (in Russian).

Loucks, D.P., J.R. Stedinger, and D.A. Haith (1981). *Water Resources Systems Planning and Analysis*. Englewood Cliffs, NJ: Prentice-Hall.

Loucks, D.P., P.N. French, and M.R. Taylor (1985). "Interactive Data Management for Resource Planning and Analysis." *Water Resources Research* 21, 131-142.

Moiseev, N.N. (1982). *Mathematical Problems of Systems Analysis*. Moscow: Nauka Publishers, (in Russian).

Thiessen, E.T. and D.P. Loucks (1992). "Computer Assisted Negotiation of Multiobjective Water Resource Conflicts." *Water Resource Bulletin*, 28/1.

Williams, F. and J.V. Pavlik (eds.) (1994). *The Peoples Right to Know: Media, Democracy and the Information Highway*. Hillsdale, NJ: Erlbaum.

Wierzbicki A. (1981). "A Mathematical Basis for Satisficing Decision Making." in J. Morse (Ed.), *Organizations: Multiple Agents with Multiple Criteria*. Berlin: Springer, 465-485.

8 INTEGRATED RURAL ENERGY DECISION SUPPORT SYSTEMS

Shaligram Pokharel and
Muthu Chandrashekar

1. Introduction

Rural populations in most developing countries have to meet their energy requirements mainly with locally available biomass because of the lack of other energy alternatives. Local energy supplies are decreasing, however, because of indiscriminate use
of local resources in the past. Some efforts have been undertaken to augment these
supplies by installing new technologies, subsidizing fuel costs, and promoting interfuel substitutions. However, these programs are usually initiated at the national level
(Tingsabadh, 1988) and often overlook local needs because they do not involve local
participation. The absence of local participation in rural development programs causes
conflicts that often lead to the failure of such programs (Chambers, 1993).

The Asian and Pacific Development Centre (APDC, 1985) has suggested that, in
order to develop consistent energy policies, energy plans should be integrated with
national economic plans. The Food and Agriculture Organization (FAO, 1993) suggests that rural energy plans would be more realistic if they were coupled with rural
economic planning activities. In this chapter, we propose a rural energy decision support system (REDSS) to integrate the data generation, storage, and management capabilities of a geographic information system (GIS) and the analysis capability of multiobjective programing (MOP) methods. The results obtained from the system could
be integrated later with rural development plans.

In a REDSS, energy data are generated from a GIS database. A collection of such
data is referred to as an energy information system (EIS). The data are then analyzed
with a suitable MOP method through a consideration of conflicting issues such as
cost, employment, and environment. We expect that REDSS will provide a more realistic rural energy program and, because of local participation, the chosen program
would have a greater chance of sustainability. The planners would include some im-

portant objectives from the local participants to analyze in REDSS. In this study, employment generation is considered as the main objective of local people.

2. Rural energy flow

The energy flow path from sources to end uses in a typical rural area is shown in Figure 1.

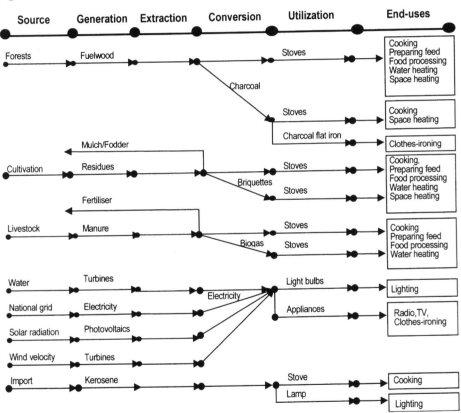

Figure 1. Energy flow diagram for a typical rural area.

The three main stages of energy transformation in an energy path are generation, conversion, and utilization. Some of the energy resources might have to be extracted for nonenergy use. In Figure 1, crop residues and animal manure have reverse arrows to show that they could be extracted for fodder and fertilizer. Consideration of these factors would allow planners to have more realistic data on resource availability for energy use.

Cooking, lighting, heating, feed preparation, and food processing are the main energy end uses in rural areas. Since more than two-thirds of the energy consumed in rural areas is for household use, a small alteration in the pattern of household energy consumption can bring significant changes in total energy consumption. Our analysis, therefore, focuses on energy consumption by rural households.

3. Energy information system

GISs are one of the tools used in REDSS. In their simplest form, GISs can be described as systems to store and manage location-specific information such as population, forests, livestock, and stream type, called *information layers* (as shown in Figure 2). In spatial analysis these information layers are processed at various stages and in different combinations to obtain application-specific information. The collection of energy information generated by the analysis of GIS data is referred to in this paper as EIS. For example, a GIS database might contain forest information by tree species, density, and area. From this, the quantity of available fuelwood from a patch of forest can be calculated through the use of established conversion factors. These patches can also be regrouped to show high and low fuelwood-yielding areas. In this example, the forest database is stored in a GIS and fuelwood information is stored in an EIS.

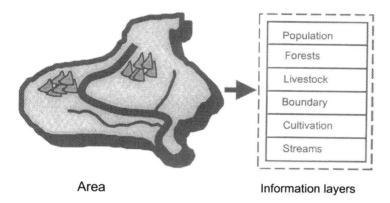

Area   Information layers

Figure 2. Information layers in geographic information systems.

Reviews of different GIS software are found in Peuquet and Marble (1990). These software programs have been used in applications like natural resource management, transportation planning, and environmental impact analysis. GIS applications are being introduced in many developing countries (Yeh, 1991). The availability of spatial data and the need to address rural energy problems in many developing countries will make the implementation of REDSS easier and its analysis more realistic.

4. Multiobjective programing methods

Energy planning should consider aspects like economy and environment, which are often conflicting and competitive. A suitable MOP method can be used to analyze such conflicting aspects. Various MOP methods have been used in the analysis of route selection, resource allocation, and location of radioactive disposal sites. Chetty and Subramanian (1988) and Ramanathan and Ganesh (1993) used goal programing, a MOP method, to analyse energy resource allocation.

For a set of objectives, $f_i(x)$, with decision variable x, and set X of feasible constraints, a MOP problem could be written as

$$Min [f_i(x)], \quad i = 1, ..., q$$
$$s.t. \qquad x \in X \subset \mathbb{R}^n$$
$$where\ X = \{ x \mid g_j(x) \geq 0, j = 1, ..., n; x_j \geq 0 \}. \tag{1}$$

When equation (1) is solved for the kth objective, an optimum value f_k^* is obtained. In a single-objective programing (SOP) problem, this optimum value forms a basis for decision-making. In a MOP problem, for each objective i, ($i = 1, ... , q$), there may be a different optimal solution. The adoption of an option (solution) for which all objectives take their optimum values is impossible, as they lie outside the feasible objective space for a two-objective minimization problem, as shown in Figure 3. A point that represents all optimum values is called an *ideal point* (or *ideal solution*). Therefore, in MOP problems, a compromise between the objectives should be worked out.

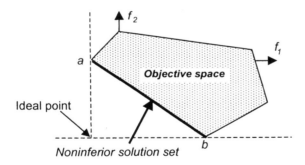

Figure 3. Ideal point and noninferior solution in MOP.

A feasible solution x^* to a MOP problem is called a *non-inferior solution*. A solution is non-inferior if there are no other feasible solutions in which the value of one objective function is improved without degrading the values of other objective functions. A collection of non-inferior solutions is called a noninferior solution set. In Figure 3, line ab of objective space is the non-inferior set. Therefore, optimal solutions a and b are also non-inferior solutions – that is, one of them could be adopted as the decision for implementation. It is assumed here that decision-makers (DMs), a group of analyst(s), planners and local representatives, will seek a planning option (a non-inferior solution) by interacting with a suitable MOP algorithm.

Various methods have been developed to analyze problems with more than one objective. A review of such methods is given in Shin and Ravindran (1991). For rural energy planning, we have chosen the STEP-Method (STEM) proposed by Benayoun et al. (1971). This method generates one non-inferior solution in one iteration and provides DMs with energy allocations for each solution. This method is relatively simple to understand, code, and implement in a generic SOP software; therefore, it has a greater chance of adoption in rural energy analysis. Applications of STEM for planning and resource allocation are discussed in Johnson and Loucks (1980), Ramirez-Rosado and Adams (1991), and Benson, Lee, and McClure (1997).

A brief review of STEM is given below. Each objective of the DMs is optimized first in STEM. The ideal solution obtained this way is arranged in a matrix and used to calculate relative weights, π_i, for each objective. These weights are used from the second iteration on. For the second iteration, the problem is formulated as

$$
\begin{aligned}
&\textit{Minimize } \delta \\
&\text{s.t.} x \in X \\
&\delta \geq \pi_i * [f_i^* - f_i], \quad i \in \mathbb{R}^n ,
\end{aligned}
\tag{2}
$$

where δ is the deviation between the ideal solution and the preferred non-inferior solution.

The non-inferior solution x^* to equation (2) is presented to the DMs, who, if satisfied, accept it as a decision and terminate the analysis. Otherwise, the aspiration level for one (or more) objective $f_m \in f_i^*$ is changed by some amount Δf_m and treated as a constraint X. The modified problem is solved to obtain a new solution. This process continues until the DMs have obtained a satisfactory non-inferior solution.

5. Energy decision support system

We have proposed that an energy decision would be better represented if it could be analysed with GIS and MOP. The use of GIS and MOP is complementary because GIS provides storage of data and MOP provides a multifaceted analysis capability on the database to arrive at a decision. The hierarchical structure of REDSS in relation to the national planning system is shown in Figure 4. It is expected that the outputs obtained from the proposed system could be used to formulate better rural energy plans, which would lead to a better national energy plan.

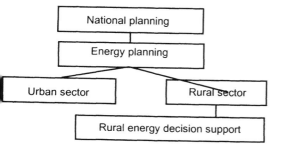

Figure 4. Rural energy decision support and national planning.

6. System architecture

The process to obtain an energy database in REDSS is shown in Figure 5. Thematic and remote sensing maps, aerial photographs, and digitized information are first collected and converted to digital format in GIS software. This information is analyzed by means of appropriate attributes to generate an EIS. Information layers in an EIS can be grouped into energy-source layers and energy-consumption layers. Energy-source layers, for example, are combinations of each energy source available in a rural area.

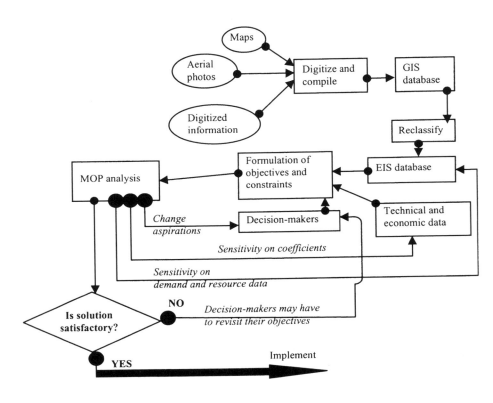

Figure 5. The system architecture of REDSS.

Spatial analysis of energy-source and energy-consumption layers helps to generate an energy-balance map which shows the regions with an energy supply/demand balance, supply deficit, and supply surplus. This energy-balance map helps DMs understand in more detail the existing relation between consumption centres and available resources. It also assists in designing location-specific energy options such as installation of solar photovoltaic modules in a particular village to supply energy for lighting. Traditional methods of compiling energy data for a planning area dilute such information, which is very important to microlevel energy planning. Energy-balance information as well as technical and economic data are required to formulate objectives and constraints, as specified by DMs. These formulations are then analyzed to obtain a satisfactory non-inferior solution. Sensitivity analysis on data and changes in aspiration levels on objectives might be necessary to arrive at a solution. DMs might have to revisit their original objectives and constraints if they are not satisfied with all of the possible noninferior solutions generated by STEM.

Both GIS and EIS modules are implemented in ARC/INFO[1] software. The ARC Macro Language (AML) has been used for display and structural queries. MOP analysis is implemented in LINDO[2] software.

7. System application

The decision support system has been applied to examine the energy situation of the Phewatal watershed, which covers about 120 km^2 in western Nepal. There are six administrative zones, called village development councils (VDCs). With help from the government, each VDC can develop small-scale natural resource-management plans such as community forestry and development of hydropower projects.

The map of the study area is shown in Figure 6. High annual precipitation there creates numerous seasonal streams, but there are only three large ones, whose confluence at an altitude of about 900 m forms Harpan Khola (stream), which are capable of producing hydropower. Harpan Khola drains to Lake Phewa at the eastern end of the watershed.

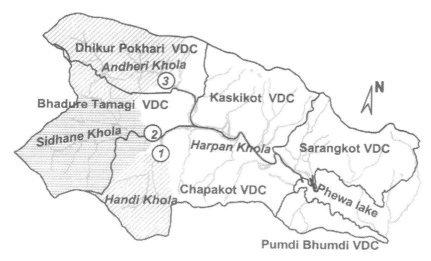

Figure 6. Map of the study area showing VDCs, streams, and major catchment areas.

Fuelwood is the main energy source in the watershed and its use is beyond the sustainable supply capability of the forests. Overextraction has degraded some of the forest patches severely, causing soil erosion and reducing the water-holding capacity of the soil. If an energy plan could be designed and implemented to alter the local energy consumption patterns, fuelwood consumption might be reduced to a sustainable production level. This might help in forestry management and interfuel substitution.

Forests (mostly hardwood) cover almost 46% of the watershed. As shown in Figure 7, most forested areas lie in the southern part. The distribution of forest by tree species and annual sustainable fuelwood production as obtained from EIS is shown in Table 1. Accessible forest area is estimated to be about 75% of the total (IWMP, 1992); therefore, the accessible sustainable fuelwood supply is about 9 900 tons (t).

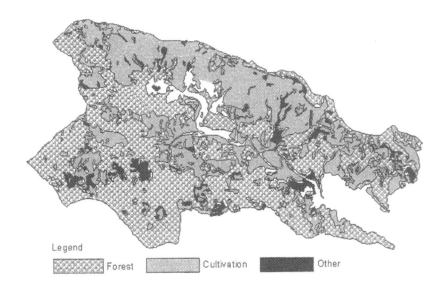

Figure 7. Major land-use pattern in the study area.

Table 1. Annual sustainable fuelwood supply.

Forest types	Area (km^2)	Fuelwood supply (t)
Shrub	3.3	210
Hardwood (<10%)	8.6	74
Hardwood (10%-40%)	10.0	1 246
Hardwood (40%-70%)	21.0	5 809
Hardwood (> 70%)	12.0	5 787
Plantation	0.7	45
Total	55.6	13 162

Note: % refers to crown density.

Cultivated land, which covers about 37% of the watershed area, is located mainly in the north and in the Harpan Khola valley. Crops such as rice, maize, wheat, millet, potato, mustard, and soybean are grown in the watershed. While rice is the main crop in the valley, maize and millet are the main crops in the uplands. The cropping intensity varies between 260% in uplands and 150% in the valley. Crop residues are used mainly as fodder; as a result, only about 4 670 t of crop residues are available for energy purposes (Table 2).

Table 2. Annual crop residue production.

Cultivated land	Area (km^2)	Crop residue produced (t)	Crop residues for energy (t)
Terrace (25-50%)	1.8	245	67
Terrace (50-75%)	20.8	4 842	1 503
Terrace (> 75%)	16.8	5 700	1 817
Slope	0.4	65	18
Fan (25-50%)	0.4	77	35
Fan (50-75%)	0.7	218	102
Fan (> 75%)	2.1	897	429
Valley	3.3	1 497	706
Total	46.3	13 541	4 677

Note: % refers to the cultivated portion of the land.

Of an estimated population of 31 000, 58% are economically active and 8% are unemployed on an annual basis (DECORE, 1991). The average population density is 258 persons/km^2, although it varies from 92 in the Chapakot VDC to 363 in the Dhikur Pokhari VDC. The annual population growth is estimated at 1.6%.

The people keep about 20 000 head of large livestock (cattle, buffalo, sheep and goats), and the manure they produce is used as fertilizer. Some biogas plants have been installed in the valley and biogas is used for cooking.

Currently, electricity is supplied through the national grid in some parts and is used mainly for lighting. There is a possibility of producing electricity from solar radiation and water. The watershed receives about 1.7 million watt-hours (MWH)/km^2 of solar radiation on an annual basis. Photovoltaic modules could be installed in barren lands to generate solar electricity. The three potential hydropower sites are indicated by the three circles in Figure 6. Altogether, they can produce about 260 gigajoules (GJ) of electricity for lighting. One of the three sites (in the Dhikur Pokhari VDC) lies on the northern part of the watershed.

Cooking is the main energy end use in the watershed and is done mainly with fuelwood. Lighting is the second most important, with both electric lights and kerosene lamps in use.

Figure 8 shows the relative shares of energy resources while Figure 9 maps the energy demand patterns for the watershed. The energy balance map obtained by overlaying these two maps is shown in Figure 10.

The energy balance map shows that if all resources could be used, and all combinations of interfuel substitutions (for example, the use of fuelwood for lighting) were possible, then the watershed would have a supply surplus. If only sustainable energy supplies in VDCs were considered, the northern VDCs would have a deficit of about 42 000 GJ (2 500 t) in fuelwood supply. Therefore, DMs might have to adopt a plan to reduce deforestation in the northern part. Such a plan requires development and supply of alternate energy sources. The installation of biogas, efficient fuelwood stoves (EFS), and switching to kerosene could be some of the options to consider. If alternate energy sources were not made available, the cross-VDC flow of fuelwood from southern to northern VDCs would have to continue. However, with decentralization of resource use, southern VDCs could stop such a practice at anytime, thereby creating an energy crisis.

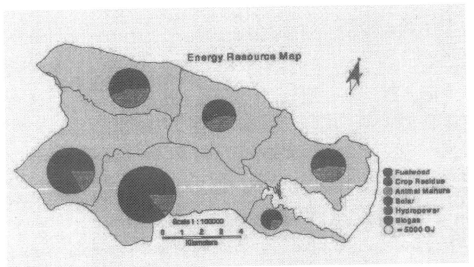

Figure 8. Energy resource map.

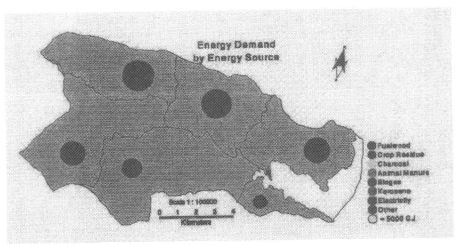

Figure 9. Energy demand map.

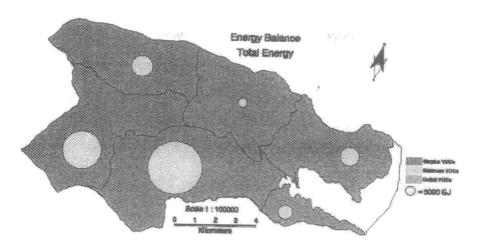

Figure 10. Energy balance map.

8. The multiobjective model

Three planning objectives have been considered in this paper for the evaluation of energy options. Tor illustrate MOP-based energy analysis, only fuelwood-deficit VDCs have been considered. The energy data for these three VDCs have been combined and analyzed. However, analysis could be done for each VDC or watershed as a whole.

Let x_{ijk} be the decision variable representing energy supplied by fuel type i (for example, fuelwood) in end-use device j (for example, EFS) to provide an energy service k (for example, cooking). Let the subscripts represent the following: $i = 1 =$ fuelwood, $2 =$ crop residue, $3 =$ kerosene, $4 =$ hydroelectricity, and $5 =$ biogas.

Let $j = 1 =$ tripod stove, $2 =$ traditional stove, $3 =$ heating stove (open), $4 =$ EFS, $5 =$ biogas stove, $6 =$ kerosene lamp, $7 =$ kerosene stove, and $8 =$ electric light bulb. Similarly, let $k = 1 =$ cooking (including feed preparation), $2 =$ heating, $3 =$ food processing, and $4 =$ lighting. The subscript 'N' refers to the new installation of a technology that is already present in the planning area. The use of solar photovoltaics is not considered, mainly because of the lack of expertise in the installation of such modules in and around the watershed.

The objectives and constraints are formulated in the following sections. The coefficients, such as end-use device efficiency, costs, and employment used in the analysis are estimated on the basis of the work of Pokharel and Chandrashekar (1995).

8.1. Formulation of objectives

8.1.1. Minimize costs to the government for implementation of proposed solution (coefficients are in 000 Rupees[Rs]/GJ). The cost indicated below refers to the first cost because the government should be ready to spend that amount during the first year of program implementation. The fuelwood variable coefficients include the cost of forest management. The crop residue coefficients refer to the collection cost. The EFS (x_{141N}) coefficient includes the costs for EFS, human labour, and forest management. The kerosene coefficients (x_{371} and x_{364}) refer to the cost to the government in terms of subsidy (Rs 2/L). In the case of hydropower, the coefficient refers to the subsidy provided for power generation from a small-scale hydropower turbine. The biogas coefficient (x_{551N}) refers to the subsidy (Rs 10 000/plant). We have assumed the installation of a 10-m^3 biogas plant, which can produce about two m^3 of biogas on a daily basis and is the most popular design in Nepal. The objective function for cost minimization is formulated as

$$0.003\,(x_{111} + x_{121} + x_{132} + x_{123} + x_{141}) + 0.06\,x_{141N} + 0.002\,(x_{211} + x_{232}) +$$
$$+\,0.005\,(x_{361} + x_{374}) + 0.25\,x_{484} + 0.7\,x_{551N}\,.$$

8.1.2. Maximize rural employment (coefficients are in employable person-years/GJ). The generation of employment through energy projects can motivate the local people to accept the proposed changes. The objective function for maximization of employment is formulated as

$$0.01\,x_{141N} + 0.003\,(x_{361} + x_{374}) + 0.09\,x_{484} + 0.009\,x_{551N}\,.$$

8.1.3. Maximize system efficiency (coefficients are end-use efficiencies). The end-use device efficiency used for lighting represents end-use utilization efficiency. In rural areas, for example, if kerosene lamps are to be replaced by electric light bulbs, the replacement would take place on the basis of the number of lamps and not on the basis of candlelights of illumination. The objective system efficiency function is formulated as

$$0.03\,x_{111} + 0.1\,x_{121} + 0.2\,(x_{141} + x_{141N}) + x_{132} + 0.1\,x_{123} + 0.03\,x_{211} + x_{232} +$$
$$+\,0.4\,x_{361} + x_{374} + x_{484} + 0.4\,x_{551} + 0.4\,x_{551N})\,.$$

8.2. Formulation of constraints

The constraints are formulated below.
1. For energy demands to be met :
$$0.03\,x_{111} + 0.1\,x_{121} + 0.2\,(x_{141} + x_{141N}) + 0.03\,x_{211} + 0.4\,x_{361} + 0.4\,(x_{551} + x_{551N}) \geq$$
6 595;
$$x_{132} + x_{232} \geq 2\ 800;$$
$$0.5\,x_{484} + x_{374} \geq 2476 \text{ (for lighting, 1 GJ of kerosene} \approx 0.5 \text{ GJ of electricity); and}$$
$$x_{123} \geq 9\ 960.$$

2. Potential for generation of hydro power:
$$x_{484} \leq 35 \text{ (10 kW installation).}$$

3. Limit to biogas installation in a year:
$$x_{551N} \leq 280 \text{ (20 plants).}$$

4. Limit on the use of crop residues as fuel:
$$x_{211} + x_{232} \leq 1\ 200.$$

5. Limit on fuelwood extraction on a sustainable basis:
   $x_{111} + x_{121} + x_{141} + x_{141N} + x_{132} + x_{123} \leq 68\,570$.

6. Limit on the installation of efficient fuelwood stoves:
   $x_{141} \leq 630$.

7. Limit on the supply of kerosene:
   $x_{361} + x_{374} <= 5\,000$.

8. Requirement for the existing EFS and biogas energy supply:
   $x_{141} = 1346$; and
   $x_{551} = 40$.

9. All decision variables should be non-negative:

   $x_{ijk} \geq 0$.

Individual optimization of objectives indicates that the ideal solution would be to adopt a program that costs Rs 225 000, or generates 24 person-years of employment, or requires 14 500 GJ of final energy to meet all energy demands. Any one of these solutions could be selected for implementation. The cost minimization option indicates that hydropower, EFS, and biogas plants should not be installed even if they provide much needed labour and energy locally: instead, the energy deficit should be met by kerosene. To maximize employment, however, biogas, EFS, and hydropower should be installed. The objective of maximizing system efficiency emphasizes the use of more local energy sources, and therefore promotes the use of crop residues, biogas, and hydropower. All of the above solutions indicate that the use of the tripod stove for cooking should be discouraged. If the DMs are satisfied with one of these solutions, the analysis is terminated. Otherwise, another solution is generated by changing the value (aspiration level) of one objective.

For this analysis, the cost of the project was increased from the optimal cost to Rs 245 000. With this increase, the amount of energy required was decreased, mainly because of increased use of kerosene for cooking. The employment level was decreased to 20 person-years for the same reason.

Table 3. The result of the MOP analysis: x values are in gigajoules (GJ).

x_{111}	x_{121}	x_{132}	x_{123}	x_{141}	x_{141N}	x_{211}	x_{232}	x_{371}	x_{364}	x_{484}	x_{551}	x_{551N}
0	52 025	3 098	9 960	1 346	488	0	1 200	2 524	2 476	0	40	0

The energy allocations for this iteration as shown in Table 3, which indicates that the use of tripod stoves for cooking (x_{111}) should be discouraged, while the use of traditional stoves (x_{121}) should be encouraged. Any deficit in cooking energy should be met by kerosene (x_{371}). The installation of hydropower (x_{484}) and new biogas plants (x_{551N}) should be put off for the time being. If the DMs are satisfied with these energy allocations, then the solution is adopted for implementation and the analysis is terminated. Otherwise, analysis should be carried further by changing the cost of the project again and performing sensitivity analysis, or by changing the aspiration levels of other objective functions.

9. Conclusions

We have shown that the GIS and MOP platform can be used to develop a decision support system for rural energy planning. At present, REDSS is an academic exercise and the objectives are designed by authors to simulate an energy planning process. However, the analysis shows a potential practical use of REDSS for the examination of rural energy problems.

Implementation of REDSS in the Phewatal watershed shows that biogas and fuelwood could be two major energy sources for cooking. If kerosene were affordable, progressively less fuelwood would be required for cooking. Interfuel substitution of this type can help protect degraded forests. Options like biogas, hydropower, and EFS are very attractive, as they reduce energy consumption per household. They do not merit installation, however, if only costs are considered. Such alternative energy sources and devices should be judged in terms of increased local energy supply capability and a more sustainable energy future.

Acknowledgments

This research was partially funded by the Natural Sciences and Engineering Research Council (NSERC) of Canada.

Endnotes

1.   ARC/INFO is a registered trademark of Environmental Systems Research Institute.
2.   LINDO is a trademark of LINDO Systems, Inc.

References

APDC (1985). *Integrated Energy Planning: A Manual*. R. Codoni, H. Park, K. and Ramani (Eds.), Asian and Pacific Development Centre, Kuala Lumpur, Malaysia.

Benayoun, R, J. de Montgolfier, J. Tergny, and O. Laritchev (1971). "Linear Programming with Multiple Objective Functions: STEP-Method." *Mathematical Programming* 1, 366-375.

Benson, H.P., D. Lee, J. P. and McClure (1997). "A Multiobjective Linear Programming Model for the Citrus Rootstock Selection Problem in Florida." *J. Multi-criteria Decision Analysis* 6, 283-295.

Chambers, R. (1993). *Rural Development: Putting the Last First*, Wiley, UK.

Chetty, K.M. and D.K. Subramanian (1988). "Rural Energy Consumption Patterns with Multiple Objectives." *Int. J. of Energy Research*, 12, 561-567.

DECORE (1991). *Socio-economic Baseline Survey of Phewatal Watershed*, Integrated Watershed Management Project, Development, Communications and Research Consultancy Group, Kathmandu, Nepal.

FAO (1993). *A New Approach to Energy Planning for Sustainable Rural Development*, Food and Agriculture Organization, Rome, Italy.

IWMP (1992). *Watershed Management Plan of Phewatal Watershed,* Integrated Watershed Management Project, Kathmandu, Nepal.

Johnson, L.E. and D.P. Loucks (1980). "Interactive Multiobjective Planning using Computer Graphics." *Comput. Ops. Res.* 7, 89-97.

Peuquet, D. J. and D.F. Marble (Eds.) (1990). *Introductory Readings in Geographic Information Systems*, Taylor and Francis, UK.

Pokharel, S. and M. Chandrashekar (1995). "Analysis of Cooking Energy in Developing Countries." *Natural Resources Forum* 19(4), 331-337.

Ramanathan, R. and L.S. Ganesh (1993). "A Multiobjective Programming Approach to Energy Resources Allocation Problems." *Int. J. of En. Res.* 17, 105-119.

Ramirez-Rosado, I.J. and R.N. Adams (1991). "Multiobjective Planning of Optimal Voltage Profile in Electric Power Distribution Systems.", *Compel* 10, 115-126.

Shin, W.S. and A. Ravindran (1991). "Interactive Multiple Objective Optimization Survey I - Continuous Case." *Comput. Ops. Res.* 18(1), 97-114.

Tingsabadh, C. (1988). "Decision Flows in the System" in *Rural Energy Planning: Asian and Pacific Experiences*, K. V. Ramani (Ed.), Asian and Pacific Development Centre, Kuala Lumpur, Malaysia.

Yeh, A.G. (1991). "The Development and Applications of Geographical Information Systems for Urban and Regional Planning in the Developing Countries", *Int. J. Geographical Information Systems* 5(1), 5-27.

9 DECISION SUPPORT SYSTEM FOR SUSTAINABLE LAND MANAGEMENT

A South-East Asia Case

Mohammad Rais, Samuel Gameda,
Eric T. Craswell, Adisak Sajjapongse and
Hans-Dieter Bechstedt

1. Introduction

Decision support systems (DSSs) have usually been developed in the context of a vast information technology infrastructure. Organizations that develop DSSs can draw on the necessary personnel and technology to collect, catalogue, and house large amounts of data, resource inventories, research findings, and scientific knowledge. The DSSs developed tend to be data- and information-intensive and usually require highly skilled personnel to manage and run them.

In contrast, the information technology infrastructure in developing regions is traditionally very inadequate, lacking in sufficient data, natural resource inventories, and trained personnel. Existing data are unreliable and full of discrepancies. In fact, weak national information infrastructures are thought to be the main hindrances to the creation and use of DSSs in developing regions (see Chapter 19). There is, therefore, a need for a new generation of DSSs that can bridge the gap between conventional information technology tools and the needs of decision-makers in developing regions. These needs require contextually and operationally relevant, as well as technologically simple and accessible DSSs that facilitate the dissemination of appropriate techniques and technologies, the identification of constraints to sustainability, and the practice of sustainable production systems.

A range of initiatives necessary for the development of DSSs for developing countries is provided in Chapter 19. DSSs for sustainable development should address decision-making contexts and processes in developing regions; development of software workbenches or toolkits for DSSs including knowledge-based systems geographic information systems (GISs) and process models; development of databases that include local practitioner or indigenous knowledge; and appropriate training and capacity building for DSS developers and users.

This chapter explores the possibility of developing a DSS for sustainable land management (SLM) with a focus on data and knowledge-base issues. This is done in the context of regions where, although data is often scarce and the use of information technology rare, the impact of land management decisions at the local level could profit from informed decision-making. This would have far-reaching consequences on the sustainability of natural resources in surrounding regions. Sustainable development and the definition of SLM, including a methodology to evaluate the sustainability of land management systems and practices, are discussed. Next, the challenge and opportunities for information gathering in data poor regions are explored. Finally, the methodology to elicit a knowledge base from land users, extension officers, and agricultural scientists in three regions in South-East Asia, and the resulting prototype DSSs for assessing the sustainability of different land management systems, are presented.

2. Sustainable land management

Sustainable development consists of objectives that are economically viable, environmentally sound, and socially acceptable. It requires the consideration of disparities in spatial and temporal scales, and an understanding of the intricate interdependence of economic, social, and environmental factors (Campbell, 1995). An evaluation methodology that encompasses these issues is central to the assessment of SLM systems.

An international working group, headed by the International Board for Soil Research and Management (IBSRAM) and the Research Branch of Agriculture and Agri-Food Canada (AAFC) in collaboration with the Food and Agriculture Organization (FAO) of the United Nations, and others, has developed a scientifically based, international Framework for the Evaluation of Sustainable Land Management (FESLM) (Smyth and Dumanski, 1993). The FESLM is based on the definition of SLM as a system that combines technologies, policies, and activities aimed at integrating socioeconomic principles with environmental concerns so as to maintain or enhance production and services simultaneously; reduce the level of production risk; protect the potential of natural resources; be economically viable; and be socially acceptable (Dumanski and Smyth, 1994). The five basic pillars of SLM are its five objectives: productivity, security, protection, viability, and acceptability.

The FESLM consists of a logical pathway analysis procedure for guiding the evaluation of land-use sustainability through a series of scientifically sound steps (Smyth and Dumanski, 1993). It comprises three main stages: i) identification of the purpose of evaluation, specifically land-use systems and management practices; ii) definition of the process of analysis, consisting of evaluation factors, diagnostic criteria, indicators, and thresholds to be utilized; and iii) an assessment endpoint that identifies the sustainability status of the land-use system under evaluation.

2.1. Local knowledge

In many developing countries, gaps and/or inaccuracies in information limit the quality of SLM evaluation, even in cases where exhaustive efforts are made to amass the necessary data. In such instances, it is essential to mobilize all data and information resources available. Where quantitative cause-and-effect relationships are absent or scarce, particularly across biophysical and socioeconomic domains, qualitative local or indigenous knowledge can be an important complement to decision support for sustainable development. Local perceptions of the operational issues and strategies for achieving sustainability are key components that should be integrated into DSSs.

Local or indigenous knowledge encompasses implicitly aggregated facts and information, incorporates uncertainty, and draws from experience, resulting in intuitive, general relationships or correlations between different factors affecting sustainability. The importance of the integration of local farmer knowledge with scientific principles has been recognized in many areas.

Some researchers have shown the merits and efficiency of using local knowledge for soil classification in developing regions (Sandor and Furbee, 1996; Zimmerer, 1994; Pawluk et al., 1992; Furbee, 1989). Others have shown that farmers can provide reliable assessments of soil health and quality (Romig et al., 1995), and that the more experienced local land users implement effective and economical soil conservation methods (Rajasekaran and Warren, 1995). The World Bank has recognized the importance of incorporating farmer-provided production estimates as a source of valid crop-production data, and has developed methodologies for including these estimates in assessments of regional crop production figures (Murphy et al., 1991).

2.2. DSSs for sustainable land management

The flexibility of FESLM methodologies and the characteristics of expert system technologies lend themselves to the integration of local knowledge with scientific research for the development of DSSs for SLM. The FESLM accommodates the use of different levels of available data and information, ranging from rigourous quantitative relationships to local or indigenous knowledge. Expert systems provide a framework for capturing non-algorithmic knowledge (Plant and Stone, 1991), can make use of qualitative knowledge, incorporate uncertainty, and provide solutions using disparate and/or incomplete information.

Several studies have demonstrated the feasibility of using expert systems for integrating local knowledge in decision-making (Furbee, 1989; Gameda and Dumanski, 1998). Moreover, researchers and practitioners meeting at an expert group workshop have identified the importance of using indigenous knowledge as an integral information source for the development of DSS for sustainable development (see Chapter 19).

Researchers at IBSRAM, in collaboration with the Land Resources Program, Research Branch, AAFC, are implementing the FESLM within a computerized DSS for sustainable land management on sloping lands in South-East Asia. The SLM-DSS development process targets two types of decision-maker: local farmers who need to know the sustainability of their land-use systems; and extension personnel, NGOs and agribusinesses that need to assess the sustainability of the packages of technologies and practices they promote.

3. Methodology

The knowledge base for expert systems is usually developed by eliciting the expertise of a domain expert. In local or indigenous knowledge, however, agricultural expertise does not reside in one expert but is shared among many experienced farmers. The procedures used by knowledge engineers to extract expertise from domain experts and structure the knowledge base are suitable for eliciting the knowledge of agricultural researchers and extension specialists.

It is relatively difficult to tap and represent the expertise of experienced farmers, because the knowledge base may overlap or conflict within the expert group. This implies the need for large-scale extraction of knowledge, coupled with resolution of any conflicting information that may arise. In particular, there is greater need for interpretation of farmer-provided information in order to transform it into a form suitable to an expert system framework.

The procedure for developing the knowledge base for the SLM-DSS entailed several stages:

- knowledge acquisition by means of case studies and in-depth interviews;

- structuring the information into relationships between pertinent FESLM parameters; and

- the establishment of correspondences between qualitative local knowledge and quantitative scientific research.

3.1. Knowledge acquisition and representation

To provide reliable SLM decision support, the characteristics that constitute sustainable land use within given agro-environmental and socio-economic conditions need to be known. An efficient way to achieve this is by identifying sustainable land-use systems, characterizing the attributes that make them sustainable, and identifying criteria by which to measure the status of these attributes in other land-use systems.

Three case studies, based on guidelines derived from FESLM methodology (Bechstedt and Renaud, 1996), were conducted in Thailand, Indonesia, and Vietnam (Wattanasarn *et al.*, 1997; Santoso *et al.*, 1997; Phien *et al.*, 1997) to identify factors affecting all dimensions of sustainability, critical indicators that can serve for monitoring the status of identified factors, and thresholds beyond which land management systems were considered unsustainable. Questionnaires were prepared with the cooperation of agricultural scientists, soil scientists, rural anthropologists, and economists for use in the case studies. Data and information were solicited from farmer cooperators in IBSRAM's ASIALAND Management of Sloping Lands (ASL/MSL) Network. Information and knowledge from participating farmers were elicited through the use of participatory rural appraisal principles (Chambers, 1994). In addition to the questionnaires, in-depth interviews were carried out with extension workers and regional agricultural scientists. Information from the case studies was supplemented with data from long-term research on IBSRAM research sites, and from pertinent scientific literature.

The information provided by farmers tended to be of a sensory, symbolic and qualitative nature. For example, farmers would describe their land as exhausted, and that this could be observed from the colour of their soils. They would infer soil fertility and nutrient availability from plant growth and leaf colour, and would associate this with crop yield reductions across soil textures, farm fields, and seasons to obtain heuristic relationships between crop growth, leaf colour, and crop response. Similarly, they would describe their access to markets in terms of the time required to reach them by walking or on a drawn cart. In contrast, researchers and regional specialists would identify the status of soil organic matter as a measure of soil quality and productivity, relationships between crop yields and nutrient availability as a measure of soil fertility, and the types of transportation and roads, and actual distances as a measure of market accessibility.

It is necessary to establish correspondences between the types of local understanding that farmers have and the more quantitative ones from standard research methodologies in order to use the information across regions and translate it into structured relationships more suited to DSSs.

The concurrence between local knowledge expressed in farmers' symbolic understanding and scientific research was achieved through extensive consultations and in-depth interviews with regional agricultural scientists, rural anthropologists, socioeconomists, and extension workers. These specialists were provided with the suite of indicators obtained from farmers and were asked to identify, on the basis of their experience or from the scientific literature, what quantitative characteristics best corresponded with each of the indicators. Sufficient consensus on the correspondence between a local indicator and its scientific equivalent indicated the establishment of an acceptable relationship between qualitative and quantitative parameters. Examples of correspondences between such parameters are given in Table 1.

Table 1. Examples of correspondences between quantitative and qualitative parameters for SLM.

SLM pillar	Quantitative parameter	Qualitative equivalent
Productivity	Soil organic matter	Soil colour
	Availability of nutrient N	Colour of whole leaves; vigour of crop growth
	Availability of nutrient K	Colour of leaf edges; vigour of crop growth
Security	Long-term precipitation trends & statistical variability	Tactile level of soil dampness at seeding
Protection	Amount of topsoil loss	Depth of rills & gullies; extent of crops covered by deposition
Economic viability	Annual income of farmers; number of all-season roads in region	Economic status of farmer; walking distance to market

Symbolic relationships obtained from combined farmer knowledge and scientific research were used as the basis for structuring the DSS. The range of information provided by experienced farmers identified indicators for each of the pillars of SLM. Initial identification of these indicators did not differentiate among them in terms of their relative analytic importance, although it was clear that they had implicit criteria for selective use of the indicators. To signify the relative importance, rank indicators were obtained.

It was important to associate a relative score with each indicator in order to categorize the indicators obtained from local knowledge for use in the knowledge base for expert system development. A two-stage weighting system was used to achieve this. Initially, indicators were classified as strategic, cumulative, or conditional, on the basis of the following criteria : srategic indicators provide definite information on the sustainability of land use systems under evaluation; cumulative indicators identify sustainability status additively; conditional indicators imply sustainability status but need to be in combination with cumulative indicators to provide definitive information on land management sustainability. After a classification based on these criteria, the indicators were ranked according to their relative analytical merit. Class scores were multiplied by rank values to give a weighting, or certainty factor, for each indicator.

Two examples of indicators, thresholds, qualitative-quantitative correspondences, and scores based on SLM pillars and used for developing the DSS knowledge base, are given in Tables 2 and 3. A description of each indicator is given in terms of its local representation.

For the yield indicator presented in Table 2, the measures are given in terms of the observed crop yields in comparison to average yields obtained for the community. The threshold indicates the value beyond which the system is considered unsustainable on the basis of FESLM criteria. For example, the threshold for the yield indicator shows that any values less than 25% of the community average are not considered sustainable.

The next two columns of each table provide the qualitative or local knowledge representation, and its quantitative equivalent, respectively. A score of 10 for the yield indicator identifies it as a strategic indicator, indicating that the yield status on a given farm definitively reflects its sustainability with respect to productivity.

While scores were applied at the indicator level, ranks were used to differentiate analytical merits within indicators. For example, a yield rating of "high" (qualitative), or greater than 25% of the community average (quantitative) would provide definite indication of sustainability. A rating of "low", or less than 10% of the community average, would give reliable indication of the lack of sustainability in the system, but this indication would be less definitive than a high rating. A yield rating of "medium" would give indications that are less definitive than either high or low yield ratings.

Values given in the last column of each table are a product of scores and ranks. They represent the certainty factor to be associated with rules or assertions related to the status of indicators. If we again consider the yield indicator, it can be said with 100% certainty that a yield rating of high indicates sustainable productivity.

Table 2. Productivity criteria: indicators, thresholds, and scores.

Indicators	Type	Threshold	Qualitative ranking	Quantitative ranking	Score	Rank	Value
Yield	1	>25% yield reduction rel. to average	Yield reduction: High Medium Low	> 25 % 10 - 25 % < 10 %	10 10 10	10 5 7	100 50 70
Soil colour: Organic C	1	< 1.2 %	High: dark soil Medium: brown soil Low: yellowish	> 1.2 % (reduction 0%) 1-1.2% (reduction 0-20%) < 1 % (reduction > 20%)	10 10 10	7 5 7	70 50 70
Plant growth, leaf colour: Nutrient N	2	< 0.5 %	High: dark green leaves, healthy growth Medium: colour normal, moderate growth Low: yellowish leaves, stunted growth	> 0.5 % 0.2 - 0.5 % < 0.2	7 7 7	7 5 7	49 35 49
P	2	> 15 ppm	High: growth normal, colour normal Medium: growth normal Low: older leaves purple, stunted growth	> 15 ppm 8-15 ppm < 8 ppm	7 7 7	7 5 7	49 35 49
K	2	> 90 ppm	High: normal growth Medium: normal growth Low: leaves yellowish, older leaves show symptoms first	> 90 ppm 60 - 90 ppm < 60 ppm	7 7 7	5 5 10	35 35 70

Indicators type and their score : strategic (1)=10; cumulative (2) =7 ; conditional (3)=3; relative ranking: 1 to 10. Value = score x rank.

Table 3. Economic viability criteria: indicators, thresholds and scores.

Indicators	Type*	Threshold	Qualitative ranking	Quantitative ranking	Score	Rank	Value
Benefit cost ratio	1	B:C ratio 1.00 or more	Rising	> 1.25	10	10	100
			Constant	1.0	10	7	70
			Declining/fluctuating	< 1.0	10	5	50
Percentage of off-farm income	2	25 % or more	High	> 25 %	7	7	49
			Medium	10-25 %	7	5	35
			Low/none	< 10 %	7	7	47
Farm gate price vs. nearest market price	2	> 15 %	High	> 50 %	7	7	49
			Medium	15 – 50 %	7	5	35
			Low	< 15 %	7	7	49
Availability of farm labour	2	2 full-time adults	High	> 2 full-time	7	7	49
			Medium	1-2 full-time	7	5	35
			Low	1 full-time	7	7	49
Size of farm holding	3	1 ha	High	> 2 ha	3	7	21
			Medium	1 - 2 ha	3	3	9
			Low	< 1 ha	3	5	15
Availability of farm credit	3	> 50 % of requirements	High	> 50 %	3	5	15
			Medium	25 – 50 %	3	3	9
			Low	< 25 %	3	3	9
Percentage of farm produce sold in market	2	> 50 %	High	> 50 %	7	5	35
			Medium	25 – 50 %	7	3	21
			Low	< 25 %	7	3	21

*Indicators type and their score : strategic (1)=10; cumulative (2) =7 ; conditional (3)=3; relative ranking: 1 to 10. Value = score x rank.

For indicators falling under a classification of cumulative or conditional, the ability to draw conclusions on the basis of a few indicators is substantially lower. For example, when evaluating the percentage of off-farm income (Table 3), it is possible to be only 35% certain that an off-farm income rating of medium (10 – 25% of income comes from off-farm sources) indicates the economic sustainability of the production system. In other words, the status of other economic viability indicators is required to make more conclusive economic sustainability evaluations.

3.2. DSSs and sustainability evaluation

A rule-based approach was used to develop the SLM-DSS. The rule base was developed from SLM indicators and their associated scores. It reflects qualitative-quantitative correspondences established in the knowledge representation process. A schematic of the SLM-DSS is given in Figure 1.

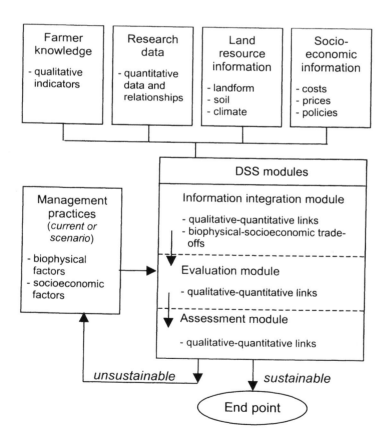

Figure 1. Schematic of the DSS for sustainable land management.

The user is prompted to respond to a set of questions based on qualitative indicators reflecting local knowledge. The rule base was structured in such a way that the quantitative correspondences to user responses were determined, and sustainability assessed, for each SLM pillar. Two examples of DSS-SLM screens consisting of selections for user response and associated rules are given in Figure 2. Conclusions on the sustainability of a given land management system are reached from an accumulation of the evidence based on user response.

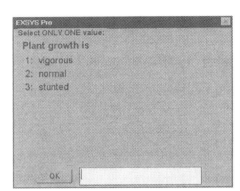
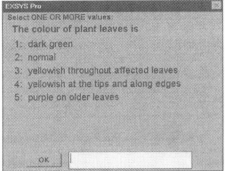

Figure 2. Screens of selections for user response.

Associated rules

IF:	Plant growth is vigorous
THEN:	Soil fertility is good
AND:	Land management practices meet sustainability requirements, on the basis of FESLM productivity criteria

IF:	Plant growth is normal
THEN:	Soil fertility is sufficient
AND:	Land management practices are marginally above the threshold for sustainability, on the basis of FESLM productivity criteria

IF:	Plant growth is stunted
THEN:	Soil fertility is poor
AND:	Land management practices do not meet sustainability requirements, on the basis of FESLM productivity criteria

IF:	The colour of plant leaves is dark green
THEN:	Soil nutrient availability is good
AND:	Land management practices meet sustainability requirements, on the basis of FESLM productivity criteria

IF:	The colour of plant leaves is yellowish throughout affected leaves
THEN:	The availability of soil nutrient - N is limited
AND:	Land management practices are marginally below the threshold for sustainability, on the basis of FESLM productivity criteria

IF: The colour of plant leaves is yellowish at the tips and along edges
THEN: The availability of soil nutrient - K is limited
AND: Land management practices are marginally below the threshold for
           sustainability, on the basis of FESLM productivity criteria

IF: The colour of plant leaves is purple on older leaves
THEN: The availability of soil nutrient - P is limited
AND: Land management practices are marginally below the threshold for
           sustainability, on the basis of FESLM productivity criteria

The DSS is designed to determine the sustainability of land management systems on the basis of the SLM definition. It is envisioned that regional extension officers or local NGO workers, with knowledge of the local agroclimatic conditions and farming practices, will facilitate the input of information from individual farmers. The sustainability status of a particular farm is provided as one of four possible scenarios:

- land management practices meet sustainability requirements;

- land management practices are marginally above the threshold for sustainability;

- land management practices are marginally below the threshold for sustainability; or

- land management practices do not meet sustainability requirements.

3.3. DSS system design and development

A protocol based on the system development life cycle was used, as much as possible, to develop the SLM-DSS. It consists of four main stages: i) requirements definition; ii) system specification; iii) design and development of the system and program; and iv) system testing. The requirements definition stage consisted of the establishment of a comprehensive set of user requirements on the basis of a series of interviews with potential users, mainly extension officers and regional agricultural scientists. The tasks undertaken during the system specification stage consisted of the translation of user requirements into a description, and a schematic of the proposed system and its data requirements. The design and development of the system and program stages consisted of the technical description and actual coding of the SLM-DSS. The resulting prototype underwent system testing by the developers, knowledge-base providers and potential users.

There were various strengths and weaknesses observed from experience in the development of the SLM-DSS. The main developers of the DSS were researchers with substantial domain knowledge in SLM, and experienced in the dissemination of SLM methodologies and techniques to local agricultural producers. Consequently, there was a good definition of the user requirement, so that the first stage of the system development life cycle could be satisfactorily completed. System specification and design and development stages were simplified by the choice of an expert system shell running under a PC/Windows environment as the main tool for the knowledge-base implementation.

Nevertheless, the development process could have been significantly enhanced by the participation of computer scientists and information technology specialists. This emerged during the system testing stage when the strength of the knowledge base was acknowledged, but where the lack of in-depth reasoning and explanatory capacity, and the excessive simplicity of the user prompts were highlighted. Collaboration between domain knowledge experts, computer scientists, and information technology specialists is necessary to incorporate the full capability of artificial intelligence and knowledge-based systems into a DSS. The testing stage also indicated several capabilities that need to be included during future enhancements of the DSS. Since the DSS is targeted to land management systems, one of the main recommendations was for the incorporation of spatial data, or GIS capabilities. Another recommendation was that the DSS should be programed with the capacity to link with, or draw data from, various crop growth and socioeconomic models or algorithms.

The collaboration and enhancements mentioned above are necessary if the DSS is not to remain a research exercise, but is to be deployed as an effective tool in land management systems and sustainable development. In making the recommended enhancements, care should be taken to maintain the simplicity of the user interface while building powerful analytical, reasoning, and explanatory capabilities into the DSS.

4. Conclusions

The knowledge base development process for the SLM-DSS was successful because of a number of key components. The criteria for the building of the knowledge base were based on a sound methodology, the FESLM. Development of the surveys for knowledge acquisition involved the collaboration of biophysical scientists, socioeconomists, rural anthropologists, and regional agricultural scientists. Finally, local knowledge was identified and integrated into the DSS as an important and necessary source of information and decision-making.

The FESLM-based case studies and the corresponding knowledge representation efforts have provided a means for establishing explicit criteria and correspondences between local knowledge and scientific research for a comprehensive evaluation of the sustainability of different land management systems. The inclusion of local farmer knowledge was also useful for addressing data and information gaps, and for determining farm-level SLM indicators.

The surveys and interviews conducted to develop the knowledge base for DSS development were prepared and carried out at a reasonable cost over a period of one-and-a-half years. Obtaining the information across the three regions by conventional means through the use of methods such as site characterization, sampling, field testing, data analysis, and compilation would have taken considerably more time and resources, and would have been too expensive to carry out.

It was obvious that this first series of knowledge base development did not fully capture the depth of knowledge involved in decision-making for achieving sustainability. This attempt could benefit from a second series of knowledge acquisition and representation exercises aimed at addressing the gaps and weaknesses in knowledge from the first effort at DSS development. An enhancement of the knowledge base with information on regional- and national-level parameters that impact on local sustainability also would be useful in giving it greater depth and breadth. This could be accomplished through surveys and interviews of regional and national planners and policymakers as well as through the use of demographic and census data. Additional information suitable for monitoring the natural resource base can also be obtained by remote sensing, thereby enhancing the integration of local knowledge with data acquired through the use of information technology tools.

The DSS development process requires the collaboration of domain knowledge experts with computer scientists and information technology specialists for successful implementation and deployment. Systems developed primarily by domain knowledge experts will have a strong knowledge base, but will be poor in the implementation of artificial intelligence and information technology capabilities. Conversely, ones developed by computer scientists will have good depth and complexity in analytical, reasoning, and explanatory capabilities, but may be poor in the depth of domain knowledge incorporated into the DSSs.

The SLM-DSS currently remains in the prototype stage. Plans are underway to enhance the system, on the basis of testing results and user feedback. These include modification of the DSS structure to incorporate greater reasoning and explanatory capabilities, and to develop linkages for GIS and modeling functionality. The opportunity is there for the developer of the SLM-DSS, IBSRAM, to collaborate with a computer science faculty to bring about these enhancements. The collaboration would be beneficial to IBSRAM because its capacity to include information technology tools for the dissemination of SLM techniques would be enhanced.

The DSS development process is useful for cases of poor data availability, dissimilar knowledge types, and dispersed sources of knowledge and expertise. This methodology shows the feasibility of mobilizing local knowledge for the development of DSSs in most developing countries, where data is often scarce or unreliable, but where a wealth of knowledge resides.

References

Bechstedt, H.-D. and F. Renaud (1996). "Protocol for Conducting Case Studies under the Framework for Evaluation of Sustainable Land Management", Singapore: IBSRAM.

Campbell, C.A. (1995). "Landcare: Participative Australian Approaches to Inquiry and Learning for Sustainability", *J. Soil Water Conservation* 50(2), 125-131.

Chambers, R. (1994). "The Origins and Practice of Participatory Rural Appraisal", *World Development* 22(7), 953-969.

Dumanski, J. and A. J. Smyth (1994). "The Issues and Challenges of Sustainable Land Management", *Proc. International Workshop on Sustainable Land Management for the 21st Century*, in R.C. Wood and J. Dumanski (eds.). Vol. 2: Ottawa: Agricultural Institute of Canada, 11-22.

Furbee, L. (1989). "A Folk Expert System: Soils Classification in the Colca Valley, Peru", *Anthropol. Quarterly* 62(2), 83-102.

Gameda, S. and J. Dumanski. 1998. SOILCROP: A Prototype Decision Support System for Soil Degradation. Crop Productivity Relationships", in A. El-Swaify (Ed.), *Multiple Objective Decision Making for Land, Water, and Environmental Management*, Boca Raton, FL: Lewis Publishers, .349-362

Murphy, J., D. J. Casley and J. J. Curry (1991). "Farmers' Estimations as a Source of Production Data: Methodological Guidelines for Cereals in Africa", World Bank Technical Paper No. 132, 71pp.

Pawluk, R.R., J. A. Sandor. and J. Tabor (1992). "The Role of Indigenous Soil Knowledge in Agricultural Development", *J. Soil Water Conservation* 47(4), 298-302.

Phien, T., T.D. Toan, et al. (1997). "Framework for evaluation of sustainable land management: A case study of rainfed cropping systems on sloping lands in northern Vietnam", Singapore: IBSRAM/NISF.

Plant, R.E. and N. D. Stone. (1991). *Knowledge-Based Systems in Agriculture,* New York: McGraw Hill.

Rajasekaran, B. and D. M. Warren (1995). "Role of Indigenous Soil Health Care Practices in Improving Soil Fertility: Evidence from South India", *J. Soil Water Conservation* 50(2), 146-149.

Romig, D.E., M. J. Garlynd, R. F. Harris, and K. McSweeney (1995). "How Farmers Assess Soil Health and Quality", *J. Soil Water Conservation* 50(3), 229-236.

Sandor, J.A. and L. Furbee, (1996). "Indigenous Knowledge and Classification of Soils in the Andes of Southern Peru", *Soil Sci. Soc. Am. J.* 60, 1502-1512.

Santoso, D., A. Rachman, et al. (1997). "Framework for evaluation of sustainable land management: A case study at Pauh Menang Village, Indonesia" Singapore: IBSRAM/CSAR.

Smyth, A.J. and J. Dumanski (1993). FESLM: An international framework for evaluating sustainable land management. World Soil Resources Report 73. FAO, Rome.

Wattanasarn, C., B. Tansiri, and others (1997). Framework for evaluation of sustainable land management: A case study at Pha Duea, Maechan District, Chiang Rai Province, Thailand. Singapore: IBSRAM/DLD.

Zimmerer, K.S. (1994). "Local Soil Knowledge: Answering Basic Questions in Highland Bolivia", *J. Soil Water Conservation* 49(1), 29-34.

10 MAPPING CRITICAL FACTORS FOR THE SURVIVAL OF FIRMS
A Case Study in the Brazilian Textile Industry

Carlos A. Bana e Costa, Émerson C. Corrêa, Leonardo Ensslin and Jean-Claude Vansnick

1. Introduction

The textile industry is one of the most important economic sectors in the Brazilian State of Santa Catarina. It comprises more than 3 000 registered firms, most of which are small (some very small) or medium-sized enterprises (SMEs), and employs about 100 000 in all. The Brazilian textile industry is currently in crisis, as a consequence of Brazil's entry into the process of economic globalization. According to official state data, employment in the textile sector has decreased by 25% over the last six years. Although all enterprises have been affected, SMEs in particular are facing severe commercial and financial problems or even going bankrupt. Their financial, technological, and managerial structures (inadequate equipment, old-fashioned management, and rigid — or complete lack of — planning) have kept them from being competitive on world markets.

This chapter describes a project conducted at the Federal University of Santa Catarina, aimed at helping SMEs in the textile industry to implement strategies to prevent failure, and ensure sustainable survival. An earlier study (Bana e Costa et al., 1999) has addressed this problem already, and shown how useful decision support methods and systems developed by problem structuring and multicriteria approaches can be for strategic management. Here, the discussion will focus on the cognitive mapping process of identifying critical (or key) success factors (C(K)SFs)[1]. Identifying C(K)SFs is a crucial step in the process of developing sustainable survival strategies for a firm in today's rapidly changing socioeconomic environments.

In Section 2 we present the basic concepts relevant to the study, in particular the concept of C(K)SFs, as they appear in management literature. Section 3 introduces a constructive perspective for the identification of C(K)SFs and posits the Kelly-based version of cognitive mapping as a means of putting it into practice. Finally, the process of identifying C(K)SFs for Santa Catarina textile SMEs is described in detail and a list of 11 C(K)SFs is provided. A brief conclusion ends the chapter.

2. Relevant conceptual background

The concept of C(K)SFs was first introduced in the management literature by Daniel (1961) in a discussion of the "management information crisis," almost four decades ago:

> "{...} a company's information system must be discriminating and selective. It should focus on 'success factors.' In most industries there are usually three to six factors that determine success; these key jobs must be done exceedingly well for a company to be successful."

One decade later, Anthony et al. (1972) rediscovered the usefulness of Daniel's concept while designing a management control system. According to Rockart (1979), "Daniel focused on critical success factors that are relevant for *any* company in a *particular* industry"; Anthony et al. "went a step further", suggesting that "there are other sources of CSFs than the industry alone", and thus, "CSFs differ from company to company and from manager to manager." However, whether a C(K)SF is mainly environment-dependent, industry-based, or firm-specific is not of key importance. The authors of this chapter agree with Leidecker and Bruno (1984) that all levels of analysis are important:

> "Critical Success Factors (CSFs) are those characteristics, conditions, or variables that when properly sustained, maintained, or managed can have a significant impact on the success of a firm competing in a particular industry. {...} The concept of critical success factors has been applied at three levels of analysis (firm-specific, industry, and economic, sociopolitical environment). Analysis at each level provides a source of potential critical success factors. Firm-specific analysis utilizes an internal focus to provide the link to possible factors. Industry-level analysis focuses on certain factors in the basic structure of the industry that significantly impact any company's performance operating in that industry. A third level of analysis goes beyond industry boundaries for a source of critical success factors. This school of thought argues that one needs to perpetually scan the environment (economic, sociopolitical) to provide sources that will be the determinants of a firm's and/or industry's success. We believe all three levels of analysis have merit as sources for critical success factors."

The concept of C(K)SFs was popularized in the field of management information systems by the end of the 1970s, when Rockart's MIT research team revealed the importance of identifying C(K)SFs in the design of information systems and termed their approach the "CSF method" (Rockart, 1979). Conceived as a means of "helping executives to define their significant information needs," this approach focuses on the company and works toward the market.

Another approach starts with the customers (external analysis) and works backwards (self-analysis) to the firm. This approach is advocated in the field of strategic (market) management, in which C(K)SFs are seen as a means to pursue a "competitive advantage" (Porter, 1985; Day 1986, 1990; Aaker, 1984, 1989, 1995).

The case study illustrated in this chapter adopts a strategic market-oriented perspective.

There are strategic factors other than C(K)SFs that are mandated by the particular economic environment and are beyond the control of management but nevertheless can have an impact on the success or failure of firms, especially SMEs (Covin and Slevin, 1989). We will refer to them as strategic risk factors (SRFs: Bouquin, 1986; Verstraete, 1997). SRFs are mainly "external risk factors," but managers should be aware of the fact that because C(K)SFs are likely to change, the uncertainty of projecting them into the future is itself an SRF that affects the sustainability of a survival strategy.

If the set of C(K)SFs in a particular industry is intended to include *all* "the factors that determine ultimate success or failure" (Day, 1986), then SRFs would be considered just a particular type of C(K)SF. Nevertheless, we prefer to differentiate the two concepts by limiting the notion of C(K)SFs to only those factors that can be managed at the firm level.

The identification of the relation between SRFs and C(K)SFs is of paramount importance for the management of the C(K)SFs of an industry, particularly in business climates that are affected significantly by external variables, such as in Santa Catarina. "The firm's sensitivities for unexpected changes in the risk factors can themselves be viewed as critical success factors" (Brännback and Spronk, 1997). The influence of external sources of risk can be captured through macroanalyses of the environment and the industry (Figure 1).

One example is the instability of interest rates. Moreover, Rockart (1979) has noted that "for smaller organizations within an industry dominated by one or two large companies, the actions of the major companies will often produce new and significant problems for the smaller companies." This applies directly to the textile industry in Santa Catarina. Unexpected actions on the part of the major textile enterprises are a SRF for the SMEs, as was the case when some major enterprises decided to update their production technology.

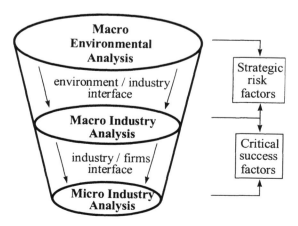

Figure 1. Frame of analysis.

3. Identifying critical factors through cognitive mapping

3.1. A constructive perspective for the identification of C(K)SFs

The identification of C(K)SFs is not an easy task. Many techniques have been proposed in the management literature (Leidecker and Bruno, 1984; Desreumaux, 1993; Verstraete, 1997). Most of them rely on a normative approach that is essentially based on some notion of "objectivity". We propose a constructive interactive approach that leaves space for actors' participation and "subjectivity". It is never certain that a particular aspect is critical/key for survival/competitiveness. In other words, there is no purely objective procedure that definitively identifies a C(K)SF.

These considerations, together with the scarcity of reliable information in developing economies (Bornstein and Rosenhead, 1990), necessitate the direct involvement of industry/business experts and problem-owners in a "soft" strategic thinking learning process (Rosenhead, 1989; van der Heijden, 1996). The articulation of objective information is required, together with the exploration of "tacit knowledge" (Brännback and Spronk, 1997). Managers and experts have to articulate issues, concepts, and consequences that are often based on influences such as tradition, intuition, values, beliefs, experience, and faith.

We stand for a market-oriented expert-based learning process to help identify C(K)SFs. This paradigm is predicated on two general methodological convictions for decision-aiding, suggested by Bana e Costa and Pirlot (1997):

> "**The conviction of the interconnection and inseparability of the objective and subjective elements of a decision context:** a decision process is a system of relations among elements of an objective nature (e.g., the characteristics of the actions) and elements of a subjective nature (stemming both from the value systems of the actors and the ill-defined decision context). These types of elements are intricately linked and neither can be neglected in decision support. Although the search for objectivity must be of great concern, one must not forget that decision-making is above all a human activity subtended by the notion of value. Subjectivity is always omnipresent in decision-making, and therefore one must recognise the limits of a pure objective approach. {…}
> **The conviction of constructivism and learning:** a decision situation is, in general, an ill-defined entity, unclear even to the actors involved in the decision process. A constructivist methodological approach is therefore most appropriate for decision-aiding, in light of a new paradigm of learning, that should replace the old paradigm of normative optimisation (prevalent in Operational Research and Systems Analysis). A direct implication of the adoption of a constructivist attitude in decision-aid is that simplicity and interaction are fundamental tools for effective participation."

This constructive perspective leads to a cognitive notion of a C(K)SF as an aspect perceived by the involved actors as being — by itself or together with other aspects — a manageable fundamental contributor to the success or failure of a firm. This dichotomy of success or failure is inherent to the cognitive aspect of C(K)SFs and presents a firm's attractiveness as a bipolar concept, or "construct" (Kelly, 1955) with a positive pole (strength) and a negative pole (weakness). In his Personal Construct Theory, Kelly (1955) stated that "a person's construction system is composed of a finite number of dichotomous constructs, each of which has a pole of affirmation and a negative pole, rather than their being categories of a unipolar type."

3.2. The Kelly-based version of cognitive mapping

There are many ways to help people to address a problem situation, from the least structured method, consisting of giving "people time and space to 'ramble' around [the] subject", to "structured interviews of a question/answer nature where the interviewer runs through a list of questions devised in advance" (Eden *et al.*, 1983). It is the authors' opinion, however, that the semistructured interactive approach called cognitive mapping is a better way to learn from managers and experts about the essential aspects that determine the success or failure of firms.

In general terms, a cognitive map can be defined as "a graphical representation of a set of discursive representations made by a subject with regards to an object in the context of a particular interaction" (Cossette and Audet, 1992). Since Tolman (1948) introduced the term, cognitive maps have been used in various domains of knowledge, including strategic decision-making, management and organizational studies (Eden, 1992). Fiol and Huff (1992) offer an overview of the many graphic maps that managerial cognition may take and discuss different map functions and dysfunctions.

This case study follows the Kelly-based version of cognitive mapping developed by Eden and his co-workers. This system is a powerful facilitative device for identifying C(K)SFs (see Section 3.1). In this style of map, each node is a concern represented by a construct, with its presented pole separated from its contrast pole by "..." (meaning "rather than"). Perceived influence relations between concerns are represented in the map by arrows between nodes. As explained by Eden (1988):

> "The linkage between constructs represents the meaning of the construct in terms of the explanations and consequences - these links are not taken to be causal in a precise way. {...} an arrow out of a construct shows a consequence and an arrow into a construct an explanation; each arrow therefore gives explanatory meaning to one construct and consequential meaning to another. A negative sign on the 'head' of an arrow implies that the first pole of the explanatory construct implies the second pole of the consequential construct."

Of particular relevance to this study is the ability of cognitive mapping to highlight key factors. In this case, Eden's Kelly-based version of cognitive mapping helped to identify concerns about the problem, and to recognize the most relevant ones later on. To facilitate this process, we used a decision support system (DSS) software developed by Eden and his colleagues once called Graphics COPE, and recently renamed Decision Explorer (Banxia, 1997).

3.3. The cognitive mapping process

This section describes in detail the interaction between two actors who were directly involved in the process of identifying C(K)SFs for textile SMEs in Santa Catarina. One is an expert in the textile industry, a consultant for several SMEs who was the key interlocutor in providing the market-oriented external perspective. The other is the owner-manager of a firm that produces 40 000 pairs of jeans per month. His participation offered an internal perspective of a typical SME.

First, the authors met with the expert twice in one day – in the morning for about two hours, and in the afternoon for one hour. The goal was to help the expert to identify specific concerns and their relationships. The first meeting was a free-thinking reflection session to discuss

- an initial statement of the expert's view of the issues: SMEs are having difficulties selling their products; and

- two broad but related questions: What factors most affect these difficulties? and How can SMEs improve their situation?

As the expert spoke, the authors drew a map, identifying the links that emerged between identified concepts. This process was quite helpful; when the expert began to understand the logic of the cognitive mapping technique, it became obvious that the map was helping him to identify new concepts and relationships. Several times he commented that the map had helped him see links he had not seen before. Other times he would say, after examining the map, "I don't think this concern is well enough explored. I believe we could go deeper into it;" or "I think this concept has some influence on that one, too. Let's link them."

Essentially, cognitive mapping was used as a semistructured device to generate ideas, facilitated by questions such as "why, or for what, do you think this aspect matters?" Making the map while the expert was speaking resulted in a messy hand-written graph. The purpose was to elicit from the expert a large number of concerns about the critical situation facing the textile firms. To illustrate the process and the movement from a messy and almost unreadable structure to a clean computer generated map we show this structure in Figure 2.

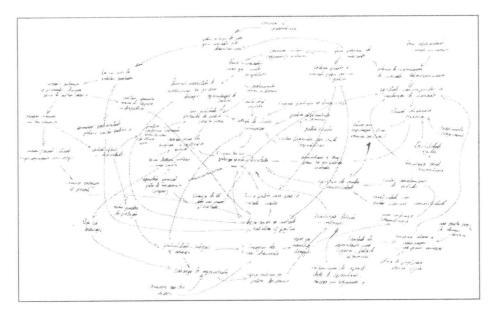

Figure 2. Cognitive map constructed with the expert.

Obviously, considerable work was necessary to clarify and arrange the original map (Figure 4 shows an English version of this map after it had been arranged and some remarks by the owner-manager had been added). As Eden and Simpson (1989) remark, "we do not expect the map of an interview to be: (i) tidy; (ii) easily readable by anyone other than the note-taker; (iii) complete; or (iv) a map with many contrasting poles." To provide a "more understandable" version of the map for a second discussion, the authors would work at "cleaning up" the map during the day. For example, they would reorganize it by joining closely related concepts, fill it in with additional concepts (raised during the interview but not immediately included), and fill in any missing contrasting poles.

The objective of the second meeting with the expert was to review the map by elaborating on concerns that had not been adequately explored that morning, and to review the information. It was important to schedule this second meeting as soon as possible after the initial one so that the expert's memory would be fresh, and valuable time would not be wasted (Belton et al., 1997).

After the meetings with the expert, the owner-manager of the firm was consulted. He was shown the expert's map and invited to highlight concerns he found fundamental and contribute any key factors he thought were missing. Figure 3 shows the map that was constructed on the basis of the notes the authors took during the one-hour interview with the owner-manager. Two important points arose during this interview.

1. In comparing the expert and the owner-manager maps, one can see that while the expert adopted a proactive attitude toward the problem, the owner-manager approached it reactively — he was dissatisfied with the current situation. According to the mapping method used, the first pole of a construct represents the descriptive dimension first proffered by the interlocutor (Eden et al., 1983). Therefore, these contrasting attitudes are directly reflected in the maps. For example, the central concern of each is stated in negative terms in the maps represented by Figures 2 and 3 (Company can't sell all its products rather than can sell everything [owner-manager], as opposed to, Company is selling all its products rather than having difficulties selling [expert]).

2. The owner-manager introduced a concern which had not been raised by the expert: SRFs. Indeed, the map in Figure 3 includes concerns (highlighted) that cannot be controlled by the firm like, for example, the competitiveness of large companies and government reaction to Asian textile imports. Both factors have a significant impact on the stability of SMEs.

The owner-manager introduced SRFs into the discussion as a justification for his firm's situation vis-à-vis some of the concerns raised by the expert that he too perceived as important for the survival his business. For example (see Figure 3), he agreed that having flexible technology would be important because it would make the firm more dynamic and enable it to produce high-quality goods. Both of these results would in turn improve his firm's capacity to sell all its products on the market. However, he immediately introduced two SRFs (high interest rates and lack of floating capital) which apparently had been preventing him from investing in flexible technology. This reasoning also highlights the reactive attitude noted above, although in this case it is perhaps realistic.

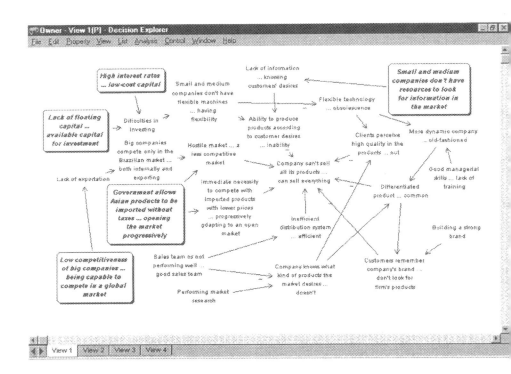

Figure 3. Cognitive map constructed with the owner-manager.

The authors moved the process forward and merged the two maps by adding the new concerns and relationships raised by the owner-manager to the expert's map. Merging maps is not usually an easy task, but in this case it was because the expert's map had been used as the basis of discussion with the owner-manager. This aggregated map is shown in Figure 4. Note that no SRFs are included, for they already had played their procedural role as a means that contributed to identifying the end-aspects that can be managed by the firm.

The map represented by Figure 4 was edited with the Decision Explorer software. Related concepts were placed as close together as possible, with as few link crossings as possible, so that the information graphics would be presented in a flexible and appealing way.

Next, to identify the most relevant aspects for the development of a sustainable survival strategy for the firm, the authors arranged for one last meeting with the expert. This meeting took place one week after the first two meetings, and the aggregated map was used as a facilitative device to identify C(K)SFs.

In preparation for the final meeting, the authors used some facilities provided by Decision Explorer to carry out several analyses that could help the expert in identifying key issues such as "consequences" of an action, "explanations" for an objective, and "cluster analysis." For example, Figure 5 shows the possible consequences of performing "market research." In this map, the dotted arrows indicate indirect influences between the concepts.

Another example is provided by Figure 6, in which the explanations for "client satisfaction rather than dissatisfaction" are presented. Note that this aspect has several components. Therefore, some of its links were considered inclusion links (indicated by the thick arrows) in the sense given by Cossette and Audet (1992), who make a relevant, though subtle, distinction between inclusion links and influence links.[2]

Cluster analysis was used to find groups of closely linked concerns. According to Eden *et al.* (1992), the "intention is to attempt the formation of clusters where the nodes in each cluster are tightly linked to one another (similar) and the number of links (or bridges) with other clusters is minimized." Each cluster can correspond to one or more key issues according to some parameters introduced into the software, namely, the target and the minimum size of the cluster.

Figure 7 shows the results of one of the many cluster analyses carried out for this study. When he saw this particular analysis, the expert immediately recognized the dynamism of the firm as a C(K)SF in the textile market. It is evident that many concerns were related to high dynamism. The identification of the elementary concerns was quite important, for it was these that were used later to make C(K)SFs operational for the evaluation of a SME's strengths and weaknesses (Bana e Costa *et al.*, 1999). On the map, the thick arrows denote strong relations, while the dotted arrows denote weak relations, as defined by the expert.

At the last meeting, the expert identified concerns which he perceived as being the C(K)SFs for textile SMEs in Santa Catarina on the basis of his experience, beliefs, and intuition, and aided by the analyses exemplified above. When asked to point out the C(K)SFs on the map, the expert surrounded each one (10 in all) with a rectangle, as shown in Figure 8.

Of the seven end-concerns directly linked to the main issue in Figure 8, four were considered to be C(K)SFs: *Efficient distribution system...*, *Good reputation...*, *Good management...*, and *Stable sales level....* Five other C(K)SFs are specifications or decompositions of the three remaining end-concerns: *High dynamism...* is a specification of *Flexible and active organization...*; *Fit costs to market reality...* specifies *Product fits the company's market...*; and *Deliveries on time...*, *Differentiated product...*, and *Clients perceive high quality in the products...* are a decomposition of *Client satisfaction...* (see the thick arrows in Figure 7). Finally, *Company knows what kind of product the market desires...*, which concerns the firm's ability to obtain information from the market (including knowledge about competitors), was considered a very important objective; for that reason, it was also perceived as a C(K)SF.

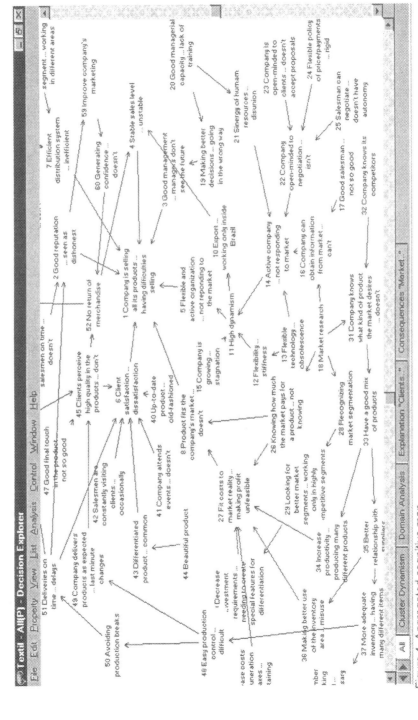

Figure 4. Aggregated cognitive map.

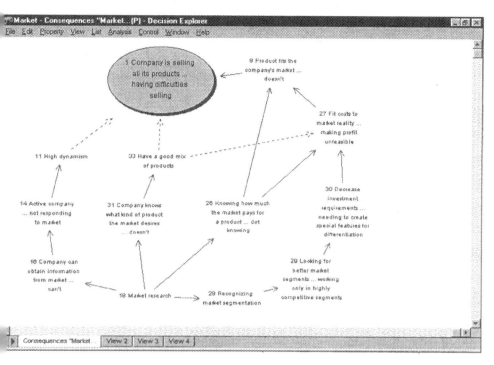

Figure 5. Consequences of performing market research mapped.

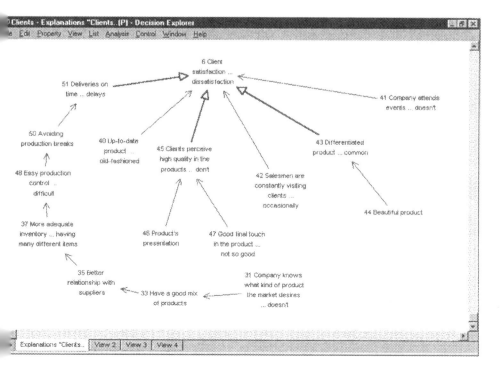

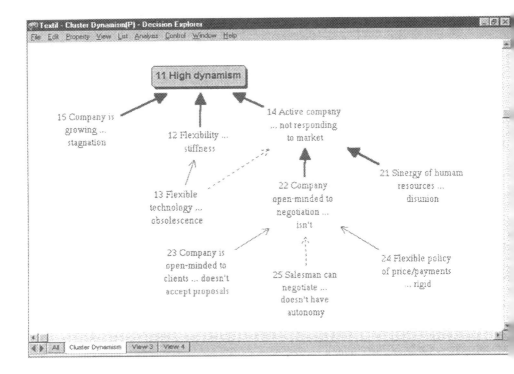

Figure 7. Cluster analysis for high dynamism on the map.

Figure 9 shows a "collapsed" view of the aggregated cognitive map. This view hides some concepts while maintaining the links (either direct or by way of other concepts) between them. It allows a clearer appraisal of the relationships among the C(K)SFs.

The final question of the meeting was to determine whether or not the set of C(K)SFs could be considered exhaustive. The answer to this question was no; for example, the expert pointed out that the concern *How competitive are the products of the firm in terms of price?* was missing. Although price is not considered in the cognitive map (probably because it was regarded as a consequence of good management of other aspects), it is indeed a C(K)SF in the market under consideration.

In the final stage of the process, each C(K)SF was described by a statement to clarify its meaning and avoid ambiguity and misunderstanding. In this way, the authors obtained the 11 factors described in section 3.4.

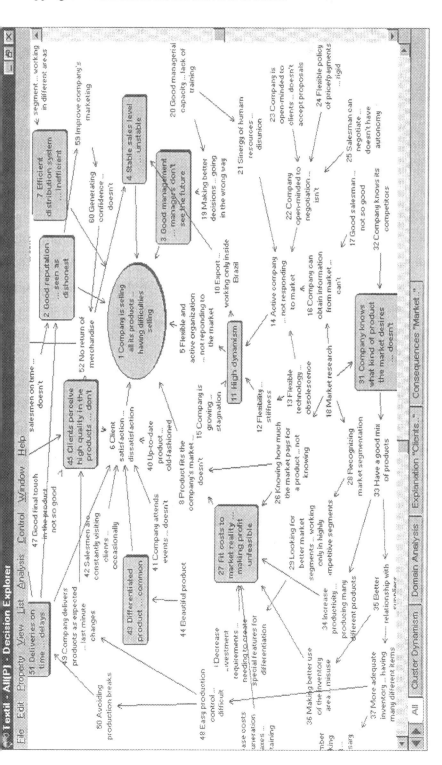

Figure 8. Identifying C(K)SFs from the cognitive map.

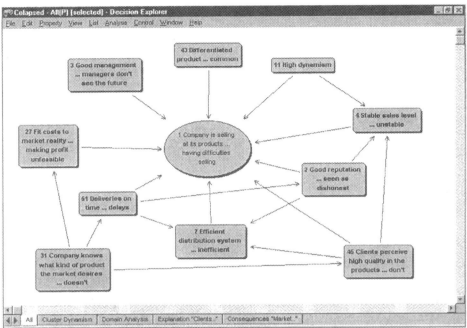

Figure 9. The "collapsed" cognitive map.

3.4. Eleven critical or key success factors

1. *Management* – the extent to which managers use available resources appropriately (both human and material) to meet client expectations. Two more elementary concerns are related to this C(K)SF:

    experience – the length of time managers have worked in the textile industry; and

    knowledge – the extent to which managers know client expectations, the available resources in the firm, the "state-of-the-art" in textile technology, and how far the firm is from it.

2. *Sales level* – the firm has a constant sales level in the Brazilian market, does not have product returns, and is able to export.

3. *Reputation* – the extent to which the firm has a good market image and is considered serious by its clients and trustworthy by its suppliers.

4. *Costs* – the extent to which production costs are known and managed, thereby improving productivity and keeping operational stock as small as possible.

5. *Dynamism* – the extent to which the firm reacts to changes in the external environment. Three more elementary concerns are related to this C(K)SF:

growth rate – the extent to which the firm is expanding its market share;
agility – the extent to which managers are flexible when negotiating with clients and staff members, or in answering client requests; and
equipment flexibility – the extent to which the firm's equipment is flexible, enabling it to retool quickly to meet fashion trends.
6. *Price* – the extent to which the firm's prices are competitive in relation to the market segment in which it is competing.
7. *Differentiation* – the extent to which the firm's products conform to the latest fashion trends and are recognized as up-to-date by the clients.
8. *Quality* – the extent to which the firm maintains a good quality level, from the client's viewpoint. Two more elementary concerns are related to this C(K)SF:
    product quality – the extent to which clients are satisfied with the final touches, conformity to standards, and the packaging of the products; and
    warranty – the extent to which clients are satisfied with the after-sales service.
9. *Delivery service* – the extent to which the firm makes sure that deliveries are made on the right dates and in the right quantities and, when there are unavoidable delays, that clients are notified.
10. *Distribution system* – the extent to which the firm competes in all potential markets, and salespeople are capable and satisfied. Three more elementary concerns are related to this C(K)SF:
    area – the extent to which the firm competes in all potential markets;
    salesperson satisfaction – the extent to which salespeople are satisfied with the financial aspects, the company's products, and their own negotiating power; and
    salesperson capability – the extent to which the firm is able to hire good salespeople.
11. *Information* – the extent to which the company can obtain relevant information about the market.

4. Conclusion

This chapter has described an application of a cognitive mapping approach to identify the C(K)SFs for textile SMEs in Santa Catarina, which the authors consider to be a fundamental step in the creation of sustainable survival strategies for those firms.

This approach was a constructive forum for discussion about strategic opportunities rather than a simple technical intervention. It allowed for tacit knowledge to be obtained directly from the actors, thus accomplishing more than would have been possible if a strictly analytical approach had been used. The DSS software Decision Explorer proved to be a user-friendly way of processing the information and an effective platform to facilitate learning and provoke debate. In the end, those involved expressed satisfaction at gaining knowledge about the capabilities and usefulness of the DSS.

Rosenhead (1996) states that "the potential of Operational Research (OR) to make a contribution to the advancement of developing countries has been recognized at least since the late 1950s. {...} Thirty or forty years on, however, the overall impact of OR is still less than was then expected." This real-world application will contribute to overcoming this drawback, by highlighting the ability of soft OR approaches to support strategic management in developing countries. In particular, this paper demonstrated the efficiency of the proposed methodology in the context of rapidly changing business environments.

Endnotes

1. The terms critical success factor (CSF) and key success factor (KSF) are used interchangeably by various authors. In this chapter, the neutral acronym C(K)SF is used, except in citations, where the original terminology is respected.

2. "It must be noted that even if the links favoured in this research are influence links, the cognitive map also contains inclusion links, that is, links characterized by the nesting of a variable in another, since participants frequently referred to these types of variables when speaking of their reality. Given that these variables can be conceived as the decomposition of a concept into its different components, even if the list is not exhaustive, these variables have, by definition, the same relationship of influence that unites the concept to other concepts. By allowing such variables to appear in a cognitive map, it is possible to obtain a more refined representation of the schema of the owner-manager." (Cossette and Audet, 1992).

References

Aaker, D. (1984). "How to Select a Business Strategy", *California Management Review* 26(3), 167-175.

Aaker, D. (1989). "Managing Assets and Skills: The Key To a Sustainable Competitive Advantage", *California Management Review* Winter, 91-106.

Aaker, D.A. (1995). *Strategic Market Management* (4th ed.), John Wiley & Sons, New York.

Anthony, R.N., J. Dearden, and R.F. Vancil (1972). "Key economic variables", in *Management Control Systems*, Irwin, Homewood, Illinois, 138-143.

Bana e Costa, C.A. and M. Pirlot (1997). "Thoughts on the future of the multicriteria field: basic convictions and outline of a general methodology", in J. Clímaco (ed.), *Multicriteria Analysis*, Springer-Verlag, Berlin, 562-568.

Bana e Costa, C.A., L. Ensslin, E.C. Corrêa, and J.-C. Vansnick (1999). "Decision Support Systems in action: integrated application in a multicriteria decision aid process", *European Journal of Operational Research* (forthcoming), 564.

Banxia Software Lda. (1997). *Decision Explorer User Manual*.

Belton, V., F. Ackermann, and I. Shepherd (1997). "Integrated support from problem structuring through to alternative evaluation using COPE and V·I·S·A" (Wiley prize paper), *Journal of Multiple Criteria Decision Analysis* 6(3), 115-130.

Bornstein, C.T. and J. Rosenhead (1990). "The role of Operational Research in less developed countries: A critical approach", *European Journal of Operational Research* 49, 156-178.

Brännback, M. and J. Spronk (1997). "A Mutidimensional Framework for Strategic Decisions", in Fandel, G. And Gal, T. (eds.), *Multiple Criteria Decision Making*, Springer-Verlag, 581-590.

Cossette, P. and M. Audet (1992). "Mapping of an idiosyncratic schema", *Journal of Management Studies* 29(3) 325-347.

Covin, J.G. and D.P. Slevin (1989). "Strategic management of small firms in hostile and benign environments", *Strategic Management Journal* 10, 75-87.

Daniel, D.R. (1961). "Management information crisis", *Harvard Business Review* September-October, 110-119.

Day, G.S. (1986). *Analysis for Strategic Market Decision*, West Publishing Co., St. Paul.

Day, G.S. (1990) *Market Driven Strategy Process for Creating Value*, The Free Press, New York.

Desreumaux, A. (1993). *Stratégie*, Dalloz, Paris.

Eden, C. (1988). "Cognitive Mapping: a Review" *European Journal of Operational Research* 36, 1-13.

Eden, C. (1992) "On the nature of cognitive maps", *Journal of Management Studies* 29(3), 261-265.

Eden, C. and P. Simpson (1989). "SODA and cognitive in practice", in J. Rosenhead (ed.), *Rational Analysis for a Problematic World: Problem Structuring Methods for Complexity, Uncertainty and Conflict*, John Wiley & Sons, Chichester, 43-70.

Eden, C., S. Jones, and D. Sims (1983). *Messing About in Problems*, Pergamon Press, 47-48.

Eden, C., F. Ackermann, and S. Cropper (1992). "The Analysis of Cause Maps", *Journal of Management Studies* 29(3) 309-324.

Fiol, C.M. and A.S. Huff (1992). "Maps for Managers: Where are we? Where do we go from here?", *Journal of Management Studies* 29(3), 267-285.

Kelly, G.A. (1955). *The Psychology of Personal Constructs; a Theory of Personality*, Norton, New York.

Leidecker, J.K. and A.V. Bruno (1984). "Identifying and Using Critical Success Factors", *Long Range Planning* 17(1), 23-32.

Porter, M.E. (1985). *Competitive Advantage, Creating and Sustaining Superior Performance*, The Free Press, New York.

Rockart, J.F. (1979). "Chief executives define their own data needs," *Harvard Business Review* March-April, 71-93.

Rosenhead, J. (ed.) (1989). *Rational Analysis for a Problematic World: Problem Structuring Methods for Complexity, Uncertainty and Conflict*, Chichester: Wiley.

Rosenhead, J. (1996). "Developing Operational Research", in J. Rosenhead and A. Tripathy (eds.), *Operational Research for Development*, New Age International, 3-14.

Tolman, E.C. (1948). "Cognitive maps in rats and men," *Psychological Review* 55.

van der Heijden, K. (1996). *Scenarios – The Art of Strategic Conversation*, Chichester: Wiley.

Verstraete, T. (1997). "Cartographie cognitive et accompagnement du créateur d'entreprise," *Revue Internationale P.M.E.* 10(1), 43-72.

11 LEARNING NEGOTIATIONS WITH WEB-BASED SYSTEMS
The Case of IIMB

T.R. Madanmohan, Gregory E. Kersten,
Sunil J. Noronha, Margaret Kersten,
David Cray

1. Introduction

Sustainable development requires a number of skills, some technical, some interpersonal and some managerial. One skill that cuts across all three of these categories is negotiations. At virtually every step in the development process separate groups and interests must be accommodated. From arranging assistance agreements between governments to persuading local groups to co-operate on specific projects, communication and negotiation are a key part of the development process. Governments, courts, businesses and labor unions, as well as individual decision makers, are all involved in these negotiations (Zartman, 1994). These issues, however, are becoming increasingly more complex and require different forms of support to help weigh decision objectives and alternatives and analyze different scenarios and offers. Teaching negotiation skills and integrating these with the ability to utilize them in the context of advanced negotiation support systems has become an important task in making development sustainable.

Policy makers and managers in developing countries face numerous challenges in such diverse fields as labor-management relations, international affairs, business partnerships, and environmental regulations. Governments, courts, businesses and labor unions, as well as individual actors are breaking new ground in decision making (Zartman, 1994). In most cases these decisions are made through negotiations, one of the most common processes for making decisions and resolving conflicts at all levels of society.

With the globalization of markets, the consequences of cross-cultural interactions have received considerable attention (Hofstede 1989; Adler 1993; Faure and Rubin

1993). Prompted by the growing economic and political roles of developing countries studies have been undertaken contrasting developing and developed countries on the process, context and form of negotiations (Graham 1993; Druckman et al., 1976; Stone, 1989; Pechter, 1992). Cultural implications impact on attitudes towards contracts, value for formality, and status in human relations in both developing and developed countries (Swierczek, 1990).

Pechter (1992), having analyzed more than fifty real-life negotiations between Western and developing countries, concluded that the ethic of trust in most Asian countries is alloyed with an appreciation of shrewdness. While compromise is considered an appropriate outcome of negotiation in the Western world, it may often be considered an indicator of failure in Asian countries. Graham (1985, 1993), in his studies of negotiation styles in various countries, observed that the Japanese offered more extreme initial offers, used the word "no" less frequently, were silent longer and used aggressive tactics only in later stages of negotiation. Brazilians' negotiation behavior was characterized by more extreme first offers (even more extreme than those made by the Japanese), fewer promises and commitments, more commands, and longer interactions than exhibited by Americans in their negotiations.

These culturally-based differences in the understanding of the negotiation process have significant implications for designing training programs for negotiators. They also have to be considered in the design of software tools which seek to promote an accurate understanding of the valuation of decision alternatives, the assessment of concessions made by both sides, and the utility of a compromise in some situations. Without an understanding of the cultural framework which interfaces with the support system, decision and negotiation aids risk making the situation worse instead of helping the parties to a fair settlement.

The existing organizational and institutional structures in a developing country often do not provide adequate support for negotiation efforts. In a developing country, negotiators may not have a past bargaining relationship with their counterparts nor a history that establishes channels of communication. This may be one factor which emphasizes less structured settings for resolving disputes in industrializing societies (Ghauri 1988).

With increases in international trade and an accelerating shift of manufacturing from developed to developing countries, there is increased pressure on managers to engage in cross-cultural negotiations. This leads to growing interest in studying the way that culture affects negotiation theory and practice. Studies have revealed that most developing countries have few negotiators capable of translating their own and their organizations' principles and general goals into concrete bargaining proposals. They also lack systems for widespread and efficient training of decision makers (Stubbs, 1984; Schermerhorn et. al, 1985; Ghauri, 1988). To bargain effectively, one must not only have the ability to articulate interests and bargaining positions, but also the skill to communicate with one's opponents and to interpret accurately their responses. This requires an appreciation of the opponents' mindset and understanding of how their actions and positions are situated in their own national and organizational cultures.

Since the ability to understand and effectively communicate with counterparts from different cultures is critical to international negotiations, its absence may be the source of a serious weakness. Feliciano (1990) argues that since developing countries do not have a long history of negotiations with foreign countries or corporations,

> "They generally lack cadres of experienced negotiators in their foreign offices, in their ministries of finance and of trade and industry, in their boards of investment, in their agencies charged with coordination and implementation of development work, and in their private sector." (Feliciano 1990, p. xxi).

Language, customs and time zones also act as barriers to effective communication between the developing and developed worlds (Xing 1995; Grindsted 1994).

Negotiators from developing countries often rely, "more or less consciously, on confused, romantic notions of 'special' or 'historic relations' or shared 'fundamental interests', and feel grievously disappointed when such counterparts refuse to sacrifice their own interests and defer to the former's claims." (Feliciano 1990, p. xxii).

Negotiators with little cross-cultural experience may focus on cultural differences while ignoring the processual and analytical aspects that are similar in any type of negotiation. These underlying concepts of negotiation technique and negotiation analysis are complex issues that are now regularly taught in universities and in executive development courses although they are not yet widely available in developing countries.

Training in negotiations at the university level was first introduced in the United States and later spread to other parts of the world. In developing countries there are a few educational organizations with highly developed infrastructures which allow this type of training. In most, however, the pedagogical infrastructure is lacking or poorly utilized due to a lack of skilled instructors and current teaching materials. Under these conditions teaching negotiations to students or managers is fraught with many problems.

The traditional tools for teaching negotiations are cases, experiments and simulations. These tools are often culture-specific, require trained instructors, and organizational support. They normally focus on the development of communication skills, situation assessment and offer evaluation. Where they are used in conjunction with formal problem solving techniques and information and decision support tools, their effectiveness is greatly enhanced. In order to emphasize these tools many negotiation courses in undergraduate and graduate management programs include the use of negotiation support systems (NSS) as part of the bargaining process.

The inclusion of NSS in training programs helps introduce students to the implications of rapid changes in communication patterns and the increasing number of organizations engaged in electronic commerce. The increasing importance of electronic forms of business transactions and the ease with which they can be accessed from anywhere in the world has led to modifications of negotiation training. One trend has been to incorporate more cultural content into training sessions. The other has resulted in expanding the analytical and technological components of such courses.

The InterNeg project, with its Web site and Web-based support systems, attempts to integrate both technically sophisticated support and culturally diverse interactions. In this paper we will discuss our experience in developing the InterNeg Support Program for International Research Experiments (INSPIRE) and using it in managerial training in India. Section 2 contrasts the different approaches to teaching negotiations in developed and developing countries. Section 3 discusses the potential of Web-based systems for expanding and enriching management education. It also outlines INSPIRE, a Web-based system designed to support international negotiation training. The system and its use in bilateral negotiations are discussed in Section 4. Section 5 describes the use of INSPIRE in post-graduate and executive training at the Indian Institute of Management Bangalore, India. A discussion of possible extensions for INSPIRE and its utilization in higher education in developing countries concludes the paper.

2. Negotiation Teaching

2.1 Traditional approaches to negotiation teaching

The first course entirely devoted to managerial negotiation was offered at Dartmouth College in 1973 (Lewicki, 1986). By the early 1980s, many business schools were offering courses in negotiation. Most of these employed an experiential learning methodology which presented students with a negotiation exercise or simulation and then encouraged them to analyze and generalize from the experience. Negotiation teaching generally includes lectures about theories, discussions of case studies, and the conduct and analysis of simple experiments. Case studies describe some elements of negotiations, for example, framing, power strategies, and negotiators' personalities (Shubik, 1971).

The study of these basic elements allows students to evaluate factors that influence the chances that a dispute may be resolved through negotiation. The analysis of events that occurred in specific instances of a negotiation provides a factual grounding for discussion. This focuses the discourse on gaining insights into negotiators' strategy and behavior. Furthermore, detailed case studies enable students to see the importance of individuals and organizations involved in the negotiations and also the effect of the broader context in which they operate (Weiss-Wik, 1983).

Negotiation experiments are also used in courses as a learning tool (Winham and Bovis 1979; Carnevale 1995). They offer an opportunity for students to participate directly. Subsequent analysis reveals the dynamic aspects of negotiation and the interplay of human biases. Some experiments are conducted in the form of pen-and-paper tests involving brief, circumscribed tasks (Francis, 1991). For practical reasons, these experiments do not normally extend beyond one or two hours; they are, therefore, typically narrowly focussed. The classroom setting, face-to-face interaction and the fact that the participants know each other, all set limits on the realism of the simulation.

Discussion of cases and experiments conducted in a classroom are useful in illustrating the principles of negotiation. They may also help demonstrate formal techniques of decision and negotiation analysis. However, it is difficult for the students to apply the principles and formal techniques they learn to negotiations that resemble real-life situations. Several limitations of these methods have been identified in the negotiation and training literatures.

Low control and arbitrary focus: students and trainees often find that they have little control over the negotiation process; the focus of the negotiation is mainly determined by instructors and trainers. Pruitt (1986), (based on his analysis of six negotiation courses in U.S. universities) observed that most cases fail to anchor the actors, as control rests with the instructor. Instructors often intervene to complete the game in time, retain students' interest, or cover specific aspects of the negotiation process.

Limited flexibility in scheduling training sessions: given that negotiation experiments are conducted in a classroom and given the restrictions of semesters, instructors find that the lack of flexibility in conducting these simulations affects the learning process. This problem is often acute in International Business courses, where the instructors would prefer to arrange groups of participants from different countries. The problem is not simply the disjuncture of time zones. The substantial organizational efforts required for such negotiations typically makes them impossible.

Low level of involvement: limited time and the restricted focus of negotiation experiments, high levels of control, and the usual lack of real-life complexity in the simulations, contributes to low involvement by the participants (Thompson 1991).

Narrow domain of simulation: since simulations allow one to analyze students' behaviors and interactions and analyze these in a classroom, it is an important vehicle for teaching (and studying) negotiations (Adler and Graham 1989; Neale and Bazerman 1992). However, the lack of realism and the restrictions imposed by the classroom limit the insights that may be generated for students or researchers.

2.2 New demands and challenges

Negotiation courses are typically designed for business students and taught by instructors with a strong behavioral background. Our review of curricula for conflict resolution and negotiation courses offered at several major American universities and a few of the consulting companies offering negotiation courses (e.g., The Management Concepts, Inc. at: http://www.mgmtconcepts.com/ and The Negotiation Skills Company at: http://negotiationskills.com/) show that the organizational behavior approach to negotiation is predominant.

The focus is on types of negotiations and conflicts, the behavior of the parties involved, planning and communication, mediation and third party intervention, and the social and organizational context of negotiations. It is obvious that these are the key issues to negotiations. There are, however, other issues that are becoming increasingly important. These result from the implementation and use of new communication and computer technologies on the one hand, and a change in roles that small and medium size organizations and countries play in the world on the other.

Electronic commerce, electronic markets and intelligent systems introduce new challenges to negotiation teaching. Negotiations are already being conducted via electronic means (e.g., email) and this may require a somewhat different approach to effective communication than in face-to-face negotiations.

Data mining and knowledge discovery tools are increasingly being used in situation assessment and process analysis and they may have an important role to play in the preparation for, and conduct of, negotiations. Decision and negotiation support systems are becoming increasingly popular. They are often used in the formulation and evaluation of alternatives, assessment of offers and counter-offers, and the organization and visualization of the negotiation process. Expert systems have been developed to support teaching cross-cultural communication and negotiation (Rangaswamy, Eliasberg et al. 1989). The impact of these developments on the negotiation process is rarely considered in negotiation courses.

Another potential application of negotiation teaching and training lies in the use of computer-based simulation models. Negotiators dealing with engineering, financial or environmental issues need to be able to construct and assess scenarios to formulate offers and evaluate the opponents' offers. The more complex these issues and the longer their implications, the more likely computer-based systems will be used to assess alternatives. The use of advanced NSS allows the assessment to be linked directly to the negotiation process.

2.3 Negotiation teaching in developing countries

In developing counties negotiation training generally follows curricula prepared in developed countries with the main difference being the emphasis on cultural factors. The focus is on the negotiation process using concepts defined by particular economic, political, and cultural environments which may not apply to the current situation (Sunshine, 1990). Cultural differences are examined through the developed country framework rather than being considered as an integral part of the process itself.

In addition to the problems caused by inappropriate models, educational institutions in developing countries are often beset with other problems. Most business schools offer their courses in the local language and may suffer from a lack of training materials. While obtaining teaching materials from outside sources is not difficult, often the material will not reflect the dominant practices or culture of the country. This not only affects the quality of course delivery, but also students' involvement.

There may also be a shortage of qualified instructors in these countries and the number and variety of available cases which are relevant to local conditions. The result is a demand for educational programs and systems that will allow for the easy development, storage and retrieval of cases and simulation models. It is expected that such systems would also facilitate access to and use of decision and negotiation techniques. Links which ease communication between instructors and students as well as links to participants in a variety of cultures are also desirable.

New technologies such as the World Wide Web offer exciting avenues for teaching and training international negotiation. Net-based systems have the capacity to respond to a variety of negotiation training needs and address many of the educational problems that developing countries face. By offering easy access to materials and computer-based support, these systems can assist in upgrading instructors' qualifications. Web-based negotiation systems can accommodate any number of users, thus many participants, be they instructors or students, can benefit from using them at any given time.

Systems such as INSPIRE offer an opportunity for direct participation in a negotiation and thus experiential learning which is deeply rooted in the theory and practice of negotiation training. The level of participants' involvement is considerably higher (comparable to that in real life negotiations) than in the traditional training sessions. This is largely due to the fact that each participant is fully responsible for making all decisions and for communicating them to their counterparts. Hence, the locus of control lies with the participants.

One of the features of the INSPIRE system that shows potential, is the possibility for users to develop their own materials (i.e. case studies) that are relevant to their particular situation and interests. This allows for the development of "custom made" materials. A pool of such materials will grow fast, providing a group of enthusiastic users from developing countries becomes involved.

What the system does not do however, is highlight cultural divisions. The system is designed to focus attention on the process and analysis of the negotiation. Since the negotiations are anonymous, any information about an opponent's cultural traits will be revealed in the course of the negotiation or inferred by the negotiator. In this way the practice of negotiation is paramount for the student and cultural considerations only surface as part of an ongoing process.

The following section describes the technological advantages offered by Web based systems such as INSPIRE, a negotiation support system that enables unconstrained inter-cultural negotiations.

3. The Web and Negotiation Teaching and Training

Technology is a critical resource that can eliminate some of the problems related to teaching of communication and negotiation in developing countries. Some of these technologies, being system independent (in terms of operation and maintenance), allow users from remote parts of the world to communicate with others and to utilize previously inaccessible resources. Widespread use of computer networks, especially Web-based systems, indicate that the information access barrier between the developing and developed worlds could be overcome.

World Wide Web provides people in different locations and time zones a communication medium that is rich in functionality and content and which offers them the ability to use previously inaccessible computational resources (for example, decision and negotiation support systems). While the Web is currently used as a powerful source for the dissemination of information, it is increasingly being used as a means for remote execution and control of complete software systems, thus adding another dimension to the value it delivers. In education, its ability to access and run remote programs and databases allows users to extend classroom and laboratory boundaries across time zones as well as geographical boundaries. This flexibility can reduce the sense of inequality between managers, policy makers and citizens in developing and developed countries by enhancing their ability to communicate, negotiate and participate in commercial, educational and cultural activities.

Computer technologies allow rich communication amongst the actors in a negotiation by virtue of computation-intensive techniques and data visualization. The users can review the negotiation and its dynamics as the process unfolds. Language and other barriers shrink or disappear since these technologies allow extreme customization. User-specific front ends can be built which are then linked to the common core of the system, thus increasing participation while retaining functionality. The cost of duplicating a technological solution is another major factor that determines whether a particular solution can reach a larger population. Web browsers allow for portability and thus increase access for users in remote countries to training and real-life negotiations with minimal computing resources. These tools and systems are accessible to everyone, lay people and experts alike, and enable them to interact more directly with persons from different cultures, thus immensely reducing the effect of distance.

The flexibility of Web based systems facilitates customization of the case material to reflect regional specifics. It is also easier to bring about a discipline-based orientation in teaching and training sessions. The systems can be tailored to reflect, say, a behavioral, decision theoretic or any other focus to suit local teaching and training needs. This is very useful for management teaching and training where different modules are often combined to reflect the particular focus of a course. Web pages are very good at representing context, and independent Web pages may be assembled by a dispatching system that determines which page to present, based on a given situation.

The InterNeg Web site and its Web-based system INSPIRE have been constructed to exploit these technologies and their use in teaching. They aim to provide people around the world with analytical knowledge and decision support techniques within the domain of negotiations. The INSPIRE system allows participants to analyze and solve real-like decision problems and conduct negotiations with people from different cultures.

The INSPIRE system is the first Web-based negotiation support system. It is based on analytical models rooted in decision and negotiation analysis (Kersten 1985; Kersten and Szapiro 1986; Rangaswamy and Shell 1994). Developed in the context of a cross-cultural study of decision making and negotiation, the system has been primarily used to conduct and study negotiation via the World Wide Web as well as in teaching information systems, management science, international management and English as a Second Language.

The system has been implemented as an application available through remote access over the Web. It is, however, conceptualized as client-side software assisting a negotiator, much like a traditional desktop application dedicated to the negotiator, and communicating over the Internet with a similar "copy" of the software belonging to the other negotiator.

INSPIRE does not act autonomously like a third party arbitrator; rather each "copy" acts solely to support a single negotiator. It supports asynchronous negotiations, thus ameliorating the time zone problem. To facilitate this type of negotiation the system saves the current state resulting from each user's actions in a form that can be retrieved when the counterpart logs some time later (Kersten and Noronha, 1999).

INSPIRE views negotiation as a process involving three stages: pre-negotiation, conduct of negotiation and post-settlement (see Fig. 1). The first stage involves understanding the negotiation problem, issues and options and the elicitation of preferences through hybrid conjoint analysis. This allows one to obtain a rating for every possible offer. The second stage involves support for offer construction and counteroffer evaluation. Finally, the last stage involves computation of possible offers that dominate the most recent compromise and re-negotiation. Details of the methodology and the system's architecture can be found in Kersten and Noronha, (1998) and at http://interneg.org/.

The system can be used to conduct multiple bilateral negotiations. The most commonly used case involves trade negotiations between two companies: Itex, a producer of bicycle parts and Cypress Cycles, which builds bicycles. To reflect the dynamics of negotiation in developing countries, we have developed cases about negotiations for international technology transfer and for the sustainable development of natural resources.

4. Negotiations via INSPIRE

4.1 Cases

At IIMB, INSPIRE has so far been used in four different courses. The first was an elective course on Technology Management, offered to post-graduate students. The primary focus of this course was to understand issues related to technology adoption, technology pricing, adaptation of a technology to local needs and fostering technological innovations at the firm level.

Two courses were long-term executive development programs: the Management Program for Technologists and the Reliance Engineers Program. The focus of the module for these programs was on international negotiations. The fourth course was an elective offered for post-graduate students concentrating in marketing. Given the variation in the focus of the programs and participant needs, different cases were used for the above programs.

For the first course, we developed a case that focuses on commonly used mechanisms for technology transfer including preparation for effective transfer at the firm level. This is the INSPIRE *Techno* case which deals with a technology purchase decision. The case involves two companies, Pegard Technology Inc. (PTI), a U.S.-based manufacturer of industrial robots, and Intelligent Tools Inc. (ITI), a small south Asian firm dedicated to manufacturing transmissions for robots and automated, guided vehicles.

The international technology market for designs and know-how for the product and processes presents a fairly wide range of technological possibilities and choices. ITI has identified the sensors market as an important element for sustaining its competitive advantage. Mastering this technology requires a thorough understanding of optics, computer science, and electronics. ITI lacks expertise in these areas and has thus decided to obtain the needed technology from an outside source. Its search process has led to PTI which has expressed interest in co-operating with ITI. The two companies need to discuss and agree on the terms of technology transfer.

There are four issues that both sides need to discuss, namely, price, collaboration content, technology restrictions and payment. Collaboration content refers to the mode of actual technology transfer. Technology can be transferred in different forms: as blueprints, through technical collaboration involving process designs and drawings, through the acquisition of key parts of the plant or the shipping of the complete plant itself. The parties also negotiatie over constraints on further development of the technology and its sale to others. Such restrictions are common in technology transfer agreements. Each party is presented with their side of the case, told that they are to represent PTI and ITI respectively and that their companies are interested in achieving a breakthrough. No indication as to the desirability of the options (issue values) either in terms of directions or specific trade-off values is made. Since classroom sessions have already focussed on the issues this would constitute an unnecessary repetition. This design also provides for instructor manipulation of the perceived trade-offs that may be desirable for either research or instructional purposes.

For the three other programs, the emphasis was more on negotiation strategies per se. Because the majority of the participants already had some expertise and interest in the area of purchase management, the "Cypress and Itex" bicycle parts procurement case was used. In this case there are four issues that both sides have to resolve, namely the price of the components, delivery times, payment arrangements and terms for the return of defective parts.

4.2 Introduction of INSPIRE to participants

The negotiation course starts with a basic introduction to negotiation and international technology negotiation. The participants are exposed to cases such as Metro Corporation (Contractor, 1995) and Brother Surgicals (Madanmohan 1997) to familiarize them with negotiation tactics, issues related to licensing in international technology negotiation, effects of sunk costs and other topics relevant to managerial decision making and negotiations.

At the end of the class, a brief presentation about InterNeg and INSPIRE is made. It is made clear at the beginning of the course that INSPIRE is an important module of the course and participation is compulsory. However, it is also stated that the final result of their negotiation (compromise or not) and its utility score are *not* used for grading purposes. This is important because the participants should be able to negotiate in as realistic a situation as possible. That is, negotiations with one company can be broken and new negotiations initiated with another, the counterparts may have their own agenda, and the sole objective of the negotiations cannot be the achievement of a high utility value. The latter is critical because sophisticated users can easily manipulate the utility so that they may achieve a compromise yielding very high utility value but which does not reflect the interests of the company they represent.

Before beginning a negotiation, participants are asked to submit a pseudonym (to ensure anonymity) and their e-mail address (which is kept confidential by the system) before a particular date. Once the list is obtained, all the participants are informed about a demonstration session of INSPIRE. This session is intended to familiarize the participants with the Web (mostly for executive participants), and INSPIRE. The participants are shown how to log on to INSPIRE, how to construct and send offers using the system, and how to incorporate changes in any of the offers or issues in subsequent visits.

The demonstration session usually lasts an hour and a half. The instructor uses hard copies of the forms used in the INSPIRE negotiation so that the participants can actually see what kinds of forms they will fill in. This activity is quite useful as a preparatory step for structuring the negotiation process.

During the demonstration session students log in and read the INSPIRE case. At this time they may conduct initial analytical activities: specify the relative importance of each issue and the options for the issues. This information is used to determine their subjective utilities for all packages.[1] In many cases the session ends with the participant making a first offer to his /her counterpart. Before the session ends, the participants are reminded that they will be notified by INSPIRE via e-mail whenever a message or offer from the counterpart is received by the system. When they receive notification the participants log into INSPIRE to read and evaluate the offer and submit a counter-offer.

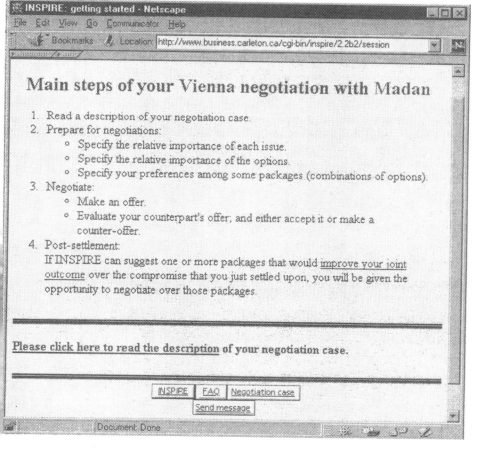

Figure 1. Initial outline of the negotiation process

---

[1] In INSPIRE each negotiation issue has several options listed a priori. A "package" or "offer" is constructed by selecting one option for each issue. For example, if there are two issues, price and quality and price options are R45, R54, R60, and the quality options are "high" and "medium", then there are 6 different packages (R45 and "high"; R45 and "medium"; R54 and "high"; R54 and "medium"; R60 and "high"; and R60 and "medium").

4.3 An example of negotiations using INSPIRE

This section describes a typical example of user activities in an INSPIRE negotiation. The description is illustrated with six screen snapshots. The first five figures are snapshots of negotiations between Gregory and Madan, two of the paper's authors.

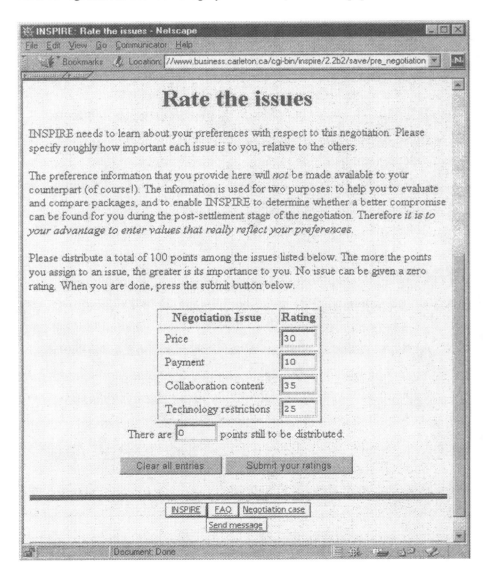

Figure 2. Rating the issues under negotiation

The INSPIRE users are assured that the information they exchange is confidential and therefore we cannot use their messages or offers for illustration purposes. The last figure presents negotiation between two users. It contains aggregated data (utility values of one side) but not any specific information that these users had exchanged.

Negotiations proceed thorough several main steps and in each step the user conducts one or more specific activities. A list of the steps denoting the steps already completed is displayed every time the user logs into the system. The initial screen that the user sees when beginning the INSPIRE supported negotiations is shown in Fig. 1. This is the screen seen by Gregory in his negotiations called Vienna, with Madan.

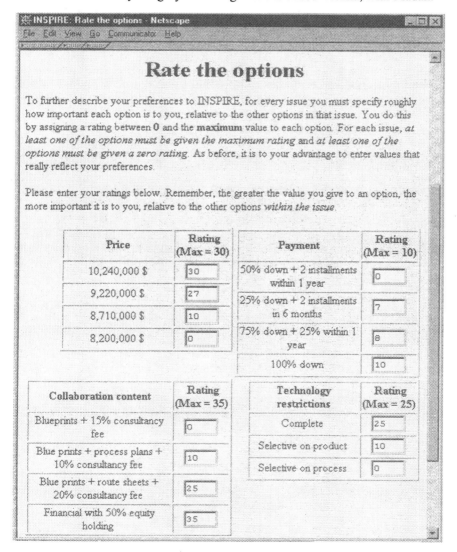

Figure 3. Option rating

At the bottom of Fig. 1 is a request to read the negotiated case that has been selected for the particular negotiations. Having read the description of the case the user can move to the next step which is preparation for negotiations. In this step the user has to evaluate the relative importance of the negotiated issues and, for each issue, their options. The issue rating activity is presented in Fig. 2 and the option rating for the four issues is illustrated in Fig. 3.

Each negotiation has a deadline. Typically, the deadline is set to expire three to four weeks from its initiation in order to allow the participants adequate time to complete their negotiation. After the deadline expires, participants can no longer send offers through the system, effectively representing a "failed" negotiation. Participants are asked to inform the faculty concerned about any difficulty encountered during the course of the negotiation. When conducting the negotiations, users construct offers, analyze counter-offers, send and receive messages, and review the negotiation's dynamics.

Figure 4 is a snapshot of the offer construction screen: it illustrates how the users can communicate either by plain messages or structured offers. It also shows how the score attached to a package helps to select an appropriate offer.

In the INSPIRE system the offer construction activity involves analysis of the counterpart's previous offer, and optionally, formulation of a message to the counterpart explaining and supporting one's position.

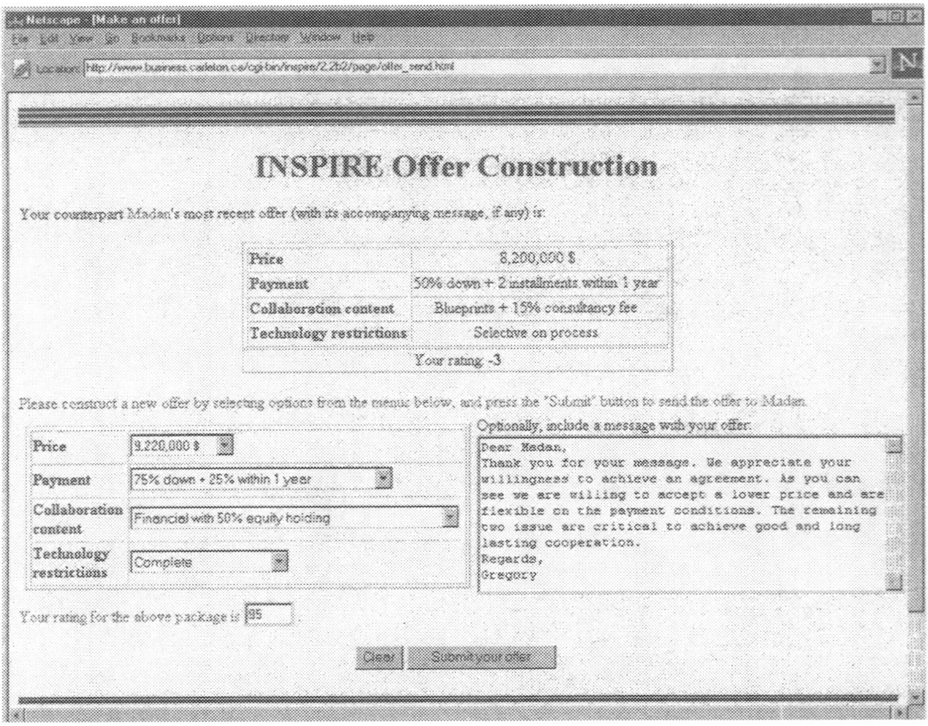

Figure 4. Offer construction.

Received offers are presented as a separate screen; the user may respond using one of four options. Figure 5 shows an offer sent by Madan. Gregory's four options are listed in Fig. 5; he may accept this offer, decide to make a counter-offer (then the offer construction screen would be displayed), send only a message, or terminate the negotiations.

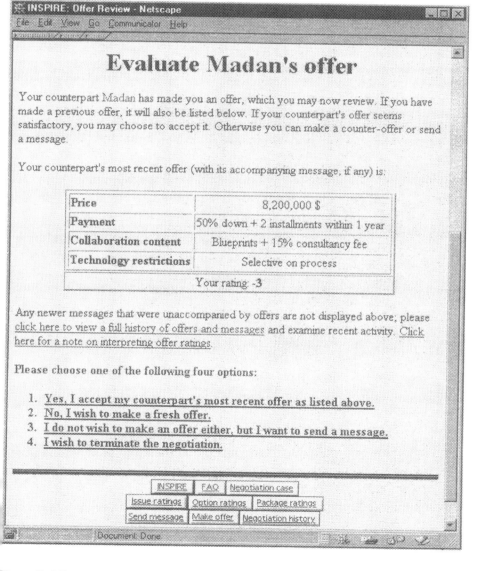

INSPIRE: Offer Review - Netscape
File Edit View Go Communicator Help

Evaluate Madan's offer

Your counterpart Madan has made you an offer, which you may now review. If you have made a previous offer, it will also be listed below. If your counterpart's offer seems satisfactory, you may choose to accept it. Otherwise you can make a counter-offer or send a message.

Your counterpart's most recent offer (with its accompanying message, if any) is:

Price	8,200,000 $
Payment	50% down + 2 installments within 1 year
Collaboration content	Blueprints + 15% consultancy fee
Technology restrictions	Selective on process
	Your rating: -3

Any newer messages that were unaccompanied by offers are not displayed above; please click here to view a full history of offers and messages and examine recent activity. Click here for a note on interpreting offer ratings.

Please choose one of the following four options:

1. Yes, I accept my counterpart's most recent offer as listed above.
2. No, I wish to make a fresh offer.
3. I do not wish to make an offer either, but I want to send a message.
4. I wish to terminate the negotiation.

INSPIRE   FAQ   Negotiation case
Issue ratings   Option ratings   Package ratings
Send message   Make offer   Negotiation history

Document Done

Figure 5. Offer evaluation

At any point the user may review the status of the negotiations by accessing a complete negotiation log that includes all offers and messages with their time stamps. This option is shown in Fig. 5, both below the table with Madan's offer and at the bottom of the screen where menu buttons are displayed.

The negotiation history contains the log of all exchanges as well as a graph that presents the dynamics of the negotiations in a simple form. An example of a negotiation graph is shown in Fig. 6.

The example in Figure 6 is an example illustrates the dynamics of the negotiation between Thomas and Andreas-Helm.[2] These names are pseudonyms that users have chosen for their negotiation. The small numbered triangles denote offers. The X axis shows the time at which each offer occurred and the Y axis represents the score associated with the offer. Note that although both parties' offers are shown, only a single utility function (that of the participant viewing the graph) has been used to evaluate all of them. This reflects the fact that INSPIRE does not expose a participant's preference function to his/her counterpart. It also stems from the fact that comparison of all offers, whether one's own or one's counterpart's, can only be meaningfully done according to one's own value system.

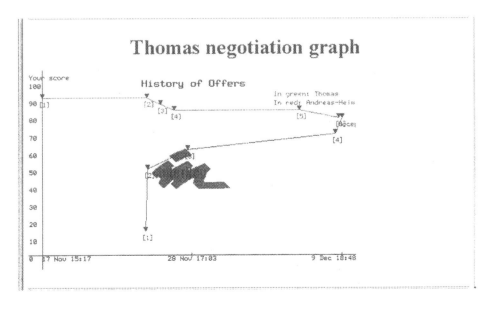

Figure 6. The negotiation graph

---

[2] Thomas (in reality Petter Westerback for the Abo Academy University in Turku, Finland) has described his negotiation experiences. His report can be accessed at: http://interneg.org/interneg/training/inspire/reports/pw/.

5. The IIMB Experience

5.1 Course offering and students' needs

The Indian Institute of Management Bangalore (IIMB), established by the Government of India in 1973, is an institution of higher learning committed to the cause of excellence in management education. The Institute offers both postgraduate and doctoral courses for students selected through a national level entrance exam and interviews. The Institute offers specialization in marketing, production and operations, finance and accounting, and human resource management. With a view to enabling practicing managers to stay current with new developments in various fields of management, IIMB offers short as well as long duration training programs for executives in general and, in particular, for functional managers. These programs can be divided into two types: (a) those that are open to managers from different firms, and (b) programs tailored to suit the requirements of a specific group or firm.

At IIMB a course on negotiation is offered as an elective, typically in the last year of the postgraduate program. Modules on negotiations are also offered in other courses. For example, in the Technology Management course there is a module on technology transfer negotiations. This module focuses on the dynamics of technology transfer between a donor and a recipient, and typically a case is discussed.

The Institute also offers a two-week executive level program on negotiation and there are several other executive programs, such as Purchasing and Supply Management in which the participants are exposed to the nuances of negotiation. The pedagogy adopted in these courses, prior to the introduction of INSPIRE, was largely lecture based, coupled with cases and games. Some of the homegrown cases were useful in helping the participants understand the behavioral part of negotiation: emphasizing mostly negotiation style and specific negotiation strategies. While these pedagogical tools were useful drivers for imparting the fundamentals of negotiation, the participants had more demands.

Feedback from postgraduate students indicated that they would actually prefer a tool that would enable them to participate in a negotiation, to understand the real motives of a human counterpart, and to see how they fared in the process. The usual restrictions of the semester and class duration also limited the effectiveness of role play and associated experiential learning techniques. A post-graduate student from the intake of '96-97 stated:

> "The international technology negotiation game should expose us to the real motives of the donors, the vulnerability of governments and the recipient. A more dynamic representation wherein we could don the role of choice and enhance our learning is needed".

The feedback from the executive program participants was even more revealing. One of the participants from the Management Program for Technologists said:

> "Given the experience we have in negotiating with the French and others, I look for the negotiation course to actually aid in understanding the dynamics of negotiation from our perspective. In a technology transfer we may be more interested in a typical arrangement, say only technical. We need a course wherein the instructor need not actually hand-hold us through negotiation, but devise programs that help us to uncover ourselves first. Well, later may be we need to know what to do better".

During the early 1990's, several departments of IIMB identified areas of research and consulting interest that would specifically address the needs of Indian industry in an increasingly multilateral and global context. Hence there has been a renewed interest in cross-cultural business, especially negotiations. Faculty teaching related courses felt a need for offering a dynamic platform from which cross-cultural research and training could be pursued along similar lines.

5.2 Users' experience

All students and participants in executive development programs at IIM Bangalore are graduates with English as their medium of instruction, hence no specific language training was required. However, for few of the executives who had had no prior computer experience, a hands-on tutorial to familiarize them with the Internet was provided. They were also guided by teaching assistants during their first few sessions with INSPIRE. Every alternate day the participants were contacted to find out whether they had experienced any snag or difficulty. The participants were requested to record every activity they undertook related to their negotiation via INSPIRE. This log was used for their personal assessment of the overall negotiations after they were completed as well as discussion in the class. It also facilitated individual discussions with the instructor about the difficulties and problems students encountered. On completion of all participants' negotiations, the results of several negotiations were discussed in subsequent classes.

Two batches of post-graduate students and three executive program participants, totaling thirty-three participants, were exposed to INSPIRE. They had registered for the Technology Management and International Management Courses, wherein either technology purchasing or managing across boundaries was the main focus. Descriptive data for the participants, who had an average negotiation experience of 2.2 years, is shown in Table 1. Despite currently low levels of access to the Web, all the participants expected a significant increase in their use of the Web. For a significant majority of the participants INSPIRE was the first DSS/NSS that they had used. Few of the participants reported any problems using the system during their negotiations.

Most participants exceeded their initial expectations for the agreement and achieved much of what they wanted. The upper limit for the utility value is 100 (i.e. the participant receives 100% of his/her most preferred package). As shown in Table 1 the IIMB participants reached an average value of 82 for their final compromises. Since the counterparts may give different weights to issues and options the total value of the final package may vary. If the negotiation is strictly competitive, that is each side assigns mirror image weights to each of the issues and options (e.g., one side wants the lowest price and the other the highest then the sum of the two sides' utility values is 100. On the other hand, if the sides had exactly the same interests, which should not happen given the case descriptions, then combined utility will be much higher.

Exceptionally high scores on expected utility (the value a participant believes he/she will achieve at the beginning of the negotiation) and actual utility (the value of the final agreement) indicate that the participants might have been more interested in maximizing their utility value rather than achieving a realistic compromise. This appears to have happened in some of the IIMB negotiations even though it was stressed that the utility value would not be considered an indicator of successful negotiations nor used for grading purposes. One might hypothesize that the very high scores reflect high competitiveness of Indian students and managers who wish to better the scores of the colleagues even though this had no impact on their success in the course. This competitive tendency had been remarked upon earlier by Druckman et al. (1976). For comparison, the expected and achieved utility values for Americans, Canadians and Finns are generally between 40 and 60 (Kersten and Noronha, 1998).

Out of the twenty two participants who reached an agreement in a negotiation, about 34 per cent achieved efficient solutions, that is the results for one side could not be improved with degrading the value for their opponent. This is one example of the usefulness of a system like INSPIRE. Despite the fact that the negotiation problem is relatively simple with only 180 potential offers, it is complex enough that most of the participants do not achieve an efficient agreement. If the original agreement is not efficient, the system displays up to five solutions that will yield a higher utility value for both negotiators.

Table 1. Descriptive statistics of users' profile and satisfaction with the experience

Current use of Internet [a]	3.8 (1.4)
Used DSS/NSS previously	19%
Expect increased Web access	100%
Satisfaction with agreement [b]	3.1 (1.3)
Satisfaction with own performance [b]	3.8 (1.4)
Agreement met expectations [c]	3.7 (1.6)
No agreement	11 (33%)
Achieved agreement	22 (67%)
Expected utility value [d]	91 (10)
Utility value of the compromise [d]	82 (22)
Efficient compromise achieved	34%

a - 1 several times a day, 6 rarely; Average (Variance)
b - 1 extremely satisfied, 7 extremely unsatisfied; Average (Variance)
c - 1 yes completely, 5 - no, not at all; Average (Variance)
d - Maximum = 100; Average (Variance)

Along with the data from the post-negotiation questionnaire presented in Table 1, additional information collected at the conclusion of the course offered further insights into the users' evaluations. Table 2 offers descriptive details of some of the feedback received. A significant majority of users said they perceived INSPIRE to have actually helped them to acquire or improve their negotiation skills. Many felt it helped them to prepare for a negotiation and focus better. For example, an executive from United Bristles and Brushes Ltd., said:

"When negotiating service contracts with a Taiwanese manufacturer we had great problem. Often, he did not understand what I wanted and I did not have a clue of what he was saying. More than language barrier, the major handicap was lack of preparation. INSPIRE prepared me for low communication negotiation and I think that does add value to my practice."

Another participant from a large public sector organization said:

"I find it extremely useful for two purposes. First as a training tool. Second as a platform for small and medium industries managers from India who can negotiate their orders through the system".

A few of the participants stated that INSPIRE helped them to see an intercultural point of view. This occurred in a situation where the participants did not know the nationality of their counterparts. A significant majority said INSPIRE did help them better to understand their counterpart's position and actually helped them to refine their own negotiation skills. Acquiring negotiation skills without direct intervention of the instructor is considered one of the biggest benefits of the INSPIRE system. Despite the participants' limited experience with the Internet, their ability to achieve expected compromises suggests that the INSPIRE system and Web-based negotiations do not introduce a significant burden or add complexity to the already complex negotiation process.

Table 2. Users' perception of their negotiations via INSPIRE

Helps in honing/development of negotiation skills [a]	3.3 (1.1)
Obtain intercultural point of view [a]	1.4 (1.3)
Understand counterpart strategies [a]	2.9 (1.6)
Will use INSPIRE for real-life negotiation [b]	2 (13%)
Will use INSPIRE for preparation of negotiation [b]	6 (40%)
Will use INSPIRE for practice [b]	8 (53%)

a - 1 no, not at all .. 5 extremely; Average (Variance)
b - based on 15 completed post-negotiation questionnaires.

5.3 Teacher's experience

Being an Internet based tool, INSPIRE required different preparation, handling and conduct of the negotiations than face-to-face methods. In contrast to instruction using cases, teaching negotiation through INSPIRE required first ascertaining the level of Internet expertise of the users. Appropriate training sessions on the nternet may need to be planned before the INSPIRE session starts. Typically, at IIMB one week orientation program was carried out to meet the requirements of the participants.

For a teacher whose class conducts a Web-based negotiation the process involves preparation and handling of three major stages:
1. introduction to the system,
2. the first exchange of offers and messages, and
3. discussion following the negotiations.

At the IIMB a formal lecture was adopted to introduce various aspects of INSPIRE and INSS. Its focus is on the specification of the environment in which the participants would negotiate, clarification of any queries regarding the sequence of activities and the submission of ratings, and the provision of certain broad guidelines about the INSPIRE system itself.

The first hands-on class was always conducted at the computer center. In this session, participants were guided through such steps as log in, reading of the case, submission of rating of issues and packages, and finally the first offer. Deft handling of varying levels of experience and expertise among the participants is of crucial importance here. We found it useful to place one skilled and less skilled participant next to each other. The role of faculty here is one of a facilitator and his presence after the submission of the first offer, was generally not necessary.

Typical problems that may arise in running an Internet based negotiation are:
1. system problems,
2. team problems, and
3. mechanisms for strong administration.

System problems include both hardware and software compatibility issues (INSPIRE requires Netscape 3, Explorer 4 or later browsers) and systemic problems (including power shutdowns, and network problems, common in India) which are often typical in a developing country.

The more acute problem in running INSPIRE negotiations was related to team dynamics, i.e. lack of a response from a counterpart at various stages of a negotiation. This poses serious difficulties in executive development programs that are normally of short duration. We worked out several strategies to address this issue. One was the expectation that students inform their instructor if they do not receive a response within two days after the submission of an offer. In such a situation, the instructor immediately e-mailed the counterpart instructor with a copy to the participant to activate the negotiations. In some cases, the instructor had to remind the participant in person about the upcoming deadline and ensure that negotiations were continuing. By design the INSPIRE negotiation exercise is a non-credit activity. To stimulate participants we posted pseudonyms of those who were active and likely to complete their negotiation. During the classes the participants were reminded of the approaching deadlines and the not-so-active participants were approached and asked if they required assistance.

To ensure successful completion of a negotiation through INSPIRE instructors need to plan and develop strong administrative mechanisms. These included identifying a module coordinator from the group of participants, who could help the group stay focused and productive during the negotiation. Administrative support also had to be planned for unintended interruptions, support that may be required during subsequent negotiations. Fortunately, once the students have begun their negotiation on INSPIRE there is very little intervention required from the instructors.

After the completion of the negotiations, most of the participants typically want to compare their analyses. The analysis of experiences can be done in many ways, i.e. instructors may comment on the process or a participant may dissect the process him or herself with or without external feedback or the instructor may conduct classroom discussions of typical negotiations. Individual introspection proved to be most useful for the executive development programs, while post-graduate students preferred classroom discussions. The instructor's role here is more to reflect on the various scores, probing the processes by which they were achieved, rather than to rank the group's scores or to evaluate specific negotiations on their basis of the outcome.

6. Conclusions

This paper outlines ongoing teaching activities at IIMB in which Web-based materials and support systems are heavily used. It is intended to share these experiences and provide directions for effective use of Web technologies for teaching and training. An important feature of our experiments is very high acceptance of the INSPIRE system and its capabilities. The system was designed for both training and research purposes and with a cross-cultural focus. However, most of the users see its practical usefulness in its analytical, presentation, and communication aspects. Participants generally said that they would use the system for training and honing negotiation skills.

Web-based solutions such as the InterNeg site and the INSPIRE negotiation support system enhance the instructors' ability to teach negotiations more effectively. Unlike other media, systems available on the Web allow for expanding the discussion beyond local borders. They create a more realistic environment by allowing communication between individuals with similar educational or professional backgrounds. Obviously, it is also possible to have students from the same group negotiate with each other and INSPIRE has also been successfully used in this mode.

At present Web-based systems and materials only allow written communication. Clearly this reduces the participants' range of tactics and strategies since non-verbal communication plays an important role in negotiations (Faure, 1993). However, this limitation may be seen as a difficulty that negotiators will have to overcome in the future as such negotiations become a more common feature of international business. Although expansion into other media are possible in computer-assisted negotiation, the communication bandwidth in many developing countries does not allow for media rich exchanges of messages (e.g., voice, video, complex images).

INSPIRE requires negotiators to define their interests, set targets, and anticipate the actions and strategies of their opponents. One of its most important features is that it allows the formulation and communication of arguments and explanations. While the system shows the value of offers it does neither flag poor negotiation outcomes nor does it highlight inadequate planning. It purposely allows users to make mistakes, employ any tactic they want, and change it whenever needed. The ability to review the verbal negotiation history and the graphical presentation of the negotiation dynamics allows users to review and assess their actions during the course of the negotiation.

The INSPIRE system has proven its usefulness in teaching and training. However, its more sophisticated users request more features and more flexibility in the use of particular decision making and negotiation techniques. Instructors have asked for more negotiation cases and for cases that can be adapted to a specific situation or teaching program. Because the system is also used for research purposes (it is a data collection tool) a decision has been made to freeze its development and instead build another system that can be continually upgraded. This second system, called INSS (InterNeg Support System) is operational and has been used in a small number of negotiations. It allows participants to select a negotiation case, define the important issues, lay out options and to modify or add options or issues during the negotiations. Further, it has mechanisms for the specification and modification of BATNA (Best Alternative to the Negotiated Agreement) and reservation prices. It also has enhanced graphic functions that can be used to display different aspects of the negotiation process.

INSPIRE and its associated systems offer an innovative, flexible tool for teaching negotiations in a variety of settings. Using INSPIRE allows managers to experience cross-cultural tensions that often arise in negotiations surrounding sustainable devel-

opment. Since the negotiations are with real human beings who have their own aspi-
rations and world views, they achieve a level of realism that is difficult to achieve
through role-playing or case studies techniques. The system also introduces its users
to the advantages of using support tools to untangle the complex issues that may arise
in development negotiations. The wide availability of INSPIRE coupled with its
flexibility, provides a means for sustaining not only technical and economic aspects of
development but the human skills for managing the trade-offs and comparisons that
must accompany any real progress in a sustainable development program

Acknowledgements

We wish to acknowledge the contribution of Anantha Mahadevan, Kumudini Pon-
nudurai and Ravi Ramsaran, the co-developers of the InterNeg site. We thank Brabara
Hauser for her help in preparing this chapter. This work has been supported by the
Social Science and Humanities Research Council of Canada, the Natural Sciences and
Engineering Research Council of Canada, and the International Institute for Applied
System Analysis.

References

Adler, N.J. (1993). "Do Cultures Vary?", in: *Societal Culture and Management*, T. D. Wein-
shall, (Ed.), Berlin: Walter de Gruyter & Co., 23-46.
Adler, N.J and J.L Graham (1989). "Cross Cultural Interaction: The International Comparison
Fallacy?", *Journal of International Business Studies*, 20 (3), 515-537.
Carnevale, P. J. (1995). "Property, Culture and Negotiation", in R.M. Kramer and D.M. Mes-
sick, (eds.) *Negotiation as a Social Process*, Thousand Oaks: Sage.
Contractor, F. J. (1995). "Metro Corporation: Technology Licensing Negotiation", in W.J.
Keegan, *Global Marketing Management*, Englewood Cliffs, N. J.: Prentice-Hall.
Druckman, D., A. A. Benton, F. Ali, and J. S. Bagur (1976). "Cultural Differences in Bargain-
ing Behaviour: India, Argentina, and the United States", *Journal of Conflict Resolution*,
20(3), 413-452.
Faure, G. O. and J. Z. Rubin (1993). "Culture and Negotiation", *Negotiation Journal*, 9(4), 380
- 381.
Faure, G. O. (1993). "Negotiation Concepts Across Cultures: Implementing Nonverbal Tools",
Negotiation Journal, 9(4), 355-359.
Feliciano, F. P. (1990). "Foreword: Process and Culture in Development Negotiations: The
management of Consent", in R.B. Sunshine, *Negotiating for International Development*,
Dordrecht: Martinus Nijhoff Publishers.
Francis, J. N. P. (1991). "When in Rome? The effects of Cultural Adaptation on Inter-cultural
Business Negotiations", *Journal of International Business Studies*, 22, 403-428.
Ghauri, P. N. (1988). "Negotiating with Firms in Developing Countries: Two Case Studies",
Industrial Marketing Management, 17(1), 49-53.
Graham, J. L. (1985). "The Influence of Culture on the Process of Business Negotiations: An
Exploratory Study", *Journal of International Business Studies*, 16(1), 81-96.
Graham, J. L. (1993). "The Japanese Negotiation style: Characteristics of a Distinct Approach",
Negotiation Journal, April, 123-140.
Grindsted, A. (1994). "The Impact of Cultural Styles on Negotiation: A Case Study of Span-
iards and Danes, *IEEE Transactions on Professional Communications*, 37(1), 34-38.

Hofstede, G. (1989). "Cultural Predictors of Negotiation Styles", in K. Mauter-Markhof, *Process of International Negotiations*, Westview Press, 193-201.

Kersten, G. E. (1985). "NEGO – Group Decision Support System", *Information and Management*, 8(5), 237-246.

Kersten, G.E. and S.J. Noronha, 1998. "Negotiations via the World Wide Web: A Cross-cultural Study of Decision Making", *Group Decision and Negotiations*, (to appear).

Kersten, G.E. and S.J. Noronha, 1999. "Supporting Negotiations with a WWW-based System". *Decision Support Systems*, 8(3), 251-279.

Kersten, G.E. and T. Szapiro (1986). "Generalized Approach to Modeling Negotiations", *European Journal of Operational Research*, 26 (1), 142-149.

Kolb, D. A. (1974). "Management and the learning process", in *Organizational psychology: A book of readings*, edited by D.A. Kolb, I.M. Rubin, and J. McIntyre, Englewood Cliffs, N. J.: Prentice Hall.

Lewicki, R. J. (1975). "A Course in Bargaining and Negotiation", *The Teaching of Organizational Behavior*, 1(1), 35-40.

Lewicki, R. J. (1986). "Challenges of Teaching Negotiation", *Negotiation Journal*, 2(1), 15-27.

Madanmohan. T.R., (1997). "*Brother Surgicals: Negotiating an Exclusive License*", Indian Institute of Management Bangalore Case Studies, Bangalore, India.

Mumford, K. (1996). "Arbitration and ACAS in Britain: A Historical Perspective", *British Journal of Industrial Relations*, 34(2), 287-305.

Neale, M.A and Bazerman, M.H., 1992. *Negotiating Rationally*, New York: Free Press.

Pechter, K. (1992). "Can we make a Deal?", *International Business*, 5(3), 46-50.

Pruitt, D. G. (1986). Trends in the Study of Negotiation and Mediation, *Negotiation Journal*, 2(3), 237-244

Rangaswamy, A., J. Eliasberg et al. (1989). "Developing Marketing Expert Systems: An Application to International Negotiations", *Journal of Marketing*, 53, 24-39.

Rangaswamy, A. and G. R. Shell (1994). "Using Computers to Realize Joint Gains in Negotiations: Towards an Electronic Bargaining Table", *Computer Assisted Negotiation and Mediation Symposium*, Harvard Law School, Cambridge MA.

Schermerhorn, J. R. R. S. Bussson, H. Elsaid and H. K. Wilson, (1985). "Managing the Interorganisational Context of Management Development in a Developing Country: A Case Study", *Leadership and Organizational Development Journal*, 6(1), 27-32.

Shubik, M. (1971). "The Dollar Auction Game: A Paradox in Non-cooperative Behavior and Escalation", *Journal of Conflict Resolution*, 15, 109-111.

Sunshine, R.B. (1990). *Negotiating for International Development*, Dordrecht: Martinus Nijhoff Publisher.

Stone, R. (1989). "Negotiating in Asia", *Practicing Manager*, 9(2), 26-29.

Stubbs, R. (1984). "The International Natural Rubber Agreement: Its Negotiation and Operation", *Journal of World Trade Laws*, 18(1), 16-31.

Swierczeck, F.W. (1990). "Culture and Negotiation in the Asian Context: Key Issues in the Marketing of Technology", *Journal of Managerial Psychology*, 5(5), 17-24.

Thompson, B. L. (1991). "Negotiation Training: Win-Win a What?", *Training*, 28(6), 31-35.

Weiss-Wik, S, (1983). "Enhancing Negotiators' Successfulness", *Journal of Conflict Resolution*, 27(4), 706-739.

Winham, G. R and H. E. Bovis, 1979. "Distribution of Benefits in Negotiation", *Journal of Conflict Resolution*, 23(3), 408-424.

Xing, F. (1995). "The Chinese Cultural System: Implications for Cross-cultural Management", *SAM Advanced Management Journal*, 60(1), 29-33.

Zartman, I.W. (Ed.) (1994). *International Multilateral Negotiation*, San Francisco: Jossey-Bass.

12 NATURAL RESOURCE CONSERVATION AND CROP MANAGEMENT EXPERT SYSTEMS
Ahmed Rafea

1. Introduction

Crop management can be defined as the set of agricultural operations that produce the crop. Soil and water science experts, plant pathologists, entomologists, agricultural scientists, and breeders all contribute to these operations. Soil and water conservation are also important factors in crop management. It is important that reliable methods of crop management be transferred to growers in an efficient way. From an examination of the existing systems for knowledge and experience transfer, we found that they depend mostly on extension documents (leaflets that contain instructions to growers) and direct communication between extension agents and growers. The limitations of the conventional extension system are due mainly to the fact that these documents and agents are unable to cover all the different aspects of cultivation.

In this chapter, we discuss our experience in the development and deployment of expert systems to transfer knowledge about appropriate agricultural operations from scientists and experts to the crop growers.

The focal theme of this chapter is twofold: first, to show how expert systems work to help solve agricultural practice information transfer problems; and second, to describe our experience in developing and deploying crop management expert systems.

The relation between crop management and natural resources is discussed in Section 2. Irrigation scheduling is highly related to the conservation of water resources, while fertilization is highly related to soil conservation. Rationalizing pesticide application also helps in conserving water and soil resources. Section 3 reviews some similar systems in developed and developing countries. In developed countries, the interest in expert systems to transfer agricultural knowledge dates back to the early 1980s, while in developing countries the interest in expert systems did not start until the late 1980s and early 1990s.

In Section 4, we describe the agricultural practice information transfer problems and analyze the suitability of expert systems to assist in solving these problems. Section 5 presents a development methodology that consists mainly of knowledge engineering and software engineering. Knowledge engineering includes acquiring the knowledge, analyzing and modeling it, and testing the model. Software engineering includes requirement specifications, design, implementation, and evaluation.

Section 6 describes the components of crop management expert systems. We describe the relation between these components and the expert systems developed for five crops. The functional requirements, knowledge engineering methodology, and implementation tools are noted. The results of the evaluation of one of the developed expert systems are discussed in Section 7. The procedures for validation by external experts are shown and the results are discussed. The field validation results and the economic and environmental impact of expert systems usage are also explained. In Section 8, we describe our experience in deploying expert systems in the field. This includes the training of extension workers, systems distribution, expert systems usage monitoring, and maintenance. The discussion ends with conclusions and some notes on future directions in Section 9.

2. Effect of crop management on water and soil conservation

Crop management can be defined as all agricultural operations that produce the crop, including irrigation, fertilization, and pest control. The conservation of natural resources is an integral part of these agricultural operations. Although there are other factors to consider in water and soil conservation, they are outside of the scope of this chapter and we concentrate on the status of the water and land resources in Egypt.

Water is the scarcest resource in Egypt, as its supply is almost fixed and water demand for different sectors is continually increasing. The water supply can be classified into three categories: surface water, ground water, and reusable, treated drainage water. The Nile River is the principal source of surface water: it provides Egypt with more than 95% of its water requirements. Egypt's annual quota from the Nile is 55.5 billion m^3.

There are two sources of ground water in Egypt. The first is Nile Valley and Delta ground water. This is not considered an additional resource because the Nile River and excess irrigation water recharge it. The total storage of its aquifer is about 500 billion m^3 with an average salinity of 800 parts per million (PPM). The volume of water extracted from this source for domestic, industrial, and agricultural use was estimated at 2.6 billion m^3 in 1990. The second ground water source is nonrenewable deep-desert ground water. The major part of this source consists of the aquifer in the western desert. Preliminary estimates indicate that the total ground water storage in the western desert is about 40 000 billion m^3, with a salinity of between 200 and 700 PPM. Ground water is also available in numerous aquifers in Sinai, but there is less information about them. The volume of water extracted from deep desert ground water in 1990 was estimated at 0.5 billion m^3.

The portion of drainage water that can be reused depends on its salinity and contamination from municipal, industrial, and agricultural effluents. The volume of drainage water reused in 1990 for irrigation was estimated at 4.9 billion m^3.

The challenge for Egypt's water resource managers is to balance the limited water supply with an increasing water demand. Lack of water is the major constraint to the land expansion needed for self-sufficiency in food production. Water demand for the year 2000 is expected to be 74.0 billion m^3, while available water resources, including agricultural drainage water, treated municipal sewage water, and ground water, are expected to be just 69.4 billion m^3. Another challenge is how to reduce the water pollution due to the use of chemical fertilizers and pesticides.

Good irrigation schedules, fertilization schedules, and integrated pest management help meet these two challenges. Irrigation scheduling and integrated pest management are part of the crop management package produced by the Ministry of Agriculture and Land Reclamation (MOALR) in Egypt and disseminated to growers through extension services. Finding good tools for transferring this information will expedite its dissemination.

After water, land is the major limiting factor for sustainable agricultural development. Cultivated land in Egypt totals about 7.7 million acres, or only 3% of the total land area. Because of the arid climate, agriculture relies mainly on irrigation from the Nile River. Most of these cultivated lands are concentrated in the Nile Valley and Delta regions. The areas lying outside the Nile basin (oases of the New Valley and Mediterranean coastal plains, including Sinai) are partly rain-fed or irrigated with ground water.

Cultivated lands in Egypt are classified into two categories: Old Lands and New Lands. Old Lands are located in the Nile Valley and Delta regions that are irrigated directly from the Nile. These lands have been under irrigation for a long period of time. They are very fertile and the soil is alluvial silt and clay loam. New Lands are the less fertile desert areas outside the Nile Valley, where the soils are generally sandy and calcareous. The New Lands are distributed west of the Delta (Nubaria), east of the Delta (Salhia and along the western side of the Suez Canal), and in Sinai and the New Valleys of the Western Desert. The Old Lands total about 5.6 million acres, with about 2.1 million acres in New Lands reclaimed since 1952.

Old agricultural land losses and limited water resources for expanding the land area are the major constraints to sustainability of land resources and consequently, food self-sufficiency for the growing population. The land losses can be measured in quantity (land converted to nonagricultural use) or quality (soil degradation, mainly due to salinization and waterlogging because of an inefficient drainage system). The problems of soil degradation and soil conservation can be addressed with proper fertilization scheduling as a part of a crop management package.

3. Review of similar expert systems

3.1. Expert systems in developed countries

The expert system applied to the problem of diagnosing soybean diseases (Michalski et al., 1983) is one of the earliest expert systems developed in agriculture. A unique feature of the system is that it uses two types of decision rules: those representing expert diagnostic knowledge and those obtained through inductive learning from several hundred cases of disease.

COMAX (Lemmon, 1986) is a crop management expert system for cotton which can predict crop growth and yield in response to external weather variables, physical parameters of the soil, soil fertility, and pest damage. The system is integrated with a computer model, Gossym that simulates the growth of the cotton plant. This was the first integration of an expert system with a simulation model for daily use in farm management.

POMI is an expert system for integrated pest management of apple orchards (Gerevini et al., 1992). Integrated pest management is defined as the scientific application of agricultural methods and products, including chemicals, to allow optimum production while respecting the farm worker, the environment, and the consumer. POMI addresses the preliminary phase of the complex process of apple orchard integrated control, namely the detection of insect populations in the field and their approximate size. The system consists of two parts: classification of user findings and their explanation through abductive reasoning.

Schulthess et al. (1996) developed a picture-based expert system for weed identification. Most such systems are rule-based and use text with a large number of botanical terms. In this system, the hierarchical classification of generic tasks was used, and the text descriptions were replaced by pictures to minimize the use of technical terms. Hypotheses are established or ruled out on the basis of the user's choices among options presented as pictures.

3.2. Expert systems in developing countries

An agroforestry expert system known as UNU-AES was designed to support land-use officials, research scientists, farmers, and other individuals interested in maximizing benefits gained from applying agroforestry management techniques in developing countries (Warkentin et al., 1990). UNU-AES is a first attempt to apply expert systems technology to agroforestry. This system addresses the option of alley cropping, a promising technology that has potential applicability when used under defined conditions in the tropics and subtropics. Alley cropping involves the planting of crops in alleys, or interspaces, between repeated hedgerows of fast-growing, preferably leguminous, woody perennials. With the inclusion of more climatic and socio-economic data and improved advisory recommendations, UNU-AES can be expanded to provide advice on alley cropping in more diverse geographical and ecological conditions and eventually address other agroforestry techniques.

In India, an intelligent front-end for selecting evapotranspiration estimation methods (Mohan and Arumugam, 1995) has been integrated with methods that calculate the evapotranspiration factor (ET). The ET is an important parameter needed by water managers for the design, operation, and management of irrigation systems. Since there are many ways to compute the ET, an inexperienced engineer or hydrologist might be unsure of the appropriate method to use. An intelligent front-end expert system (ETES) has been developed to select suitable ET estimation methods under South Indian climatic conditions. Ten meteorological stations located in different climatic regions and 13 ET estimation methods have been considered in this expert system. Along with the recommended method, ETES suggests suitable correction factors for converting the resulting ET values to those of methods that result in accurate estimation.

In Indonesia, expert systems have been used for information transfer of soil and crop management (Yost et al., 1992). The work began with a prototype to determine lime requirements for managing soil acidity. This led to the development of three expert systems: FARMSYS, ADSS, and LIMEAID. The goal of FARMSYS is to integrate information about farmers and their families into a system that provides recommendations from a farm household's perspective. ADSS provides soil acidity management recommendations. LIMEAID is an enhanced version of ADSS.

4. Do expert systems help in agricultural practice transfer?

Expert systems are an excellent technology to transfer information from highly qualified experts to less qualified practitioners. In this section, we list reasons for the selection of expert system technology to assist managers in solving agricultural practice transfer problems.

4.1. Conventional agricultural practices transfer problems

On examining existing information on irrigation, fertilization, and pest control as it relates to environmental conservation, we found it to be static and not necessarily responsive to the growers' needs. Extension documents can give only general recommendations because there are too many factors to take into consideration. Most of them handle problems related to a particular factor including, plant pathology, entomology, nutrition, or irrigation. In real-life situations, however, the problem is usually due to more than one factor and may need integration of the knowledge behind the information included in the various extension documents and books.

Changes in chemicals, their dosage, and their effect on the environment need to be considered. The update of this information in documents and the subsequent distribution of the materials take time.

Sometimes, information may not be readily available in any form other than human experts, extension workers and/or experienced growers. In addition, the limitations of conventional information transfer from specialists and scientists to extension workers and farmers represents a bottleneck for the development of agriculture on the national level. We are currently living in an era that is witnessing vast developments in all fields of agriculture. Therefore, there is a need for trained experts in these new technologies to transfer all this new information to farmers.

4.2. Is expert system development possible?

To determine whether expert system development for crop production is possible, we should consider the following questions.

4.2.1. What are the key requirements for this task? The crop management task does not require "common sense" knowledge: it requires a combination of cognitive and physical skills. The expert system can deal only with cognitive skills. However, a description of the operation and the equipment to be used can be handled by the system.

4.2.2 Does genuine expertise exist? Crop management includes many different disciplines: soil science, water and irrigation, plant nutrition, plant protection, economics, and production. Expertise on all crops cultivated in Egypt has accumulated over the years. Experts in many disciplines and crops work at the Agriculture Research Center (ARC) within MOALR. ARC consists of 16 research institutes and six central laboratories. The activities of each institute are either crop- or discipline-oriented. For example, there are institutes for horticulture and field crops, plant protection, soil and water management, and more. ARC is therefore equipped to provide knowledge and expertise on any crop for which an expert system is developed. The Central Laboratory for Agricultural Expert Systems (CLAES) is one of the main laboratories within ARC.

The first interviews with domain experts reveal whether they can articulate their problem-solving methods effectively and explain their decisions and reasoning. Most ARC scientists can, because they have extensive practical experience in addition to their scientific background.

The experts who participated in the knowledge acquisition process were motivated by their eagerness to participate in a new application that would help them in their research and extension activities. They were also financially compensated for their time.

The establishment of CLAES as part of ARC shows that MOALR is committed to expert system development for Egypt's agriculture. National funds are allocated to conduct research on the use of expert systems in agriculture and to install computers in extension offices.

4.2.3. Is the task easy to manage and well understood? Examination of the crop management task revealed that it could be divided into smaller, more manageable modules. It can be broken down into irrigation, fertilization, plant care, and diagnosis tasks. Each of these tasks is not too difficult and can be understood easily.

4.3. Is expert system development justified?

The need for expert systems for information transfer in agriculture can be demonstrated by showing that they generate a high payoff where human expertise is scarce and the environment hostile.

Crop yield definitely decreases when the required expertise is not available. If the appropriate decisions are not taken during the growing season, yields can be seriously affected. In some cases, a total crop failure could be the result.

One reason for the scarcity of experts is that the solutions for some problems require many experts. The system can thus help an extension worker identify the cause of a problem much more efficiently than a single expert or document can. Another reason is that a few experts and growers are mastering the undocumented experience and/or new agricultural technology in a specific discipline or for a certain commodity. The system can make this experience available to all growers.

In Egypt, there are 26 governorates, 200 districts, and 4000 villages. At the different levels of administration, there are extension offices. However, even though there are so many agricultural experts available, it is too difficult to cover all 4000 villages when a problem occurs. Some of these villages are far from cities and difficult to reach in a short time.

4.4 Is expert system development appropriate?

The key factors in determining when it is appropriate to develop an expert system are the nature, complexity, and scope of the problem to be solved.

4.4.1. Does the task require heuristic solutions and symbolic manipulation? Our investigation indicated that the system should satisfy two main goals: provision of advice on different agricultural issues and consultation to seek solutions to a specific problem a grower may encounter. A system of that nature deals extensively with non-numeric data and heuristics. For example, advice to a user who needed to find out the best way to prepare the soil for cultivation would involve factors such as soil type and depend on available equipment and tools. Such factors are non-numeric and lend themselves well to an heuristic approach.

Another example of the heuristic nature of a problem is the certainty factors associated with a conditional clause. If there is a disease symptom, there could be numerous reasons for it, such as nitrogen deficiency or low temperature. Conventional software is incapable of working effectively with ambiguous data and inexact ("fuzzy") relationships. For problems of this nature, developers often choose to apply the technology of knowledge-based expert systems. Expert systems provide acceptable results, advice, or diagnosis even when the available data is ambiguous or incomplete. Expert systems can apply heuristic and inference control strategies to find answers to problems in less time than conventional software and often when conventional systems cannot.

4.4.2 Is the task too complex? It takes years of study and practice to become an expert in only one aspect of the production of a crop. It is not possible to generate all the recommendations suitable for each field and put them in a printed document. This is because there are too many factors that can affect the appropriate recommendations. The manual solution for this problem is to give the extension worker and growers the rules for generating recommendations and let them make inferences from these rules. Generating documents that include these rules is not easy.

4.4.3 Is the task of manageable size? Does it have practical value? The crop management task can be divided into smaller modules. Each of these modules is well defined and its size is manageable. For example, the function of the diagnosis task is to identify the cause of a certain disorder. The total number of disorders is finite, and hence acquisition of knowledge about their symptoms is possible.

Developing an expert system for the transfer of information to agricultural operations is, therefore, highly practical as it will solve all the problems found in the conventional extension system.

5. Development methodology

The methodology is divided into two main parts: knowledge engineering and software engineering. The two parts interact through a spiral model. The output of the knowledge engineering process becomes the input for the software engineering activities. The main interaction occurs when the results of the knowledge analysis phase are fed into the design activity. The methodology also preserves the design model by introducing a method to transfer the design into an expert system development shell. This methodology (Rafea *et al.*, 1994) is based on KADS (Wielinga *et al.*, 1992).

5.1. Knowledge engineering methodology

The CLAES methodology was developed after three other knowledge engineering methodologies were tried: rule-based, knowledge analysis and design structuring (KADS), and generic task (GT). The following subsections will briefly describe these approaches and then provide a brief description of our methodology. For readers who want to know more about expert systems, a recent reference explains different aspects of expert systems (Liebowitz, 1997).

5.1.1. The rule-based approach. The rule-based approach acquires heuristic knowledge from domain experts and codes this knowledge into rules so that it can be processed by means of forward and/or backward chaining inference mechanisms. This approach is the first one used when expert systems are being developed and is commonly called the first generation approach. The main problem with the rule-based approach is that not all types of knowledge can be represented as rules and not all problem-solving or reasoning mechanisms use merely forward and/or backward chaining. Consequently, the nature of expert system tasks, including diagnosis, planning, and assessment cannot be represented explicitly.

5.1.2. Knowledge analysis and design structuring. The knowledge analysis and design structuring (KADS) approach was developed within an ESPRIT project between 1985 and 1989. The goal of the KADS project was to produce a methodology for the development of knowledge-based systems. In KADS, the development of a knowledge-based system is viewed as a modeling activity. Its methodology is based on a number of principles derived from cognitive psychology, artificial intelligence, and software development, including multiple model, expertise model, reusable model elements, knowledge differentiation, and structure-preserving design (Wielinga *et al.*,1992).

The expertise model is at the core of the methodology. It consists of three layers, known as domain, inference, and task. The domain layer contains concepts, properties, relations between concepts, and relations between expressions. The inference layer consists of a set of inference steps that use the domain layer to achieve a specific output. An inference step has input and output roles. They are connected in a diagram, called the inference structure, showing the flow of reasoning. The task layer is a procedure that orders the execution of the inference steps.

The advantages of KADS include a complete methodology for building knowledge-based systems which include all development phases and help for the engineer in the knowledge acquisition process. It is important to add that KADS is a generic methodology and not limited to a particular knowledge representation or reasoning mechanism.

The disadvantage of KADS is that the model of expertise is not a computational model, and the knowledge engineer has to transform the expertise model into a computational environment.

5.1.3 Generic task. The generic task (GT) approach was evolved at the laboratory for artificial intelligence research at the University of Ohio, under the guidance of Chandrasekaran (1988). His group's work in medical decision-making led to the postulation of a framework for expert systems design where the domain knowledge is decomposed into substructures that correspond to different problems.

Each problem-solving type is called a GT. Any complex task can be broken down into these GTs, which are a "natural kind" of information-processing task and correspond to a strategy to provide basic building blocks of intelligence. To date, there are six GTs identified by Chandrasekaran and his colleagues: classification, abstraction, knowledge-directed information passing, design by plan selection and refinement, hypothesis matching, and abduction assembly of explanatory hypotheses.

The advantage of the GT approach is that it enables the knowledge engineer, once he has broken down the underlying problem into a set of GTs, to concentrate on eliciting knowledge. The disadvantage of this approach is the scarcity of GT tools to be used by knowledge engineers, and the problems encountered if the underlying system cannot be decomposed into the GTs that are available. In such a situation, the knowledge engineer might want to define his/her own task, and this would violate the main purpose of this methodology.

5.1.4. CLAES methodology. Methodology developed at CLAES includes knowledge-acquisition, analyzing and modeling the acquired knowledge, and verifying the model. KADS was selected for knowledge modeling because it is a general methodology that can be used to model crop management tasks. However, GT methodology was considered when the knowledge model could fit into one of the well-known GTs, like hierarchical classification, that is suitable for diagnosis expert systems. The rule-based approach was not considered because it does not have the facility to model knowledge at the task level and consequently, will cause user interaction problems if there are too many rules.

Knowledge was acquired through the use of structured interview, concept sorting, and protocol analysis techniques. The knowledge engineering team consisted of two knowledge engineers and five domain experts in the following specialties: production, irrigation, nutrition, plant pathology, and entomology. Directors of the appropriate research institutes were asked to nominate two experts with practical experience in addition to a scientific background.

Once a model for each crop management task had been developed, the knowledge-acquisition process became more efficient as the relations were predetermined by the domain model. The developed prototypes were used as an automated tool for refining the acquired knowledge, which was documented in different forms depending on the knowledge type. It emerged that a dependency network could be used to represent the domain knowledge relationships effectively and as a successful communication tool for domain experts. During this phase, all types of media that could be used to enhance the explanation capabilities of the systems were collected, including images, video clips, and texts.

The knowledge analysis and modeling procedure included domain analysis, inference analysis, and task analysis. Domain analysis involves documented knowledge and aims at identifying concepts, their properties, and relationships. These relationships are either between concepts or between expressions. Concepts and relationships used by more than one subsystem were identified and grouped in a common knowledge base.

The purpose of inference analysis is to determine what knowledge the expert should use to reach a conclusion from specific components in the domain layer. Therefore, inference analysis was aimed at modeling the inference layer. The project has succeeded in developing a set of inference models for the crop management domain. These models can be used later for developing any crop management expert system.

The objective of task analysis is to find the sequence or the procedure for applying the inference steps, which the experts use to reach a final conclusion. The analysis phase was skipped when the GT approach was used, as it provides a model for each generic task, such as diagnosis and routine design, for example.

Knowledge is verified at the knowledge acquisition, analysis, and implementation stages. A review of the results is conducted by the domain experts at the knowledge elicitation stages. A review of the filled forms describing the domain layer is conducted by the domain experts at the analysis and modeling stage. Since it was difficult for the domain experts to review the inference and task layers, the task layer was explained in natural language and approved by the expert as a valid way for solving the problem. At the implementation stage, early reviews were conducted by the domain experts of any prototype. To verify the acquired knowledge, a multiple-expert conflict resolution process was used. If no consensus was reached, the view of the expert recognized to be the most specialized in the area of disagreement was accepted.

Explanation also was used for knowledge verification. The explanation facility was customized to follow the reasoning process at the task, inference, and domain levels. In addition to its verification function, the explanation facility provides the end user with greater confidence in the system recommendations.

5.2. Software engineering methodology

The approach we proposed for software development was a combination of rapid prototyping, incremental, and traditional methods. Rapid prototyping was used first to complete the requirement specifications. The incremental method started by implementing the laboratory prototype and ended by implementing the production version. The software development included the following activities: requirement specifications, design, implementation, verification, and validation.

5.2.1 Requirement specifications. An initial set of requirement specifications was determined as a result of early knowledge elicitation activity. This initial set was the basis for further knowledge-acquisition efforts and the basis for the preliminary design of the research prototype. The requirement specifications were revised regularly after each prototype implementation. This activity is well settled now and does not need to be carried out for new systems.

5.2.2 Design. A preliminary design was established just after the set of initial requirement specifications were determined, and a preliminary model of knowledge layers was specified. This design was the basis for the research prototype that was used to produce the requirement specifications for the laboratory prototype. The design was revised after the implementation of each prototype. Major areas that were considered in the design were the representation of knowledge, interfaces, explanation, databases, multimedia components, and control strategies.

5.2.3 Implementation. The first decision taken after the approval of the design was the selection of the implementation tools. This decision was based on the type of knowledge representation supported by the tool, the number of interfaces to external modules, explanation facilities, the primitives provided to code the control mechanism, and the hardware and software needs for system delivery. Although using a shell speeds up the implementation process, two major constraints on using shells were identified during the course of the project. First, implementing a special explanation module and/or a special control mechanism is not feasible in less expensive tools. Second, the delivery of a developed system needs a runtime license for expensive tools that give the user the environment to customize an application, which may cost about US $1200. We therefore built our own tool, KROL, on top of the programing language PROLOG.

5.2.4 Verification and validation. The demonstration of consistency, completeness, and correctness is the only verification a software needs (Adrion *et al.*, 1982). O'Keefe *et al.* (1987) define verification as "building the system right," that is, making sure that the implemented system matches the proposed design functionally and is free of semantic and syntactic errors. This was done internally at the laboratory by matching the design document against the implementation and running cases acquired from the design.

Validation is the process whereby the system is tested to show that its performance matches the original requirements of the proposed system. It is defined by Adrion *et al.*, (1982) as "the determination of the correctness of the final program or software produced from a development project with respect to the user needs and requirements." Our procedure for internal validation was to demonstrate the system to the knowledge experts for their approval and comments by preparing typical cases. This process was repeated until the experts were completely satisfied.

6. The components of crop management expert systems

Analysis of agricultural operations has revealed that an expert system for crop management is a family of systems that work together to generate a schedule for agricultural operations. The functionality of these operations was analyzed and classified into five categories: site assessment, seedling production, cultivation preparation, agriculture practice management, and disorder diagnosis and remediation. Analysis has revealed also that knowledge about these operations is dependent on a particular crop, although the same categories stand for any crop. The five systems that were developed are crop-wise expert systems. The subsystems implemented for each of them are listed in Table 1. Each cell in the table contains the knowledge-engineering methodology used to develop the corresponding subsystem and the tool used in the implementation.

Table 1. The subsystems included in each of the five expert systems developed.

	Site assessment	Seedling production	Cultivation preparation	Agricultural management	Disorder diagnosis
Cucum-bers	Not included	Separate (RB) (EXSYS)	Included (KADS) (KROL)	Included (KADS) (KROL)	Included (KADS) (KROL)
Tomatoes	Not included	Not included	Not Included	Not included	Included (KADS) (KROL)
Oranges	Included (KADS) (KROL)	Not included	Not Included	Included (KADS) (KROL)	Included (KADS) (KROL)
Limes	Included (KADS) (CLIPS)	Not included	Not included	Included (KADS) (CLIPS)	Partially (KADS) (CLIPS)
Wheat	Not included	Not included	Included (GT) (Small talk)	Included (GT) (Small talk)	Included (GT) (Small talk)

Figure 1 depicts the structure of CUPTEX, an integrated expert system to grow cucumbers under plastic tunnels, as an example of the architecture of a crop management expert system.

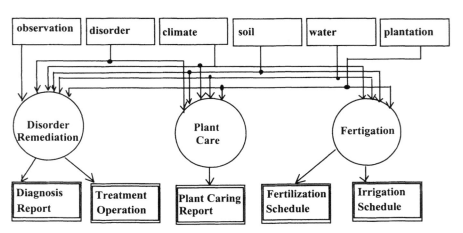

Figure 1. Overall structure of CUPTEX.

In the following subsections, a brief description is given for each of the crop management subsystems.

6.1. Site assessment

The site assessment subsystem that was developed here was for oranges (Salah *et al.*, 1992) and limes. Its function is to assess whether soil, water, and weather characteristics are suitable for orange cultivation. The decisions that can be generated from this subsystem are: the site is perfect for cultivation; a set of treatment operations has to be applied before cultivation; it is feasible to cultivate but you have to follow a set of recommendations; or the site is not suitable for cultivation. It would be useful to execute this subsystem before any consultations, because if a plantation is not suitable for cultivation, there would be no need to go to any other subsystem like diagnosis, for example.

This subsystem was developed in NEXPERT/Object shell (NEXPERT, 1988) for oranges and CLIPS (CLIPS, 1993) for limes. The oranges subsystem was transferred to a knowledge-representation object language (KROL) based on logic and object-programing and it was developed in CLAES (ESICM, 1992; Shaalan *et al.*, 1998).

6.2 Seedling production

The subsystem for seedling production was developed for cucumber seedlings (Rafea *et al.*, 1991; El-Dessouki *et al.*, 1991; El-Dessouki *et al.*, 1992). It has six functions: seed cultivation; cultivation medium preparation, which includes the determination of the ratio between vermiculite and peat moss; environmental growth factors control; diagnosis; treatment; and protection. This subsystem was developed with the EXSYS shell (EXSYS, 1989).

6.3 Cultivation preparation

The subsystem for cultivation preparation was developed for cucumbers (Rafea *et al.*, 1992) and wheat (Kamel *et al.*, 1994). Its main objective is to advise on precultivation activities at the production stage. For example, in the Wheat Expert System, the precultivation activities include variety selection, tillage, and seed cultivation. Some of these operations are unique to a specific situation while others are routine operations. However, the description of these routine operations was found to be very useful for novice growers. The subsystem was developed for cucumbers by NEXPERT/Object shell (NEXPERT, 1988) and then transferred to KROL, and for wheat by using a routine design GT that uses the programing language Small Talk (Kamel *et al.*, 1994).

6.4. Agricultural management

The subsystem for agricultural management was developed for cucumbers and oranges (Salah *et al.*, 1993). As noted above, two other subsystems have been developed: one for wheat (Kamel *et al.*, 1994), and another for limes. The main objective of such subsystems is to generate irrigation, fertilization, and preventive operation schedules. The irrigation and fertilization schedules include the water quantity, irrigation interval, nutrient quantity, and application interval. These outputs are based on quantitative rather than heuristic reasoning. The preventive operation schedule includes agricultural operations and preventive chemical spraying. General agricultural practices, such as pruning, are described in detail with the help of a video clip to show the grower how to perform them correctly. The system also gives the date when an agricultural operation should be done and helps determine the correct dosage for preventive chemical sprays. It also provides precautions to take while performing such operations. The cucumber and orange subsystem was developed with the NEXPERT/Object shell (NEXPERT, 1988); the wheat subsystem was based on SmallTalk; and the lime susbystem on CLIPS.

6.5 Disorder diagnosis and remediation

The disorder diagnosis subsystem was developed for cucumbers, oranges, tomatoes and wheat. Its main objective is to identify the cause and severity of an observed disorder and then to propose the appropriate remedy. The user can ask for the remedy at once if the cause of the disorder is known to him. In this case, the system would verify the cause given by the user before providing remediation advice. The subsystems for cucumbers, oranges and tomatoes were implemented by KROL (ESICM, 1992) those for limes by CLIPS.

Figure 2 presents a typical output screen of the diagnosis and remediation subsystem.

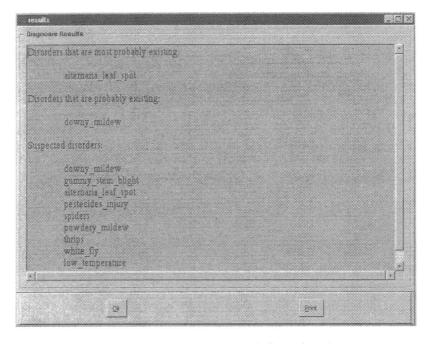

Figure-2. Output screen of the diagnosis and remediation subsystem.

7. Expert system evaluation

Expert system evaluation is the procedure by which we can be confident that the developed expert system satisfies the needs of the end users. We have mainly three types of users: researchers, extension workers, and growers. To ensure that the system is acceptable to the research community, an external validation is conducted; to ensure it is acceptable to extension workers and growers, a field validation is done.

7.1 External validation

External validation ensures the usability, quality, and utility of the expert system (Benbasat et al., 1989). The external expert validation is actually like an internal validation but their decisions are not taken for granted. Senior experts discuss the validation results with the domain experts to settle any disagreements. The rationale for conducting this external validation is to compensate for any defect due to erroneous knowledge and to ensure the acceptability of the system by the professionals in the field. Erroneous knowledge may have been provided by the selected experts, or the verification and validation steps did not locate it. The process of external validation is outlined below.

1. *Case description forms and comparison criteria preparation.* An evaluation criterion was designed to enable a formal judgement on solutions generated by human experts, and the expert system. The selected criteria provided both quantitative and qualitative evaluation bases for judgement. The qualitative criteria were excellent, good, acceptable, and unacceptable, with corresponding scores of three, two, one, and zero. The overall performance score P_i for each expert was calculated from

   $$P_i = 3*NE_i + 2*NG_i + 1*NA_i + 0*NU_i$$

   where P_i – performance score for expert no. i; Ne_i – number of cases evaluated as excellent; Ng_i – number of cases evaluated as good; NA_i – number of cases evaluated as acceptable; and NU_i – number of cases evaluated as unacceptable.

2. *Test case generation.* Test cases were prepared manually by knowledge engineers. The most import criterion for these test cases was that they had to account for both routine and difficult, rare cases.

3. *Test case solution.* Generated test cases were given to three domain experts. The same cases were introduced in the expert system. Each of the domain experts, as well as the expert system, worked out the test cases independently

4. *Test case evaluation.* Test case solutions were "double-blind" evaluated so that distinguishing between the solutions of the expert system and those of the domain experts was not possible. Two domain experts other than those who provided the knowledge acquired by the expert system were given the solved test cases for evaluation according to a previously prepared formula. A score was given to each solution, and solutions were ranked by their scores.

5. *Final observations and remarks.* A meeting was held to discuss solutions. Domain experts who provided the knowledge acquired by the expert system,

domain experts who solved the test cases, evaluators, and knowledge engineers attended this meeting, to analyze solutions and reach a final conclusion about the behavior of the expert system.

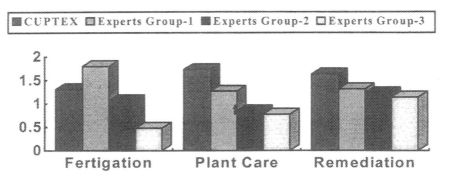

Figure 3. Evaluation results.

Figure 3 shows the scores given by groups of experts to each subsystem of CUPTEX (Rafea *et al.*, 1995). The "Experts Group-1" score is the score of the ranking expert in each area of expertise, the "Experts Group-2" score is the second-ranked, and so on.

The maximum (best) score is under 2.0. This is because the group decision was better than some individual decisions. CUPTEX outperformed human experts in two subsystems, and its score in the third one was 0.75 of the ranking experts group. Actually, CUPTEX was trained to be the best.

7.2. Field validation

Field validation is a continuous activity. Regular visits and meeting are conducted at the sites where the expert systems were deployed. The user comments are collected and discussed with the domain experts before being approved for updating the knowledge base. Another type of field validation is an experiment conducted during the growing season, with data collected from the field being managed by the expert system and a control field. The results of this experiment are used to update the knowledge base and measure the impact of the expert system. The main objective of the field validation is to determine, as part of the MOALR mission, whether the expert system is giving farmers good service at the extension offices. If it is, national production will increase and the environment will be protected. This will more than offset the cost of developing expert systems.

During 1995/96, an experiment was conducted at six sites: El-Bousily, El-Noubaria in the Behira governorate, Toukh in the Kaliobia governorate, El-Haram, El-Douki in the Giza governorate, and Mariot in the Alexandria governorate. The first objective was to validate the system in the field and measure its impact. The experiment was conducted by selecting two greenhouse tunnels: one would be cultivated by CUPTEX without any interference from the agricultural scientist and the other, the control tunnel, would be cultivated as usual. The field test results proved that CUPTEX was very successful in reducing cost and increasing yield (Figure 4). The total cost decreased from $478.95 to $430.87, a savings of approximately 10%. The yield price, on the other hand, increased from $1153.24 to $1407.65, an increase of approximately 22% (all costs are given in U.S. dollars).

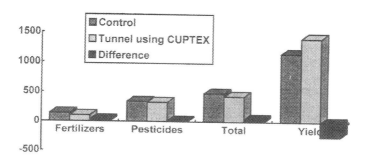

Figure 4. Average cost of chemicals used and the yield for one tunnel ($).

The second objective was to use fewer pesticides and fertilizers. Although our initial strategy was to optimize the economic aspect of the agricultural process, we ended up with an agricultural practice which is less harmful to the environment. Using cost as an indicator, we found that the overall use of fertilizers had decreased from $139.84 to $104.86, a decrease of approximately 25%. The use of pesticides decreased from $339.11 to $326.01, which represents a decrease of approximately 4%. This reduction in chemical use did not affect yield. When we took the ratio between the total cost of the chemicals used and the yield price, we found that this ratio has decreased from 0.415 to 0.306, with a percentage decrease of approximately 26%.

We noticed also that more water was being used in the expert system tunnel – 229.85 m^3 as opposed to 213.14 m^3 in the control tunnel (Figure 5). As water is not priced in Egypt, we did not add it to the cost. If this quantity is divided by the yield prices obtained from the expert system and control tunnels respectively, we get the amounts of water it takes to produce $1's worth of cucumbers. These amounts are 0.16 m^3 and 0.18 m^3, calculated for the results obtained from the expert system and control tunnels, respectively. This means that we can produce the same number of cucumbers in the plastic tunnels for approximately 11% less water than in the control tunnels.

8. Deployment

This section discusses activities related to the deployment of the developed expert systems. These activities start with training the extension workers and end with maintenance of the system.

8.1. Extension-worker training and system distribution

Once the expert system is approved for dissemination after the external validation is conducted, the training activity starts. This includes trainee selection, training sessions, and trainee evaluation. Trainees keep the system diskettes at the end of their training.

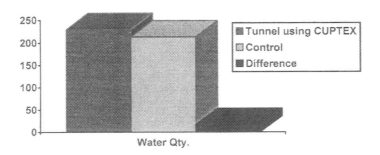

Figure 5. Average water quantity used for one tunnel (m^3).

Trainees were to be selected by age, computer literacy, and eagerness to learn. Unfortunately, computer literacy had to be dropped from the criteria as most of the extension workers were computer illiterate. As a result, a one-week training course in computer literacy was added. Eagerness to learn is measured subjectively by interviewing several candidates and selecting two from each site.

The training material was prepared and distributed on the first day of training. One week was dedicated to learning how to use the expert system. Lectures and hands-on training sessions were held. On the last day, the trainees were taught to install the system by using the distributed diskettes. Also, they were trained to report any bugs in the system and/or any comments on the recommendations generated by the expert system. A one-day training course is offered when a new version is produced.

The trainees were evaluated by presenting them with typical problems and checking whether they could come up with the correct recommendations. Hands-on testing was the main method used to assess the trainees and ensure that they could install the system successfully, handle minor computer problems, and report on bugs and the quality of recommendations generated by the system.

The computers are installed at extension offices where farmers usually come to ask extension workers for advice and recommendations.

8.2. Expert system monitoring

The developed expert systems can be used by decision-makers at both operational and planning levels.

On the operational level, extension workers in a village, district, and/or governorate used the system to support their decisions and to give appropriate advice to the growers. A survey was conducted to measure the adoption level of the expert system by trained extension workers (Nawar et al., 1997). The measurements were based on three main factors: the frequency of operation, the training of others to use the system, and the views of extension workers on system adoption. The results showed that 57% of the trained extension workers use the system at least once daily and 16% use it several times daily. Therefore, we can say that 73% of the extension workers use the system on a regular basis. It was found also that 68% of the trained extension workers had trained others to operate the system. The reasons preventing greater adoption of the system, as indicated by the extension workers, were lack of publicity (27%), problems related to the expert system itself (26%), personal (21%), and financial and administrative (10%). Only 8% responded that there was no need for the system while another 8% responded that there were no problems hindering the adoption of the system.

Another important factor to consider was the use of the explanation facility in the system. This facility was not included in the deployed system at the time of the survey; however, personal contacts and feedback from the users indicate that it gave them confidence in the system.

On the planning level, we have no experience in disseminating the expert systems to top management officials. However, we can see that the simulation models integrated with expert systems would be useful for predicting the need for water, fertilizers, and pesticides for a given crop in a given region. This generated information would be very important to the traders, exporters, and importers of these materials. The top-level management at MOALR is also interested in this type of information to plan for the crop rotation cycle and measure the inputs and outputs for each crop. Another type of application would be the prediction of yield, which could help decision-makers determine in advance how much of a crop to import or export.

8.3. Maintenance

System maintenance has two main objectives: first, to discover bugs and problems that may arise during the actual on-site running of the system; and second, to make sure that the system is up to date and possesses the most accurate and recent knowledge on the domain of application.

Users' remarks and complaints on the operation of the expert system, and the results of field validation should be collected in a field evaluation report. This is the main source of support for decisions on what is to be done to refine and maintain the system. Changes should be discussed and classified as minor or major according to the modification implementation time. Minor changes are usually implementation corrections, and can be performed immediately by the developer and transferred to the sites if they are urgently needed. Major changes are documented, studied, and discussed with domain experts if they concern the knowledge base. Approved changes are enrolled in the plan for the next version, documented, and attached to the user manual.

The second objective is achieved by arranging meetings every six months with domain experts to review the domain knowledge and discuss the latest updates. The acquired knowledge is then entered into the knowledge base.

9. Challenges and directions

The expert systems development effort contributed to the solution of agricultural practice information transfer problems. New problems arise, however, and have to be addressed. They fall into two main categories: development and deployment.

The development problems are addressed in our methodology, which includes the management of the development life cycle activities of any knowledge base system. The continuous knowledge base and software maintenance need extensive documentation and management because there is a scarcity of knowledge and software engineers. Porting systems from one platform to another platform is another maintenance problem. For example, when expert systems were first being developed in the late 1980s, the operating system on PC machines was DOS but now the common platform is Windows 95. Therefore, we had to port all expert systems developed on DOS to the new environment. Tools and methodologies should be developed to achieve the maintenance and porting procedures efficiently.

The deployment problems are specific to the agricultural environment. New activities at research stations and extension services make the introduction of the expert system as a new tool for information dissemination very relevant. Researchers should conduct field validation experiments for the deployed expert systems, analyze the results, and feed their comments back to the developers. Extension service activities include training, equipping extension offices with computers, following up on the use of expert systems, and feeding any comments back to the developers. Many of these activities are currently done on manual systems. For example, validation experiments on the extension package for specific crops, are routinely conducted at experimental research stations.

As noted above, CLAES was established at ARC to respond to these challenges. It provides extension service workers and growers with computerized packages (expert systems) in addition to the manual package provided by the research centers. CLAES is divided into three research departments: Knowledge Engineering Methodologies and Expert Systems Building Tools (KEBT), Agricultural Expert Systems Development (AESD), and Training, Evaluating, and Updating Expert Systems (TEUES).

The main goal of KEBT is to enhance the methodology continually in response to development and deployment needs, and to build tools to facilitate the development and maintenance of the systems. Another goal is to enhance the tools that researchers can use to build their own expert systems. Enhancing the explanation methodology and implementation is also being considered. Another important issue currently being investigated is the automation of knowledge-based translation. Although our expert systems are bilingual, great efforts have been made to maintain the two language versions (Arabic/English). The need for two versions is due to interaction with the international community, but in the meantime, extension workers are more comfortable when the inputs and outputs are in Arabic.

The main goal of AESD is to build expert systems for the major crops in Egypt and for animal health care. The priority for expert systems development is settled with the administration in MOALR. The sharing of previously acquired knowledge is another research goal that this department is investigating to expedite expert system development.

The main tasks for TEUES are to invent methods and tools to train extension workers, conduct verifications, participate in field validations, and update the deployed expert system. Right now, the department supports five systems deployed at 24 extension offices, seven research stations, seven research institutes, nine faculties of agriculture, and 225 private-sector farms.

References

Adrion, W., M. Branstad and J. Cherniovsky (1982). "Validation, Verification and Testing of Computer Software", ACM Computing Surveys 14(2).

Benbasat, A. and J. S. Dhaliwal (1989). "A Framework for Validation of Knowledge Acquisition", *Knowledge Acquisition* 1(2), 215-233.

Chandrasekaran, B. (1988). "Generic Task as Building Blocks for Knowledge-Based Systems: the Diagnosis and Routine Design Example," *Knowledge Engineering Review* 3, 183-211.

CLIPS (1993). CLIPS v. 6.0 Lyndon B. Johnson Space Center, Information System Directorate, Software Technology Branch,NASA, Houston, Texas.

El-Dessouki, A., A. Rafea, M. Youssef and H. Safwat (1991). "An Expert System for Seedling Production Management." *Proceedings of the First National Expert Systems and Development Workshop (ESADW-91)*, MOALR, Cairo.

El-Dessouki, A., A. Rafea and M. Youssef (1992). "Verification and Validation in Knowledge Based Systems". *Proceedings of the First International Conference on Expert Systems and Development (ICESD-92)*, MOALR, Cairo.

ESICM (1992). Specifications of Object Oriented Language on Top of Prolog (KROL). Technical Report No. TR-88-024-25. Expert Systems for Improved Crop Management Project. MOALR, Cairo, Egypt.

EXSYS (1989). EXSYS professional: Advanced Expert System Development Software. Technical Manual, EXSYS Inc., NM-USA.

Gerevini, A., A. Perini, *et al.* (1992). "POMI: An Expert System for Integrated Pest Management of Apple Orchards," *AI Applications* 6(3), 51-62.

Kamel, A., K. Schroeder, *et al.* (1994). "Integrated Wheat Crop Management System Based on Generic Task Knowledge Based Systems and CERES Numerical Simulation," *AI Applications* 9(1), 17- 27.

Lemmon, H. (1986). "COMAX: An expert system for cotton crop management," *Science* 233, 29-33.

Liebowitz, J (ed.) (1997). *The Handbook of Applied Expert Systems*. CRC Press.

Michalski, R., J. Davis, V. Visht and J. Sinclair (1983). "A computer-based advisory system for diagnosing soybean diseases in Illinois," *Plant Disease* 67, 459-463.

Mohan, S., N. Arumugam (1995). "An intelligent front-end for selecting evapotranspiration estimation methods," *Computers and electronics in agriculture* 12(4): 295-309.

Nawar,M., M. Mahmoud and I. Rashwan (1997). Social Impacts of Using Expert Systems in Field Practice, Technical Report Published by CLAES TR/CLAES/7/97.2

NEXPERT (1988). Neuron Data, NEXPERT Object Fundamentals, PC Version 1.1, Technical Manual, Neuron Data Inc. California.

O' Keefe, R.M., O. Balci and E. Smith (1987). "Validating Expert System Performance", *IEEE Expert* Winter 1987.

Rafea, A., M. Warkentin and S. Ruth (1991). "An Expert System for Cucumber Production in Plastic Tunnels", *Proceedings of the World Congress on Expert Systems*, Florida.

Rafea, A., A. El-Dessoki, *et al.* (1992). "An Expert System for Cucumber Production Management under Plastic Tunnels," *Proceedings of the First International Conference on Expert Systems and Development (ICESD-92)*, MOALR, Cairo.

Rafea, A., S. Edree *et al.* (1994). "A Development Methodology for Agricultural Expert Systems Based on KADS," *Proceedings of the Second World Congress on Expert Systems.*

Salah, A., A. Rafea and E. Mohamed (1992). "An Expert System for Citrus Cultivation Feasibility," *Proceedings of the First International Conference on Expert Systems and Development (ICESD-92)*, MOALR, Cairo.

Salah, A., H. Hassan *et al.* (1993). "CITEX: An Expert System for Citrus Crop Management," *Proceedings of the Second National Expert Systems and Development Workshop (ESADW-93)*, MOALR, Cairo.

Schulthess, U. *et.al.* (1996). NEPER-Weed: "A Picture-Based Expert System for Weed Identification," *Agron. J.* 88, 423-427.

Shaalan, K., M. Rafea and A. and Rafea (1998). "KROL: A Knowledge Representation Object Language on Top of Prolog," *Expert Systems with Applications* (4).

Warkentin, M., P. Nair *et al.* (1990). "A Knowledge-Based Expert System for Planning and Design of Agroforestry Systems," *Agroforestry Systems* 11(1), 71-83.

Wielinga, B. J., A. T. Schreiber and J. A. Breuker (1992). "KADS: A Modeling Approach to Knowledge Engineering," *Knowledge Acquisition* 4, 5-53.

Yost, R. *et al.* (1992). "Expert System for Information Transfer About Soil and Crop Management in Developing Countries," in C. Mann and S. Ruth (eds.), *Expert Systems in Developing Countries Practice and Promise*, Westview.

III RESEARCH ISSUES

13 RULE INDUCTION IN CONSTRUCTING KNOWLEDGE-BASED DECISION SUPPORT

Tu Bao Ho

1. Introduction

Decision support systems (DSSs) are computer-based systems that can help decision-makers to solve semistructured problems by allowing them to access and use data and analytic models interactively. Knowledge-based systems (KBSs) are systems that generate quality solutions to problems requiring computer-based reasoning knowledge. There are two main ways of using KBS techniques in computer-based decision-making. The first is to use KBSs directly as kinds of DSSs. Recent progress in techniques for coupling databases and knowledge bases allows the user to exploit KBSs more effectively in various decision problems such as diagnosis, design, and planning (Sprague and Watson, 1993; Blanning and King, 1993). The other is to integrate KBS technologies with conventional DSSs (El Najdawi and Stylianou, 1993; Sullivan and Fordyce, 1994; Turban and Aronson, 1998). We call knowledge-based decision support systems (KBDSSs) "software in both directions."

By assisting with the simultaneous use of data, knowledge, analytic models, and dialogue, KBDSSs have the potential to be useful in the field of sustainable development, particularly in helping developing countries meet their own unique development needs. Knowledge and advanced information technology from industrialized countries could be used to solve complex problems in fields like public administration, urban and rural development, transportation, health care, special needs, education, environment, and agriculture (Mansell and Wehn, 1998). As knowledge is central to intelligence systems, how to obtain that knowledge for decision-making is a crucial issue, and it remains the bottleneck in the development of KBDSSs. Automatic access to knowledge from external resources would be particularly beneficial to developing countries where there is a lack of expertise in providing knowledge for KBDSSs, because many large databases in the fields listed above already exist. It is clear that decision-makers in these countries need to understand and use their data to find useful knowledge that can help in their country's development.

Knowledge acquisition (KA) has been seen as a problem of transferring knowledge — in other words, extracting discourse and translating it into the implementation language constructs. Recently, however, KA is not merely being seen as an exercise in 'expertise transfer' but as a modeling process. Much of the current work focuses on systems that follow the KA paradigm of expert or *external resources* → *automatic acquisition* → *knowledge base* instead of the traditional paradigm of expert to knowledge base by way of a knowledge engineer.

Automatic KA shares the same goal as the rapidly growing interdisciplinary field of knowledge discovery and data mining — extracting useful knowledge from large databases (Fayyad *et al.*, 1996). Automatic KA is based mainly on applications of machine learning and rule induction. Machine learning (ML) — the study of computational algorithms that improve automatically through experience — can provide increasing levels of automation in the construction of knowledge bases. Rule induction is one paradigm of ML that has achieved many recent successes in real-world applications.

Section 2 deals with the problem and the main steps of the rule induction process. Section 3 presents the key ideas of two rule induction methods, CABRO and OSHAM, from supervised and unsupervised data, respectively, and their implementation as interactive-graphic systems (Ho, 1995; Ho, 1997; Ho *et al.*, 1998; Nguyen and Ho, 1999). To evaluate these systems, we carry out experimental comparative studies for CABRO, OSHAM, and some widely used systems such as C4.5 (Quinlan, 1993), CART (Breiman *et al.*, 1984), AUTOCLASS (Cheeseman and Stutz, 1996), and others. Section 4 describes the mutual use of CABRO and OSHAM in the construction of KBDSSs within the framework of the knowledge-based system generator TESOR (Ho *et al.*, 1992). Section 5 presents a case study to illustrate how diagnostic knowledge was induced from a clinical database on meningoencephalitis.

2. Problem and steps of rule induction

Two main components of a KBS are an inference engine and a knowledge base. Most available KBSs provide an inference engine and knowledge editor but not the capability to acquire knowledge automatically. Langley and Simon (1995) pointed out that "rule induction may never entirely replace the knowledge engineer in constructing KBS, but significant progress toward automation in knowledge engineering has already been made." They examined various successes in applying rule induction to real-world problems, including increasing yield in chemical process control, making credit decisions, forecasting severe thunderstorms, diagnosis of mechanical devices, improving separation of gas from oil, and preventing breakdowns in electrical transformers.

Langley and Simon called rule induction the paradigm of ML that "employs condition-action rules, decision trees, or similar logical knowledge structures. Here the performance element sorts instances down the branches of the decision trees or finds the first rule whose conditions match the instance, typically using a logical matching process. Information about classes or predictions is stored in the action sides of the rules or the leaves of the trees. Learning algorithms in the rule-induction framework usually carry out a greedy search through the space of decision trees or rule sets, typically using a statistical evaluation function to select attributes for incorporation into the knowledge structure. Most methods partition the training data recursively into disjoint sets, attempting to summarize each set as a conjunction of logical conditions."

There are five main steps in the rule-induction process which the DSS user and developer should know while applying rule induction programs.

The first step is to formulate the decision problem so that part of it can be handled by an induction method. An important distinction in ML has to do with whether one uses the learned knowledge for one-step classification or prediction, or for some form of multistep inference or problem solving. Decision tasks such as process control, design, diagnosis, planning, and scheduling are often complex, yet one can identify components that involve simple classification, a task for which there exist robust induction algorithms.

The second step is to determine the representation of knowledge to be learned. Knowledge can be represented in different ways on a computer — rules, decision lists, inference networks, decision trees, and concept hierarchies. A decision tree is a classifier in the form of a tree structure that is either a leaf that indicates a class of instances, or a decision node that specifies some test to be carried out on a single attribute value, with one branch and subtree for each possible outcome of the test. A decision tree can be used to classify an instance by starting at the root of the tree and moving up through it until a leaf is encountered. A decision tree or concept hierarchy can be converted easily into a set of decision rules.

The third step is to collect the training data. Rule induction methods start from data sets representing cases in application domains. The simplest situation concerns Boolean or binary features, with each case specifying the presence or absence of a feature. A slightly more sophisticated situation concerns symbolic attributes that consist of nominal (categorical) or ordinal ones. Numeric attributes that take on real values are also possible. Data are often viewed in the instance space, which may consist of a mix of attributes. Inductive learning methods may be classified into two groups on the basis of the degree of supervision in the training data. Early works in concept learning would focus on the simpler supervised task in which the learner is asked to characterize concepts from a given set of classified instances (members of the concept). This task continues to receive much attention in ML, where most of the research is concerned with influential methods like decision tree induction.

The unsupervised task, though less developed than the supervised, can arise just as often in real-world problems; as a result, it has attracted a considerable amount of interest. In this task, the learner is asked to determine concepts and knowledge structures from a given set of unclassified instances. In general, rule induction programs work on data sets represented in data tables where each line stands for the description of an instance and each column stands for an attribute. A special column, which exists only in supervised data, contains the label of instances. Training data are also provided offline (all instances are presented simultaneously) or online (instances are presented once at a time). This leads to nonincremental or incremental learning, respectively. After collecting the training data, the user needs to distinguish between these different methods to choose which one is most suitable.

The fourth step is to evaluate the learned knowledge. Experiments show that rules induced from training data are not always of high quality. A standard way to evaluate these rules is to divide the data into two sets, training and testing. One can repeat this process a number of times with different splits, then average the results to estimate the rules performance. This procedure is known as cross-validation.

The fifth and final step is to field the knowledge base. In some cases, discovered knowledge can be used without embedding it into a computer system. Otherwise, users might expect that induced rules could be put on computer and exploited by some programs. In particular, the results of a rule program could be considered decision knowledge for a KBDSS. It is most important to determine whether the learned knowledge in fact can be used in decision-making.

3. Two rule induction methods

3.1 Rule induction from supervised data

CABRO is a method of constructing decision trees from supervised (classified) data. The basic task of supervised learning is to extract knowledge that can be used to predict classes of classified instances correctly from a given set of classified instances. Here is an example of discovering rules from data for diagnosing two kinds of heart disease. CABRO started from a database (Cleveland Clinic Foundation) containing records of patients who belong to one of two heart disease classes, 'H' or 'S' (the last column), and discovered diagnosis rules with associated (estimated) prediction accuracy. Each record consists of information observed on 14 numeric and categorical attributes: age, sex, chest pain type, resting blood pressure, vessels colored, etc. Some patient records extracted from the database follow:

> 63, male, angina, 145.0, 233.0, true, hyp, 150.0, fal, 2.3, down, 0.0, fix, H
> 67, male, asympt, 120.0, 229.0, fal, hyp, 129.0, true, 2.6, flat, 2.0, rev, S
> 41, female, abnang, 130.0, 204.0, fal, hyp, 172.0, fal, 1.4, up, 0.0, norm, H
> 56, male, abnang, 120.0, 236.0, fal, norm, 178.0, fal, 0.8, up, 0.0, norm, H
> 62, female, asympt, 140.0, 268.0, fal, hyp, 160.0, fal, 3.6, down, 2.0, norm, S
> 57, female, asympt, 120.0, 354.0, fal, norm, 163.0, true, 0.6, up, 0.0, norm, H

The diagnosis of unpredicted patients can be obtained by matching their symptoms (records) against conditions of discovered rules such as

Rule 7:   IF   sex = male
               age <= 67
               resting blood press > 108
               max heart rate <= 171
               number of vessels colored > 0
           THEN diagnosis of heart disease S [accuracy = 88.8%]; and

Rule 35: IF   sex = female
               number of vessels colored <= 0
               thal = norm

resting blood press <= 140
THEN diagnosis of heart disease H [accuracy = 93.6%].

Among approaches to supervised concept learning, decision-tree induction (DTI) is certainly the most active and applicable one. DTI systems differ from each other in the way they deal with the problems of attribute selection (or choosing the 'best' attribute to split a decision node) and pruning (or avoiding overfitting). This chapter presents our solution to the attribute selection problem, among others, including model selection, visualization, and interactive learning (Nguyen and Ho, 1999).

It is known that the most widely used measures for attribute selection are information-theory based, such as information gain or gain-ratio (Quinlan, 1993), or statistics-based, such as Chi-square or Gini-index (Breiman *et al.*, 1984). For attribute selection, CABRO employs R-measure (Ho et al., 1998), which is inspired by the notion of dependency degree in the rough set theory (Pawlak, 1991). Rough set theory is a mathematical tool to deal with vagueness and uncertainty. The basic idea in this theory is to 'view' approximately each subset X of an object set O by its lower and upper approximations, E_* and E^* w.r.t., an equivalence relation $E \subseteq O \times O$. These approximations of X are defined, respectively, by and $E_*(X) = \{o \in O: [o]_E \subseteq X\}$ and $E_*(X) = \{o \in O: [o]_E \cap X \neq \varnothing\}$ where $[o]_E$ denotes the equivalence class of an object o in E. A key concept in the rough set theory is the degree of dependency, $\mu_P(Q)$, of an attribute set Q on an attribute set P:

$$\mu_P(Q) = \frac{\text{card}(\bigcup_{[o]_Q} P_*([o]_Q))}{\text{card}(O)} = \frac{\text{card}(\{o \in O : [o]_P \subseteq [o]_Q\})}{\text{card}(O)}.$$

The measure $\mu_P(Q)$ can be used directly in decision-tree induction for selecting attributes, where Q stands for the class attribute and P stands for each descriptive attribute. Our analysis and experiments have shown, however, that it is not robust enough with noisy data or sensitive enough when partitions of O generated by Q and P are nearly identified. Inspired by $\mu_P(Q)$, we proposed a new measure, called R-measure, for the dependency of Q on P:

$$\tilde{\mu}_P(Q) = \sum_{[o]_P} \frac{1}{\text{card}(O)} \max_{[o]_Q} \frac{\text{card}([o]_P \cap [o]_Q)^2}{\text{card}([o]_P)}.$$

We show that R-measure can overcome the limitations of $\mu_P(Q)$ in different situations (Nguyen and Ho, 1999). Using this measure for selecting attributes, we developed the CABRO system in the following common framework of decision-tree induction:

1. choose the 'best' attribute that maximizes the chosen attribute selection measure;

2. extend the tree by adding a new branch for each attribute value;

3. sort training examples to leaf nodes; and

4.   if examples are unambiguously classified, then stop, or else repeat steps 1-4 for
     leaf nodes.

There are three main criteria for evaluating discovered decision trees: predictive
accuracy, size, and understandability. Predictive accuracy refers to the ability of the
decision tree to predict unknown instances into learned classes. The predictive accuracy is measured also in terms of error rates (the proportion of incorrect predictions
that a tree makes on the test data). Size relates to the generally accepted Occam's Ra-
zor principle — the fewer nodes in the tree, the better. Understandability relates to the
knowledge representation.

In order to obtain a reliable evaluation of R-measure, we have carried out an experimental comparative study. Recently, k-fold stratified cross validation has been
recommended as an appropriate method for evaluating techniques with real-world
datasets similar to those of the University of California, Irvine (UCI) repository of ML
databases. We used this benchmark to evaluate three other well-known measures —
gain-ratio (Quinlan, 1993), Gini-index (Breiman *et al.*, 1984), Chi-square in statistics
and R-measure — on 18 datasets from the UCI repository. Table 1 includes information on the datasets such as their name, dimension (number of attributes X number of
instances), and type ([s] for symbolic, [n] for numeric and [m] for mixture), as well as
the experimental comparative results of the error rates of pruned trees for the four
measures on these datasets. The lowest error rate attained at each dataset is in bold.

Table 1. Experimental comparative error rates of four measures.

Datasets	Gain-ratio	Gini-index	Chi-square	R-measure
Vote, 16X300 (s)	**5.0 ± 2.8**	5.9 ± 2.7	5.9 ± 2.7	5.7 ± 2.7
Cancer, 9X700 (s)	7.4 ± 2.9	7.4 ± 3.5	7.4 ± 3.5	**7.1 ± 3.4**
Promoters, 45X105 (s)	24.5 ± 7.5	**22.7 ± 10.0**	**22.7 ± 10.0**	**22.7 ± 10.0**
Shuttle, 9X956 (s)	**0.2 ± 0.1**	**0.2 ± 0.1**	0.3 ± 0.1	0.3 ± 0.1
Solar fare, 9X956 (s)	**25.3 ± 1.5**	27.8 ± 1.3	26.6 ± 2.0	25.5 ± 1.0
Diabetes, 8X768 (n)	25.3 ± 2.6	25.6 ± 2.5	25.5 ± 2.5	**25.3 ± 2.6**
Spice, 45X3189 (n)	**8.0 ± 1.7**	8.4 ± 1.8	8.8 ± 1.7	8.6 ± 1.9
Glass, 9X214 (n)	**34.5 ± 8.2**	36.8 ± 6.8	37.3 ± 6.4	35.9 ± 6.9
Waveform, 36X3195 (s)	25.7 ± 1.1	**24.4 ± 1.8**	26.8 ± 1.3	25.1 ± 1.1
Heart disease, 13X270 (m)	25.6 ± 4.1	25.6 ± 5.6	26.3 ± 4.9	**25.2 ± 4.6**
Vehicle, 18X846 (n)	32.7 ± 5.1	32.0 ± 3.7	31.9 ± 3.2	**31.8 ± 3.5**
Hypothyroid, 25X3163 (n)	0.9 ± 0.4	0.9 ± 0.4	0.9 ± 0.4	0.9 ± 0.4
Audiology, 70X226 (s)	30.9 ± 11.0	30.9 ± 11.9	45.2 ± 8.7	**29.1 ± 11.7**
Cars, 8X392 (n)	26.0 ± 2.0	26.8 ± 5.2	26.5 ± 5.2	**25.2 ± 4.8**
Horse colic, 28X368 (n)	**14.3 ± 5.1**	16.8 ± 3.5	17.0 ± 3.3	15.9 ± 4.2
Pima diabetes, 8X768 (n)	**23.4 ± 3.6**	23.5 ± 3.5	23.5 ± 3.5	23.9 ± 3.2
Segmentation, 19X2310 (n)	6.2 ± 1.6	**6.1 ± 2.0**	7.6 ± 2.0	**6.1 ± 2.1**
Iris, 4X150 (n)	3.3 ± 3.3	**2.7 ± 3.2**	4.0 ± 4.0	4.0 ± 4.0

Two observations can be drawn from these experiments:
•   The gain-ratio and R-measure are somehow comparable. They had the lowest
     values of error rates in nine datasets. The Gini-index and Chi-square have average
     and high error rates and attained those in five and two datasets, respectively.

Moreover, CABRO offers various advantages over other systems on model selection, visualization, and interactive learning (Nguyen and Ho, 1999).

- As for tree size, the gain-ratio demonstrated its advantage of smaller trees when datasets were small or mid-size, but not when datasets were large. However, for datasets with big trees, Chi-square often had noticeably bigger ones.

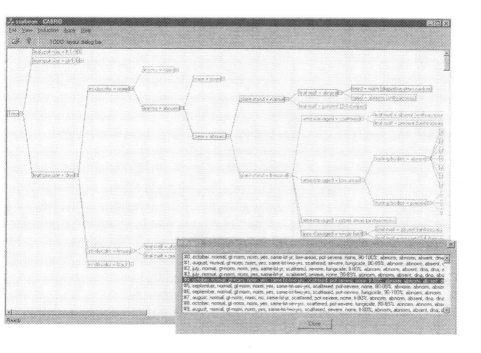

Figure 1. Generation and interpretation of decision trees from soybean data by CABRO.

3.2 Rule induction from unsupervised data

Here, the basic KA task is to find simultaneously, from a given set of unsupervised (unclassified) instances, a hierarchical clustering of subsets of instances and intensional descriptions of these subsets that satisfy certain conditions. Essentially, unsupervised induction methods differ from each other in two ways: views on concepts and constraints of categorization. Among views on concepts, the classical, prototype and exemplar views are widely known and used. Among categorization constraints, the similarity, feature correlation, and structure of the concept hierarchy are widely known and used. OSHAM employs the classical view on concepts and is able to form an effective concept hierarchy from unclassified data (Ho, 1995). Essentially, OSHAM searches to extract a good concept hierarchy by exploiting the lattice structure of Galois concepts as the hypothesis space. Recently, OSHAM has been extended to a hybrid system that allows for a higher performance by combining its original view on concepts with the prototype and exemplar views (Ho, 1996; Ho, 1997). The essential ideas of the hybrid OSHAM are

1. to extend the classical representation of a concept from a pair of extent, intent and its hierarchical structure information to a hybrid representation of a 10-tuple with additional components (these include the probability of the concept occurrence, homogeneity, typical instances, conditional probability for occurrence and dispersion of typical instances, and the quality of the concept evaluated in terms of the concept hierarchy structure); and

2. to find sufficiently general and discriminant concepts at each level of the concept hierarchy while determining their typical instances on the basis of a quality criterion that combines the similarity and structural constraints.

OSHAM has been implemented in the X Window on a Sparcstation with the direct manipulation style of interaction that allows the user to participate actively in the discovery process. The user can initialize parameters to cluster data, visualize the concept hierarchy gradually, observe the results and the quality estimation, manually modify the parameters when necessary before the system continues to go further to cluster subsequent data, or backtrack to regrow branches of the concept hierarchy with respect to the categorization scheme. Figure 2 shows a main screen of the interactive OSHAM with a hierarchy of overlapping concepts learned from breast cancer data collected at a Wisconsin hospital. A full description of a concept in Figure 2 is given below:

```
CONCEPT 43
Level = 5
Super_Concepts = {29},  Sub_Concepts = {52, 53}
Features = (Uniformity of Cell Size, 1) AND (Bare Nuclei, 1)  AND
           (Bland Chromatin, 1)  AND (Uniformity of Cell Shape, 2)
Local_instances/Covered_instances = 6/25
Local_instances = {8, 127, 221, 236, 415, 661}
Concept_probability = 0.041666
Local_instance_conditional_probability = 0.240000
Concept_dispersion = 0.258848
Local_instance_dispersion = 0.055556
Subconcept_partition_quality = 0.519719.
```

The interpretation of induced results is commonly understood as the process of matching an unknown case e to discovered classes (concepts). As the generality decreases along branches of a hierarchical structure, we say that a concept C_k matches the unknown instance e if C_k is the most specific concept in a branch that matches e intensionally (though all superconcepts of C_k match e). Naturally, there are three types of outcomes when one is logically matching an unknown instance e with the learned concepts: only one concept that matches e (single-match), many concepts that match e (multiple-match), and no concept that matches e (no-match). We developed an interpretation procedure for concept hierarchies that uses the concept intent, the hierarchical structure information, the probabilistic estimations, and the nearest neighbors of unknown instances (Ho and Luong, 1997). This interpretation procedure consists of two stages: (1) find all concepts on the concept hierarchy that match *e* intensionally, and (2) decide which one among these concepts matches *e* best.

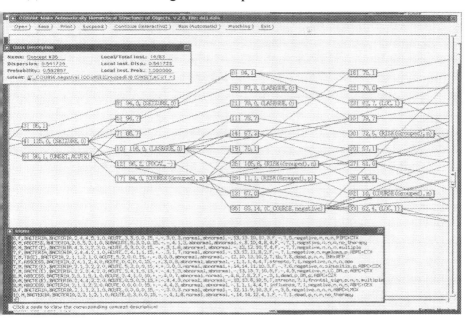

Figure 2. Generation of a concept hierarchy from medical data by OSHAM.

Comparative experiments were carried out for the well-known system AUTOCLASS (Cheeseman, 1996) and OSHAM on 10 datasets from the UCI repository of ML databases.

Table 2. Experimental comparative results of OSHAM.

Datasets	num.att.	sym.att.	instances	classes	AUTOC	OSHAM
Wisconsin breast cancer	9 -	699	2	96.6	92.6	
Congressional voting	17	-	435	2	91.2	93.7
Mushroom	23	-	8125	2	86.5	88.2
Tic-tac-toe	9	-	862	9	82.3	92.6
Glass identification	-	9	214	6	55.7	65.3
Ionosphere	-	35	351	2	91.5	84.6
Waveform	-	21	300	3	59.2	73.0
Pima diabetes	-	8	768	2	68.2	72.7
Thyroid (new) disease	-	6	215	3	89.3	84.6
Cleveland heart disease	8	5	303	2	49.2	60.8

The number of attributes, instances, and 'natural' classes of these datasets are given in Table 2. All experiments on these datasets were carried out with 10-fold cross validation by AUTOCLASS (AUTOC) and OSHAM in the same condition (the same datasets randomly divided into subsets). For AUTOC, we used the public version AUTOCLASS-C implemented in C and ran three steps of search, report, and predict with the default parameters. We obtained the predicted name and predictive accuracy of AUTOCLASS and OSHAM. In order to avoid a biased evaluation of OSHAM (although with each dataset, parameters can be adjusted to obtain the most suitable concept hierarchy), we fixed the values $\alpha = 1\%$ of the size of the training set, $\beta = 15\%$ and $\sigma = 10\%$ of the number of attributes, and the beam size $\eta = 3$ in common to all datasets.

The predictive accuracies of OSHAM and AUTOCLASS in these first trials were only slightly different. Each system performs better in different datasets; therefore, these two systems can be considered to have comparable performance. The main advantage of OSHAM is that its concept hierarchies can be more easily understood by its extended classical view on concepts and its graphic support.

4. Fielding knowledge bases in decision-making

Among knowledge representation models, the object/rule model, developed by the knowledge-based system generator TESOR (Ho *et al.*, 1992), provides an understandable structure of knowledge that allows final decisions from an initial state to be reached efficiently in orderly stages.

The process of constructing a knowledge base in TESOR is shown in Figure 3 (Ho, 1996). The Graphic Knowledge Editor (GKE) is preferred if the knowledge is overt and available, while CABRO and OSHAM are used when expertise is being described in terms of relevant attributes or correct instances, although experts cannot directly formulate rules or enter a knowledge structure either. These tools also can be used mutually in order to obtain a higher-quality knowledge base:

- The expert/knowledge engineer describes the domain knowledge by means of the GKE if the knowledge is already available; for example, in the form of a set of decision rules or an object hierarchy.

- Otherwise, CABRO can generate decision rules/decision trees from supervised data. It can handle large data tables and noisy data effectively.

- OSHAM can generate hierarchy/decision rules from unsupervised data. The output of OSHAM is compiled directly by the TESOR GKE into the internal representation form. The object hierarchy is displayed graphically to enable an expert to give his judgment until a satisfactory hierarchical object knowledge base is obtained.

- The visualization of knowledge bases by the TESOR GKE offers the user an intuitive understanding about the structure of the domain knowledge. It helps the expert/knowledge engineer in clustering data, observing the results and quality evaluations, modifying parameters, and adjusting results at each level before the system continues further.

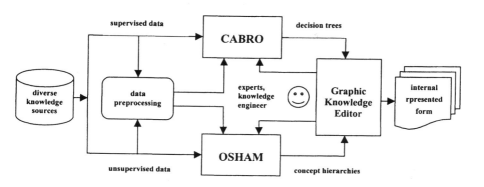

Figure 3. Mutual use of CABRO and OSHAM in constructing knowledge bases.

5. A Case Study

In this section we illustrate discovered knowledge from a medical database by two DTI systems, CABRO and C4.5. This is a database on meningoencepahalitis collected at the Medical Research Institute, Tokyo Medical and Dental University (Tsumoto, 1999). Each record contains 35 attributes. Their names and values are given in Table 3. Part of the database is shown below. The third field represents the class attribute DIAG with nine values (VIRUS, ABSCESS, BACTERIA, BACTE (E), BACTERI, VIRUS (E), TB(E), CRYPT, and Rubella), which are classified into two groups: 'VIRUS' with the values VIRUS and VIRUS (E), and 'BACTERIA' with the other values.

Table 3. Attributes in the database.

Category	Type	Number of attributes
Present history	Numerical and categorical	7
Physical examination	Numerical and categorical	8
Laboratory examination	Numerical	11
Diagnosis	Categorical	1
Therapy	Categorical	1
Clinical course	Categorical	4
Final status	Categorical	1
Risk factor	Categorical	1

58,M,VIRUS,0,12,11,0,0,0,acute,37.6,2,0,0,0,-,-,6000,0.0,7,n.p.,a,-,660,86,44,F,-,-,0,-,0,244,-,-
29,F,VIRUS,0,2,4,0,0,0,acute,38.5,4,0,1,15,-,-,8300,0.0,14,n.p.,a,-,464,88,59,F,-,-,0,-,0,423,-,-
40,M,VIRUS,0,5,3,3,0,0,ACUTE,37.5,0,0,0,0,-,-,8200,0.0,0,n.p.,a,-,208,348,0,F,-,-,0,-,0,353,-,-

16,M,VIRUS,0,2,4,2,0,0,ACUTE,37.2,3,1,0,0,-,-,1070,2.0,42,n.p.,a,-,219,33,57,F,-,-,0,-,0,90,-,-
42,M,VIRUS,0,3,0,0,0,0,acute,37.5,2,0,0,15,-,-,8300,0.0,6,n.p.,a,-,117,37,63,F,-,-,0,-,0,53,-,-
30,F,VIRUS,6,3,0,2,0,0,ACUTE,37.0,0,0,0,0,-,-,3800,1.0,0,n.p.,a,-,484,42,57,F,-,-,0,-,0,156,-,-
24,F,VIRUS,4,2,0,1,0,0,ACUTE,37.8,1,0,0,0,-,-,6000,2.2,25,n.p.,a,-,151,23,67,F,-,-,0,-,0,88,-,-
...

We did an experimental comparative evaluation of C4.5 and CABRO. The evaluation was carried out as follows:

- all instances and 35 attributes in the original data set where DIAG is the class attribute were used in order to discover diagnostic knowledge for two classes, 'BACTERIA' and 'VIRUS';

- an automatic random shuffle of the dataset then divided the dataset into 10 stratified subsets; and

- an automatic cross-validation was done for both systems with the same division of the dataset into 10 folds.

This process was repeated many times and the results were found to be stable (varying only slightly with each shuffle of the dataset). Our experiments showed that on the testing data, the accuracy of CABRO is higher than that of C4.5, but the size of pruned trees from C4.5 is smaller than that from CABRO. From small datasets, C4.5 was able to produce very small trees. From many experiments with different datasets, we observed that this difference in tree size does not exist when learning from large datasets (Ho et al., 1998). Below are some simplified diagnostic rules discovered by CABRO:

```
If THERAPY2 = ope Then BACTERIA (2.0/1.0)
If THERAPY2 = Zobirax Then VIRUS (17.0/1.3)
If THERAPY2 = PIPC+CTX Then BACTERIA (1.0/0.8)
If THERAPY2 = ABPC+LMOX Then BACTERIA (2.0/1.0)
If THERAPY2 = change Then BACTERIA (8.0/2.4)
If THERAPY2 = ABPC+FMOX Then BACTERIA (3.0/1.1)
If THERAPY2 = ARA-A Then VIRUS (8.0/1.3)
If THERAPY2 = Dara-P Then BACTERIA (1.0/0.8)
If THERAPY2 = ABPC+CZX Then BACTERIA (8.0/1.3),
...
If THERAPY2 = globulin Then VIRUS (1.0/0.8)
If THERAPY2 = ALA-A Then VIRUS (1.0/0.8)
If THERAPY2 = ABPC+CTX Then BACTERIA (2.0/1.0)
If THERAPY2 = ABPC+CEX Then BACTERIA (1.0/0.8)
If THERAPY2 = -:
|   CRP = 0 Then VIRUS (51.0/2.6)
|   CRP = 1 Then BACTERIA (4.0/2.2)
|   CRP = 2 Then VIRUS (0.0)
|   CRP = 3 Then VIRUS (0.0).
```

Conclusion

As decision-making is a process of choosing among alternative courses of action for the purpose of attaining a goal, knowledge about these choices plays a crucial role in any decision. Finding knowledge in data is the ultimate goal of rule induction, a sub-field of ML and KA that can be used effectively in decision-making theory and practice.

In this chapter, we introduced the main stages of the rule-induction process that the DSS user/developer needs to know when finding automatic decision knowledge in data. We also presented two rule-induction methods for supervised and unsupervised data. These methods have been implemented successfully in the CABRO and OSHAM systems and have several significant advantages in comparison with other systems doing the same tasks. We have demonstrated how to use CABRO and OSHAM to construct knowledge bases within the knowledge-based system generator TESOR. It is possible that the user may consider rules/concept hierarchies induced by CABRO and OSHAM as available knowledge in many other KBS/expert system tools. Since finding decision rules is one of the most difficult and expensive tasks in the development of KBDSSs, we believe that rule induction techniques in general, and CABRO and OSHAM in particular, will make a significant contribution to DSSs for sustainable development.

Acknowledgements

The author would like to thank professors Masayuki Kimura and Setsuo Ohsuga for their support, and Nguyen Trong Dung, the codeveloper of CABRO. This work has been supported by Kokusai Electric Co., Ltd., Japan Advanced Institute of Science and Technology (Japan), and the National Research Programme on Information Technology KC01-RD08 (Vietnam).

References

Blanning, R.W. and D.R. King (eds.) (1993). *Current Research in Decision Support Technology*. IEEE Computer Society Press.

Breiman, L., J. Friedman, R. Olshen, and C. Stone, (1984). *Classification and Regression Trees*. Belmont, CA: Wadsworth.

Cheeseman, P. and J. Stutz, (1996), "Bayesian classification (AutoClass): Theory and results", in: *Advances in Knowledge Discovery and Data Mining*, U.M. Fayyad et al. (Eds.), AAAI Press/MIT Press, 153-180.

El-Najdawi, M.K. and A.C. Stylianou, (1993). "Expert Support Systems: Integrating AI Technology", *Communication of the ACM*, 36(12), 55-65.

Fayyad, U.M., G. Piatetsky-Shapiro, P. Smyth, and R. Uthurusamy, (1996). "From DataMining to Knowledge Discovery: An Overview", in: *Advances in Knowledge Discovery and Data Mining*, U.M. Fayyad et al. (eds.), AAAI Press/MIT Press, 1-36.

Ho, T.B., N.K. Pham, Bach, et al. (1992). "Development and Applications of the Expert System Generator TESOR", *Proceedings of NCSR of Vietnam* 2(2), 3-14.

Ho, T.B. (1995). "An Approach to Concept Formation Based on Formal Concept Analysis", *IEICE Trans. Information and Systems* E78-D(5), 553-559.

Ho, T.B. (1996). "Integrating Inductive Learning and Knowledge Acquisition in the Expert System Generator TESOR", *3rd World Congress on Expert Systems*, Seoul, Cognizant Communication Corporation, 925-932.

Ho, T.B. (1997). "Discovering and Using Knowledge From Unsupervised Data", *Decision Support Systems*, Elsevier Science 21(1), 27-41.

Ho, T.B. and C.M. Luong, (1997). "Using Case-Based Reasoning in Interpreting Unsupervised Inductive Learning Results", *Int. Joint Conf. on Artificial Intelligence IJCAI'97*, Nagoya, 258-263.

Ho, T.B., T.D. Nguyen, and M. Kimura, (1998). "Induction of Decision Trees Based on the Rough Set Theory", in: *Data Science, Classification and Related Methods*, C. Hayashi et al. (eds.), Springer, 215-222.

Langley, P. and H.A. Simon, (1995). "Applications of Machine Learning and Rule Induction", *Communication of the ACM* 38, 55-64.

Mansell, R. and U. Wehn, (1998). *Knowledge Societies: Information Technology for Sustainable Development*, Oxford Press.

Nguyen, T.D. and T.B. Ho, (1999). "An Interactive-Graphic System for Decision Tree Induction", *Journal of Japanese Society for Artificial Intelligence* 14(1), 131-138.

Pawlak, Z., (1991). *Rough Sets: Theoretical Aspects of Reasoning about Data*, Kluwer.

Quinlan, J.R., (1993). *C4.5: Programs for Machine Learning*, Morgan Kaufmann.

Sprague, R.H and H.J. Watson, (1993). *Decision Support Systems: Putting Theory in Practice* Prentice Hall, Third Edition.

Sullivan, G. and K. Fordyce, (1994). "Decision Simulation (DSIM): One Outcome of Combining Artificial Intelligence and Decision Support Systems", in Gray P. (ed.), *Decision Support and Executive Information Systems*, Prentice Hall, 409-419.

Tsumoto, S. (1999). "Information about Clinical Databases on Miningoencepahalitis", *KBS'42 Symposium on Knowledge Discovery and Data Mining*, JSAI Press, 1-5.

Turban, E. and J.E. Aronson, (1998). *Decision Support Systems and Intelligent Systems*. Prentice-Hall, Fifth Edition.

14 ORGANIZATIONAL MEMORY INFORMATION SYSTEMS: A Case-based Approach to Decision Support

Helen G. Smith, Frada V. Burstein, Ramita Sharma, Dayo Sowunmi

1. Introduction

Good decision-makers learn from their own experiences as well as those of others; they also learn by studying the past. Decisions on development — particularly in developing countries — are complex. Information about past decisions and their outcomes can help decision-makers identify potential successes and failures; however, this information is poorly structured, and difficult to store and retrieve. Conventional databases are poor at recording the complexities of past events in a form that can be reused in the future to inform decision-makers.

Agenda 21 of the United Nations Conference on Environment & Development (item 34.16) states that "disseminated information [from international and regional clearing-houses] would highlight and detail concrete cases where environmentally sound technologies were successfully developed and implemented" (United Nations Environment Programme, 1992). There is a need for a technology that will support the dissemination and sharing of such knowledge.

Organizational memory information systems (OMISs) have been proposed as a means of retaining past experience within an organization (Walsh and Ungson, 1991; Stein, 1995). Some views on organizational memory assume that it would include all recorded information. The authors of this chapter believe that only some information needs to be recorded, and that some structure must be applied to that information in order for it to be of future use.

Organizational memory can be useful for many purposes but the focus of this chapter is its application in the domain of decision support. In decisions on the environment, lessons already learned can be captured, and both successes and failures of the past can be used to inform future decision-makers who were not involved in the original events. Where knowledge of past decisions and their outcomes is likely to be beneficial to decision-makers and their organization, an OMIS can be a useful component of a decision support system (DSS). This component must contain the shared knowledge from which decision-makers will benefit in order to make consistent decisions. Once a number of successes and failures are known, the OMIS discussed in this chapter can support the decision-maker's memory and encourage improved decision-making.

The authors propose an OMIS architecture designed to support decision-making which combines a case base, a database, and a rule base into an "intelligent" advisory system. The aim of this architecture is to support decision-making processes by its capacity to recall past decisions and adapt similar cases to new decision scenarios. Case-based reasoning (CBR) is suggested as a suitable technology for representing previous decisions since it retains past experience as cases, presenting the most similar case for the decision-maker's consideration. The DSS also can "learn" from new situations by means of the CBR adaptation mechanism. Rules are used within the architecture to describe existing policies and regulations.

This proposed architecture has been implemented within a practical decision-making situation (Sharma, 1996). A prototype is being used as the basis on which to evaluate the quality of decision-making and the perceptions of the decision-makers of such a memory aid (Burstein et al., 1998).

2. Organizational memory and sustainable development

A DSS is most beneficial in semistructured or unstructured situations when the decision-maker's exploration of the problem domain, with the assistance of the DSS, enables him/her to develop a structure within which to make a decision. The exploration of past related decisions with an OMIS can extend the value of the DSS in structuring the problem domain.

2.1. Organizational memory information systems

Stein and Zwass (1995) define organizational memory as "the means by which knowledge from the past is brought to bear on present activities, thus resulting in higher or lower levels of organizational effectiveness." According to Yates (1990), organizational memory is what allows a firm to store and retrieve facts, processes, or experience. An OMIS is a computerized system that supports organizational memory (Stein and Zwass, 1995), and can provide support to decision-makers who are geographically dispersed, or who are dispersed in time (assuming they can share the same OMIS). Decisions (and their outcomes) made by other people, at other times and in other places, can then be used to inform the current decision-maker. If the technology is available, this knowledge can be shared around the world and will provide a consistency of expertise that is difficult to achieve in other ways.

An OMIS of this nature needs to be descriptive, providing illustrations of the successes and failures in similar decision situations. It cannot be normative since this would imply that the decision situations are well structured and therefore outside the scope of decision support. There is little room for generalization in the stored knowledge since this would preempt the analysis of the current decision situation; however, the OMIS can be prescriptive in more repetitive situations where rules for decision-making already exist.

An OMIS of the type described in this chapter deals mainly with concrete, rather than abstract, descriptions of actual decision situations. A concrete description of an event avoids, to some extent, the problems of anticipating future decision situations.

The memory contained in the OMIS may be either communal (shared) or idiosyncratic (representing only one individual's experience). A communal organizational memory describes information elicited from groups of people who make decisions that affect their organization and is derived from a consensus of the decision-makers' opinions. Idiosyncratic organizational memory contains information elicited from each individual decision-maker. In this instance, differing (and perhaps conflicting) opinions can be preserved and presented as alternative approaches to addressing decision situations. Idiosyncratic organizational memory contains the valuable private knowledge of previous decision-makers, which will benefit decision-makers now and in the future.

The proposed case-based OMIS contains specific information on a particular work activity. Ackerman and Mandel (1995) refer to systems which contain task-based organizational memory as "memory in the small." Case based OMISs belong in this category since their main purpose is to capture specific important past decision situations (represented as cases) and other relevant information, in order to provide effective assistance in performing problem-specific tasks.

To summarize, for an OMIS to be useful in decision support it should be task-oriented (concrete) and should contain both private and shared knowledge about the activity. Therefore, the most important type of knowledge that should be incorporated into a case-based OMIS can be described as *"concrete + abstract / descriptive + prescriptive / private / idiosyncratic"*.

2.2. The role of OMISs in sustainable development

Many different decisions affect the environment in unpredictable ways. If both the decision and its outcomes can be described, an OMIS offers the potential for developing a cohesive memory of past decisions that could provide a basis for an analysis of the impact of these decisions and a prediction of the outcome of future decisions.

The "organizational" component of an OMIS can be interpreted generously to apply to regional, national, and international bodies. An organization can be any community of individuals who share a common goal and structures to achieve that goal. An OMIS requires a corporate entity that values the stored knowledge and is prepared to provide financial and political support for the collection, maintenance, and use of that knowledge. Such knowledge may be elicited from sanctioned decision-makers, but even more exciting possibilities exist if the knowledge of local communities and indigenous people can be identified and valued.

The descriptive nature of the OMIS simplifies the capture of knowledge, particularly that of indigenous people. Avoiding the need to abstract and categorize minimizes the impact of the differences in perception that may arise across cultures. The descriptive nature of the OMIS also promotes understanding: once the memory has attained a minimum size, a range of tools are available to analyze the stored descriptions and highlight relations that may not be obvious in individual cases.

The idiosyncratic knowledge of individuals can be pooled to produce a communal OMIS. Since the knowledge is maintained in descriptive form, past decisions of various people can be combined as long as the types of described characteristics intersect to some degree. Sharing idiosyncratic knowledge optimizes the usefulness of an OMIS, extends the range of decisions it can address, and provides more useful analyses of hidden relations.

3. Case-based technology for OMIS

The proposed DSS combines a case base with a data and a rule base to perform quantitative information processing and qualitative reasoning, depending on the complexity of the situation.

3.1. Case-based reasoning

CBR is a problem-solving technique that applies past experiences to future similar decision situations. It is "an analogical reasoning method ... reasoning from old cases or experiences in an effort to solve problems, critique solutions, explain anomalous situations, or interpret situations" (Kolodner, 1991). The experience or knowledge of past decision-makers is represented as cases that that are indexed by a range of significant characteristics (including outcome) to ensure appropriate retrieval. Each case represents a past event and is described by a number of characteristics, including the context of the event, the decision made, and its outcome, along with the significant factors that led to the decision. This collection of cases is called the case base.

CBR involves the retrieval of relevant cases from the case base. Unlike retrieval from a database, retrieval from a case base does not require an exact match. Rather, various measures of similarity are used to identify the "closest" cases, and some of these are retrieved for consideration.

This technology is said to reflect some of the features of human intelligence such as reasoning, remembering, and learning (Kolodner, 1993). In addition to their ability to store and retrieve cases, CBR systems can "learn" by adapting known cases to new situations and simulate natural language expressions by using symbolic rather than numeric representations.

CBR is seen as valuable for decision support because "the computer augments the person's memory by providing cases (analogs) for a person to use in solving a problem. The person does the actual decision-making, using these cases as guidelines ... computer augmentation of a person's memory allows this person to make better case-based decisions because it makes more cases (and perhaps better ones) available to the person than would be available without the machine." (Kolodner, 1991)

CBR systems can improve their performance (expertise) over time through abstractions induced from the collection of past cases (experiences). According to Wilcox (1972), "by observing the outcomes of past decisions ... we obtain the knowledge to make present decisions with some intelligence." This statement recommends the application of CBR technology (and similar analogical reasoning models) in decision support, which results in a system that provides situation-specific examples or illustrations to improve the processes and outcomes of decision-making.

A case-based approach to DSS offers a number of advantages:

- a decision-maker can explore a current situation in more detail by adapting the decision taken in a similar past case;

- a decision-maker is not obliged to retain a mental record of all past cases;

- novice decision-makers can "reuse" the experience and knowledge of an expert;

- the valuable experience of expert decision-makers is not lost through staffing changes;

- the quality of decisions can be improved by ensuring that relevant past cases are considered; and

- a more robust system results from representing decision-making strategies and heuristics as meta-rules which reflect the relevant decision-making knowledge.

The case-based approach is better suited for addressing open-ended, semistructured decision-making domains than well-structured domains where rules for decision-making already exist. This is because when there are no predetermined rules, judgment is often based on past experience. It is another reason to consider this technology for DSS development, as DSSs traditionally are targeted for semistructured decision situations.

3.2. Case-based reasoning

A DSS with a CBR component addresses a different class of decisions from those addressed by traditional DSSs. If DSSs are intended for unstructured or semi-structured decision-making situations, then by definition there is little domain-specific knowledge readily available. The proposed system should be regarded as a knowledge-based intelligent DSS (IDSS) (Burstein and Smith, 1995). An IDSS, as defined by Gottinger and Weimann (1992), "is an interactive tool for decision making for well-structured (or well-structurable) decision and planning situations that use expert system techniques as well as specific decision models to make it a model-based expert system (integration of information systems and decision models for decision support)." This indicates that decision-making situations appropriate for IDSSs are more structured than those appropriate for DSSs. The proposed architecture addresses these more structured decisions and is suitable for decision situations in sustainable development where lessons from the past can be applied to the future.

The proposed architecture for the IDSS is an interactive tool that uses CBR for decision-making within a class of structurable decision situations. It encodes specific problem-solving experiences as cases that can be retrieved through CBR. The IDSS is then appropriate for the purpose of repetitive decision-making within related decision scenarios. As the nature of these decisions is dynamic, human judgment is often required.

The proposed DSS is a tool specifically designed for a constrained problem area. It is not a general-purpose tool or an enterprise-wide system; the knowledge it contains should be specific to the problem domain, and thus it can act as a task-based OMIS. Although the architecture is generic, the components are domain-specific.

4. An architecture for an OMIS

This section proposes an architecture for an OMIS that consists of a range of components designed to capture any available information about a current situation and to provide appropriate reasoning mechanisms to handle decision situations with various levels of structure. It uses rules, historical data, or analogical reasoning as required, either to resolve the problem, suggest a solution, or present relevant past cases to support decision-making.

4.1. The proposed architecture

The architecture of the proposed DSS comprises a rule base, a database, a case base, an interface, and a decision and justification module. These components perform quantitative information processing and qualitative reasoning, depending on the complexity of the situation. Together they form an intelligent decision support tool.

This DSS is an interactive system that communicates information to and from the decision-maker, accepting feedback. Figure 1 illustrates the components that are essential to the basic design of the proposed architecture.

There are three storage components in which a match for the current situation can be found: a rule base, a database, and a case base, each of which are explored by the system in turn until a match is found.

4.1.1. The rule base. The rule base contains the well-structured (programmable) decision rules which correspond to the most frequent, or best-known, problems in the domain. The rules, which are only a handful in a semistructured domain, are usually explicitly known and expressed as public domain knowledge. These few decision rules are easily maintained and well documented.

4.1.2. The database. The database contains explicit historical records of all past decision situations. The system can retrieve details of past situations where the decision-making scenario is considered "identical" to the current situation, and may be a precedent for the current situation.

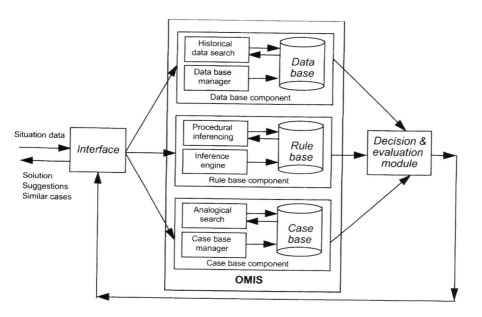

Figure 1. The OMIS architecture.

4.1.3. The case base. The case base incorporates episodic knowledge and heuristics on the basis of the decision-maker's idiosyncratic knowledge, encoded in cases. This idiosyncratic knowledge, stored in the OMIS, can be shared with future decision-makers within the organization. The case base differs from the database in that for each class of situation only one case is stored. That is, exact duplicates are not recorded, so the data in the case represents an abstraction of that class of case rather than the historical data in the database.

The case base contains selected instances of difficult cases (specific cases) and generalized episodes (generic cases). The system retrieves a number of cases that are "similar" to the current situation on the basis of several indexed attributes. These may be assigned different degrees of relevance to the search. CBR provides judgmental knowledge in the form of advice, explanations, and heuristics to guide the decision-maker through less structured situations.

4.1.4. The interface. The interface helps the decision-maker classify the current decision situation and controls the search through the database, rule base, and case base. The system may prompt the user to classify the situation at hand. The data selected as classification information depend on the case base indexing, which is a result of the knowledge acquisition and modeling phase of systems design. The classification data correlates with the particular design of the case base and indexing mechanism. The specific design of the interface is strongly determined by the application domain itself and requires domain-specific analysis.

4.1.5. The decision and justification module. The decision and justification module presents the recommended action to resolve the situation. It may deduce a solution from procedural inferencing automatically if a rule applies, it may recommend one on the basis of precedents in the database, or it may present alternative analogies of relevant past cases through the DSS interface.

4.2. Using the proposed architecture for decision support

The proposed DSS augments human memory and can perform analogical reasoning to recall similar situations in a way similar to human reasoning. It serves as a memory aid to the decision-maker, provides new insights on past experiences, and supports learning from significant past decisions. The proposed decision support process is illustrated in Figure 2. The system provides three levels of decision support by performing procedural inferencing, a historical data search, and an analogical search.

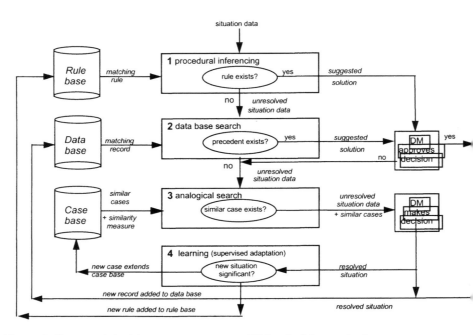

Figure 2. Proposed decision support process *(DM = decision-maker)*.

4.2.1. Procedural inferencing. If the specific details of the situation match a rule in the rule base, the situation is resolved at once and a recommendation is given to the decision-maker for ratification. At this point, the database is updated. The rules represent well-structured situations that represent the shared knowledge of the organization explicitly.

4.2.2. Historical data search. Problems that cannot be solved by rules are matched against the database. If an identical situation has been encountered in the past, it will be recorded in the database and should be used as a precedent. The decision-maker again will be asked to ratify the solution before it is stored as a precedent in the database.

The situations described in the database are less structured than those for which there are rules, in the sense that, for most of them, general patterns of decision-making have yet to be identified. The collection of precedents represents the shared memory of past actions but without any significant classification structure.

4.2.3. Analogical search. In situations where neither rules nor database access are successful, CBR is used to perform an analogical search on the case base for past similar cases. These situations are relatively unstructured, and CBR is used to recall the most similar cases to guide the decision-maker. Ideally, a range of cases with varying decision outcomes will be identified, thus prompting the decision-maker to consider several alternative solutions. These cases are presented to the decision-maker with pointers to additional relevant documentation. He or she may need some time to consider the most appropriate decision but once the decision is made and justified, the relevant data is stored in the database as another precedent.

4.2.4. "Learning" in the proposed architecture. As new situations are considered and decisions made, the knowledge of the decision-maker increases within the domain. This knowledge is more than just a memory of past events. The decision-maker develops abstractions with which to classify future situations; these are evidence that the decision-maker is "learning." A significant feature of case-based systems is that they can be said to learn. Kolodner (1993) describes the learning features of case bases in which the system adapts previous case descriptions to address new situations and extends the case base accordingly. In our architecture, whenever new decisions are made on the basis of similarity to past cases, the decision-maker will identify the significance of the new decision. When a decision is considered to be significant, a summary will be added to the case base through a process called supervised adaptation. Since not every new decision will be significant, the case base will expand only at the rate at which significant cases occur. The database component, on the other hand, expands with every decision whether it is significant or not. In this way, the case base imposes a structure on the database that is somewhat similar to human learning structures. While case bases expands slowly, the larger ones are slow to search; for this reason at least, a case base also should be able to "forget", where appropriate. This learning aspect of the architecture is being explored currently.

4.3. Knowledge acquisition and representation for this architecture

Knowledge acquisition and representation for DSSs is a relatively new area. Although artificial intelligence has been identified for many years as being useful for DSSs (Sprague, 1980), little attention has been paid to knowledge acquisition for DSSs. With regard to the proposed architecture, there are several issues related to getting appropriate information from several sources in order to populate the various data storage components of the system. Since knowledge acquisition for rule bases and data collection for databases are relatively routine, this paper will focus on knowledge acquisition for the case base component.

A significant difficulty in building CBR systems is eliciting knowledge in the form of cases. However, the theory of CBR does not suggest any generally accepted approach to the acquisition and representation of such knowledge. Moreover, it has been argued that knowledge modeling for case-based systems is an area of major importance (Simoudis *et al.*, 1992). Simoudis and his colleagues indicated that a considerable knowledge acquisition effort is still required in CBR system development. Kolodner (1993) also has acknowledged the effort involved and stressed the need for the development of CBR knowledge acquisition methodologies.

In the context of sustainable development, the challenges of knowledge acquisition include capturing indigenous knowledge in a form that can be translated into cases for cross-cultural use. Moreover, for technology to be appropriate in a given culture, it needs to be incorporated into that culture's approach to evaluating the sustainability of development efforts rather than imported as a high technology product. The benefit of a case base component is that because of its declarative nature, it is a more "natural" form of electronic representation than others.

The authors have proposed a method of knowledge acquisition as a modeling activity for case-based systems (Sowunmi *et al.,*1996), which combines a number of techniques: conventional knowledge acquisition approaches for expert systems development (interviews, teach back, and others); a personal construct psychology approach (repertory grid); and techniques for stimulating creative thinking in the expert in order to collect a set of cases and their characteristics as indexes (Sowunmi, 1996). This method provides an initial set of significant cases as a starting point for building a seed case base.

Once the initial case base has been developed, it can expand through the process of supervised adaptation mentioned above. In this process, decision-makers identify decision situations which are sufficiently new that no useful cases exist to give insight into the alternative decisions. The decision-maker will have to identify these situations with little or no support. However, once the decision is made, the decision-maker should add a description of the case to the case base for future reference. In this way, the case base learns through the increased knowledge of each of the decision-makers involved.

5. Benefits of the proposed architecture

The case base component of the proposed DSS has been implemented at Melbourne's Monash University to help faculty members make decisions about the admission of students who have undertaken prior tertiary studies elsewhere (Burstein, *et al.*, 1995; Sharma, 1996). The system provides organizational knowledge so that faculty decision-makers can determine credit equivalence in converting a student's prior studies into credit points towards the relevant Monash degree.

The case base component of the system contains a set of significant past examples of student applications. For each new application, a decision-maker is able to access relevant historical data to maintain consistency in the decision-making process. This example showed reasonable efficiency of the proposed knowledge acquisition method in the case base construction and feasibility of the proposed approach to case base organizational memory construction (Sowunmi, 1996). A prototype seed case base was built by means of the tool CBR Express™. The prototype system is being evaluated on an experimental basis from the point of view of a case base decision aid's impact on the quality of the decision-making process (Burstein *et al*, 1998).

The proposed architecture aims to assist decision-makers rather than make the decision for them. That no attempt is made to replace the decision-maker is consonant with the philosophy of DSSs, and it also distinguishes this approach from other "intelligent" approaches such as expert systems. This approach is "extended decision support" that "endeavours to influence and guide decision-making while respecting the primacy of judgment and focusing very carefully on how decision- makers think." (Keen, 1987). As human reasoning is weak on algorithms but strong on heuristics (Hollnagel, 1987), it is anticipated that the limitations of algorithmic reasoning can be overcome through the provision of better conditions for analogical reasoning by the case base component of the architecture.

Decisions are often made with uncertain information in nondeterministic situations and as such are inherently unstructured. The proposed approach is appropriate for situations which do not have strong underlying domain knowledge. It combines qualitative reasoning incorporated in a case base with rule integration and database access.

A combination of quantitative information processing and qualitative reasoning embodied in a knowledge base has been claimed to offer better support to decision-makers faced with insufficient information about a problem (Hollnagel, 1987). This approach exploits CBR technology to provide a "memory" of past decisions, which are justified and stored in a case base.

Decision-makers use analogous experiences when solving problems but find it difficult to remember the right ones. CBR performs an analogical retrieval of relevant past cases to support decisions of a similar nature when they occur. This is particularly important in handling decision-making situations where the solutions are ambiguous.

The CBR component may assist in handling unique situations by retrieving the most relevant cases to provide the decision-maker with new insights into the situation and an understanding of past relevant decisions that can be adapted to resolve the existing problem. Therefore, the CBR component of the DSS provides more consistency in decision-making. When encountering unique situations, the proposed DSS is able to learn from the decision-maker's problem-solving experience by incorporating new significant cases in the case base. The proposed approach ensures that experience is not lost and that best practice is followed.

6. Conclusion

Sustainable development will not be achieved solely by automating decision-making. In order to improve the overall quality of decision-making, we need to find ways to support human decision-making processes with extensions to the human memory. An organizational memory aid offers support throughout the process of exploring decision alternatives. A decision-maker who is aware of the history of decision-making in the domain in question can identify more alternatives and their potential impact and is in a position to make more informed — and better — decisions.

This chapter has examined the use of an OMIS as a decision support component. Capturing knowledge about significant past decision situations and making this knowledge available to decision-makers for future similar cases offers improved knowledge sharing and the opportunity for greater consistency in decision-making.

While the illustration of the proposed architecture was concerned with supporting decision-makers in an academic environment, the principles of this form of decision support can be extrapolated to a wide range of decision-making situations. In particular, the use of a case base of previous experiences offers decision-makers in environmentally sensitive areas the ability to remember the effects of past decisions and, potentially, to avoid past mistakes.

A system of this type might be employed in a number of environmentally sensitive areas where it could use the case base to record past decisions and their effects. Over time, the case base would become a form of memory, which would enable the decision-maker to explore the impact of previous decisions in similar situations; this is particularly important in situations where the cost of a bad decision can be very high. A case base is more powerful than a database since it retrieves situations that are analogous (but not identical) to the situation in hand. In addition, the case base can enhance organizational memory by recording analytical comments in addition to the details of the situation itself.

The architecture proposed in this chapter, particularly the case base component, offers an approach that enables the knowledge developed by decision-makers to become shared expertise, without the need for external experts. Decision-makers themselves provide an initial seed case base with case descriptions of further decisions. The need to elicit specific knowledge is minimized, and the case representation can be more natural than formal techniques for representing knowledge. As a result, this approach aids in the capture and dissemination of traditional and indigenous knowledge that does not lend itself to conventional knowledge elicitation. It therefore may play a very important role in preserving, storing, and applying old, natural, and well-known ways to make decisions. This would be extremely important in forming a bridge between the modern and the traditional. Furthermore, the use of such systems can provide empowerment to developing countries. The OMIS resulting from this process is a valuable resource for organizational learning.

References

Ackerman, M. and E. Mandel (1995). "Memory in the Small: An Application to Provide Task-based Organisational Memory for a Scientific Community," *Proceedings of the 28th Hawaii International Conference on System Sciences*, IEEE, 323-332.

Burstein, F. V. and H. G. Smith (1995). "Case-Based Reasoning for Intelligent Decision Support," *Proceedings of the Third International Conference of the International Society of Decision Support Systems*, Hong Kong, 603 -610.

Burstein, F.V., H. G. Smith and S. M. Fung (1998). "Experimental Evaluation of the Efficiency of a Case-Based Organisational Memory Information System Used as a Decision Aid," *Proceedings of the 31st. Annual Hawaii Int. Conference on Systems Science*, 209-219.

Burstein, F.V., A. Sowunmi, H.G. Smith, McMillan, and Cole (1995). "Building an Intelligent DSS: A Case-Based Approach," *Proceedings of the Pan Pacific Information Systems Conference (PACIS 95)*, Singapore, 60 - 70.

Hollnagel, E. (1987). "Information and Reasoning in Intelligent Decision Support Systems." *Man-Machine Studies* 275(5-6), 665-678.

Gottinger, H.W. and P. Weimann (1992). "Intelligent Decision Support Systems," *Decision Support Systems* 8, "Information and Reasoning in Intelligent Decision Support Systems". *Man-Machine Studies* 275(5-6), 665-678.

Keen, P.G.W. (1987). "Decision Support Systems: The Next Decade," in Arnott and O'Donnell (eds.) *Readings in DSS*, 1994, 29-44.

Kolodner, J.L. (1991). "Improving Human Decision Making through Case-Based Decision Aiding," *AI Magazine* 12(2), 52-68.

Kolodner, J.L. (1993). Case Based Reasoning. California: Morgan Kaufmann.

Sharma, R. (1996). "Hierarchical Classification Approach to Knowledge Acquisition for Case-based Decision Support Systems", unpublished Masters Thesis, Monash University.

Simoudis, E., K. Ford, and A. Cañas (1992). "Knowledge Acquisition in Case-based Reasoning: '...and then a miracle happens'", *Proceedings of the 5th Florida Artificial Intelligence Research Symposium (FLAIRS 1992)*.

Sowunmi, A. (1996). "A Knowledge Acquisition Method for Building a Case-Base for Intelligent Decision Support," Unpublished Masters thesis, Monash University, Melbourne.

Sowunmi, A., F. V. Burstein and H. G. Smith (1996). "Knowledge Acquisition for an Organisational Memory System," *Proceedings of the 29th Hawaii International Conference on System Sciences, IEEE* 4, 168 - 177.

Sprague, R.H., Jr. (1980). "A Framework for Research on Decision Support Systems," in G. Fick and R. H. Sprague (eds.), *Decision Support Systems: Issues and Challenges*, International Institute for Applied Systems Analysis, 5-22.

Stein, E.W. (1995). "Organisational Memory: Review of Concepts and Recommendations for Management," *International Journal of Information Management* 15(1), 17-32.

Stein, E. W and V. Zwass (1995). "Actualizing Organizational Memory with Information Systems," *Information Systems Research* 6(2), 85-117.

United Nations Environment Programme (1992). Agenda 21, United Nations Conference on Environment & Development (UNCED) available at gopher://unephq.unep.org:70/00/un/unced/agenda21

Walsh, J.P. and G.R. Ungson (1991). "Organisational Memory," *Academy of Management Review* 16(1), 57-91.

Wilcox, J.W. (1972). A Method for Measuring Decision Assumptions. Cambridge, MA: MIT Press.

Yates, J. (1990). "For the Record: The Embodiment of Organizational Memory, 1850-1920," *Business and Economic History* 19, 172-182.

15 SOFTWARE INTERNATIONALIZATION ARCHITECTURES FOR DECISION SUPPORT SYSTEMS
Patrick A.V. Hall

1. Introduction

As can be seen in the many other papers in this book, decision making using DSS occurs in all countries and is undertaken by people who are not necessarily fluent in any language other than their mother tongue. Thus the decision support software must be made available in this mother tongue. But the situation is more complex than that, for some decisions may need to be taken using information gathered from across a complete region from data sources in different languages and cultures, and the decision making process may also involve many people from similarly diverse languages and cultures. The processes of collaborative decision making are covered in Chapter 12 on business negotiations, while in this chapter we will focus on how to make software available for a range of languages and cultures.

Globalization of commodity software products like word processors, spreadsheets, and their underlying operating systems, is becoming widespread. The suppliers of these systems, like Microsoft, Lotus, Claris, Apple, the main platform suppliers, and many others, have recognized since the start of the 1990s that more than half their revenues must come from outside the US and the English speaking markets. The US market is saturating.

This has meant that the manufacturers of software have developed methods for translating their products from their original target market, typically the US, to new markets. There are a wide range of issues to be addressed – these are described in Section 2.

The methods of attack have been based on relatively simple methods, on characterizing the essential features of a market within a "locale", using a suitable character coding standard, and using resource files to factor out the locale dependent data like

The methods of attack have been based on relatively simple methods, on characterizing the essential features of a market within a "locale", using a suitable character coding standard, and using resource files to factor out the locale dependent data like messages so that they can be easily replaced. These current approaches are described in Section 3.

Nevertheless, translation of software remains expensive, and not all markets are large enough to warrant translation. Thus, for example, the Same language of Lapland in Northern Finland, Sweden, and Norway, will not have the interfaces of office products translated, though the ability to store and manipulate data in the Same language will be enabled. Same is closely related to Finnish.

New approaches are needed, and I have been involved with some of these within the EU-funded Glossasoft project. These new approaches aim to exploit software architectures to factor out the locale dependent aspects behind Application Programmer Interfaces, and linguistics to enable this and avoid internationalization imposing unnatural interfaces upon the software. These approaches are described in Section 4.

Finally, in Section 5, we look forward to the kind of work that will be necessary to enable Decision Support Systems to succeed across countries and regions of the developing and developed world. All current approaches are aimed at single languages, though the same principles would work for a single system accessed through multiple languages, such as a decision support system spanning several countries or locales. There are two levels of linguistic issue: diverse interfaces in multiple languages accessing a single common repository of data which is in some manner language neutral; and systems where the shared data is in multiple languages intermingled. But behind all this there are deeper cultural issues about how decisions are made: social relations, and the perceptions of space and time, are critical. This area is only partially understood, and the issues and some indication of how to handle these technically will be discussed.

2. The localization problem

In moving any piece of software from one part of the world to another, it is important that the different needs of the users at these different locations are taken into account. These differences are:

1. the language of the user interface,

2. the language and number representations and measurement units of the data stored within the system,

3. the language used for product support

4. "cultural" factors like representation of currency, dates and colours

5. local practices such as legal requirements

6. deeper cultural issues like those characterised by Hofstede (1991).

These factors are also equally important when a single system is used through multiple languages, as when sharing information between different countries. Even when the language is the same, some of these factors will be important because the user community is different.

2.1 Language

The user interface of software should feel "natural" to the user of that software. All aspects of the computer and its communication with the user should be in the language of that person, or some second language in which the person is very comfortable.

For example, in European countries, nationalities differ markedly in their ability to understand one or more foreign languages. In some countries as few as 20% of the population are able to understand a language other than their own, while in others like Luxembourg the proportion can be as high as 90%. In South Asia computer systems are delivered almost exclusively in English as a consequence of their colonial heritage, but English is only spoken by the 5% of the population who are the elite. However, the real ability to understand subtleties and shades of meaning of the second language, and the meaning of specific terminology used in software applications, may be substantially lower. English or some other dominant "world" language such as Arabic, Russian, Spanish, Chinese, or French just cannot be used.

The aspects of languages that are important are:

Writing systems: different languages use different scripts, which may be written in different directions. While most scripts will be alphabetic, using a small set of symbols (say less that 100) with which to write their language, by no means all are. The important languages of the Eastern Asian countries are ideographic and use a repertoire of many thousands of symbols. Many languages are written from left to right in lines which progress from top to bottom of the page, but the Arabic script is written right to left top to bottom, and some ideographic languages are written top to bottom right to left. Some scripts are composed of characters that can take a number of forms, like the capital letters of the Roman script and the letter variants in Arabic, or have compound characters as in Devanagari. Diacritics may be important, effectively extending the alphabet. There are different conventions regarding line breaks and hyphenation.

Word structure: the individual units of language used to express the meaning and typically separated in some manner in the orthography (eg by blank space or a break in the cursive flow) may be richly structured. For example, in the Finno-Ugric languages of Finalnd, Hungary and central Asia, and in some South-Asian language like Bengali, a single word may take many thousands of forms as it plays different roles within a sentence – we say that it is richly inflected. Other languages allow the free construction of compound words, like German. The general rules regarding the formation of words is known as morphology.

Collation and sorting: different languages using the same (or largely the same) alphabet and script system may arrange words in different orders in dictionaries. Of particular significance here are the accents, and "compound" letters like the Spanish "ll" which sorts between "l" and "m". Some Devanagari letters seen as compounds in Hindi and treated as distinct letter in Nepali.

Spelling: different regions using the "same" language may expect different spellings for a wide range of words – a well-known example of this is the difference between America and UK English. In South Asia the spelling (orthography) of the northern Indo-Aryan languages may be varied with font size. From this arises the equivalence between words that have the same meaning but differ only in their spelling.

These aspects of language may be of importance wherever language appears in an application. The most obvious and visible part of a system is its user interface, where all these issues are important.

When text data is stored and manipulated by the computer, these linguistic issues may also become important, particularly where searches and matches are made. For example, a word processor searching for a whole word must understand what is meant by a word in the language and may need to take into account the morphology of the language.

Some of the documentation supporting a product may be specialized in its usage, and not require translation outside a set of international technical languages. Other material of a training nature may need more than just translation since examples may be cultural specific and may need to be changed – an example here is the use of baseball in examples for the US market being changed to football for other markets.

2.2 Culture

Social anthropologists define culture as "those socially transmitted patterns for behavior characteristic of a particular social group" and "the organized system of knowledge and belief whereby a people structure their experience and perception." (Keesing 1973: page 68). So it includes the non-linguistic human conventions that distinguish one group of people from some other group, their patterns of thinking, feeling, interacting, and ways of recording this in their art and artefacts.

These cultural aspects include the way time and dates and currency are denoted. Here there can be subtle variation, such as the use of "," and "." in the representation of numbers, through to the particular symbols used to denote currency. Typically there may be a variety of ways of writing equivalent information, such as we see in the writing of dates. One particular example is the way western number systems have used thousands and millions as major groupings, while in South Asia lakhs and crores (100 thousand and ten million) are used.

Calendar systems can vary: the western calendar works with a solar year with a Christian orientation for its origin for counting years: it uses various devices to synchronize days with years, through leap years, all worked out by calculation based on scientific theories of the earth's orbit around the sun (see Duncan 1998 for a history of the Gregorian system). By contrast many parts of the Islamic world use the Hijira calendar with is lunar months and yearly cycle of 12 lunar months with no attempt to synchronize with the solar year, while synchronization between days and months is not done by calculation as in the west, but by direct observation of the moon. A person could be 30 years old in the western calendar and 33 years old in the Islamic calendar. The Bikram calendar system of Nepal (historically widely used in South Asia before the colonial imposition of the Gregorian calendar) uses a solar year but lunar months, and synchronizes these by having an extra month every three years and other devices (Pillai 1911). These calendar systems embody within them the deeper cultural values, so that for the West knowing the calendar for years in advance through the calculations of the Gregorian system is important – for Nepal it has always been sufficient to have the calendar for the next year determined a few months before the start of the next year.

Cultural areas are full of pitfalls for the unwary. For example, an icon depicting a hand held up with palm towards the viewer, commonly used to indicate 'Stop!' or 'Danger!' in the Apple Macintosh and Windows interface, has a completely different meaning in Greece, where it is extremely rude. The product name "Nova" could denote "new" to readers in one locations, but "does not go" in some other location. Colors may also have radically different connotations in different cultures, so for example red is a warning color in Western countries, but in China it illustrates joy, while white is the color of mourning and black is lucky: this could be important in the coloring of maps in GIS systems, since in the west red might well be used to denote danger zones, but lose that connotation elsewhere. Any of the many books on cross-cultural communication will have numerous examples (e.g. Jandt1995).

2.3 Local conventions and practices

Local practices are often specified by legal and professional bodies. These consist of factors such as market-specific computing practices: communication system interfaces, law and regulations, and financial accounting rules. For example, most Western European and America countries express salary in monetary units per year, but in Greece salary is expressed in monetary units per month and the annual salary consists of fourteen monthly payments.

Also included here are differences in tax regimes, and more generally legal constraints, that can influence the way software must work.

3. Current approaches to software globalization

Current approaches to the localization of software focuses on three topics:
1. the choice of character codes
2. use of locales
3. use of resource files.

Facilities for these are available in all the popular operating systems.

3.1 Character codes

The traditional 7 or 8 bit code ASCII (more properly ISO 646) for the Roman script permitted limited national variation across European languages, but has proved inadequate for more general use. Various extensions to ISO 646 have been proposed, from registration of national variants through to complete new coding systems that would handle all alphabetic languages. These were based on single byte encodings which necessarily limited the total number of codes to 256 and the alphabet size to around 100. In multilingual applications there has to be a shift between coding tables, and the meaning of a single code is context dependent.

Ideographic scripts just cannot be handled in this way, and multiple bytes are necessary. The coding standard that has emerged as the best available is Unicode, which requires two bytes per character, and its ISO extension 10646 which uses four bytes per character. It is claimed that this gives the ability to store all known scripts within the single system, with room for expansion – however inspection of the tables (Unicode Consortium 1992 and 1997) shows that many scripts and languages are as yet not included. In multilingual applications there is now no need to shift between coding tables, and the meaning of a single code is unambiguous.

On the surface of it, the use of two bytes per character by Unicode may seem wasteful, but in most applications text is stored with considerable volumes of management information and the extra byte does not double the storage requirement, perhaps only increasing it by half as much again.

No single character encoding scheme has found favor, though there does appear to be a drift towards Unicode and ISO10646. Microsoft's Windows NT has adopted Unicode, Windows 95 has not done so, though Windows 98 does and the projected Windows 2000 arising from the convergence of NT and 95/98 technologies will do so fully. Other software suppliers are looking to Unicode for support in multilingual applications, and so for example the database supplier Sybase has adopted Unicode as part of its distributed multilingual strategy. It does seem that Unicode is the way of the future.

3.2 Locales

A "locale" is a collection of all the conventions that characterize a single user community:

1. the Script, a reference to the code tables to be used, and any special rendering software

2. the Language, leading to hyphenation rules, morphological rules, and similat

3. Number, time and date system and conventions, the format for input and output and associated software to transforming these to and form the internal representations.

4. Monetary system symbols and rules

5. Messages to be used, for error messages, help, and similar in the language and terminology of the locale

and so on – the exact capabilities depends upon the operating system offering the locale facility. Kano (1995) lists 94 locales for Windows platforms, but many of these are for variants of dominant European languages with there being only 38 different languages, plus some 16 others defined but not offered. A lot of the world is unsupported through locales though they are supported through Unicode tables.

When a user buys a system the operating system's locale may already have been set, so that the system interacts with them appropriately. Where several different communities are served, the locale may be set at time of sale. For example, the Regional Settings in the Control Panel in Windows 95 gives some user control over the local details. Application programmes may also be able to set their own locale, and can enable locales to change dynamically.

3.3 Resource files and Dynamic linking

To be able to select a locale, and change it dynamically, you need the two basic facilities from the operating system: a level of indirection where the locale sensitive elements can be named but not bound into the application code, and the ability to link in these locale sensitive elements either statically or dynamically.

Most operating systems have suitable facilities. There will be "resource files" or "message catalogues" in which the locale sensitive elements can be placed, such as:

- user interface text for window titles, menus and prompts, error and information messages

- interface and report layout information like position, size, font, colour, intensity, text orientation

- help text

- symbols, graphics and icons

- sound and video

- software to print or display numerical values.

The application is designed to access these resource files using locale parameters to select the appropriate resource for a given locale. Screen layout is designed to reserve the space needed for possible text expansion during localization (Finnish requires double the space of English, while Arabic requires a lot less space than English), where dialogues and other elements may need to be resized.

4. New approaches – software architecture plus language generation

Existing practice can be taken forward by drawing upon current and recent research in software architectures and linguistics. The work reported here arises from the Glossasoft project reported by Hall and Hudson (1997), particularly involving the work of a team in Finland at VTT and a team in Greece at Demokritos.

4.1 An internationalization architecture

The idea of locales backed up by resource files and dynamic linking is readily generalized to an application programmer interface (API) and this is clearly the next step. OS standards like Posix have made steps in this direction, but what is appropriate must be determined by global user need. All the culturally and linguistically sensitive software components need to be separated from the core of the application. This leads to the kind of architecture shown in Figure 1. The locale sensitive elements are shown at the top and right of the figure, separated by the heavy line from the internationalized locale independent core of the application. Many of the local sensitive elements are associated with the human computer interface, but some may be associated with the application. Whenever the application manages data that relates to the external environment, this is potentially locale sensitive.

The locale dependent components at the top and right should be designed to be plug-compatible so that they can be easily replaced with others performing the same function but for different locales. The heavy line denotes an Application Programmer Interface (API).

Contrast this approach with that currently supplied by operating systems using locales and resource files. Some level of standardization is necessary to identify the locale data structure and its components, but no standardization is used for the formats and protocols for resource files. The interface in current practice is very low level, whereas the interfaces of Figure 1 work at the level of linguistic rules and the meaning of messages.

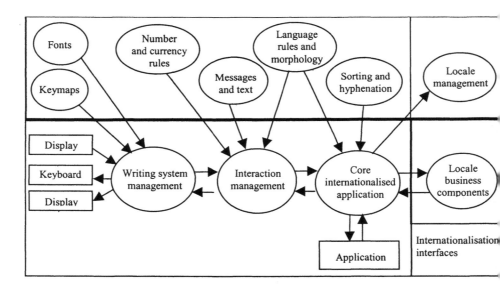

Figure 1. Internationalized interaction architecture.

Current methods would simply give messages numbers, and at the most leave a few slots in these messages for variable data like file names. What applications need to work with are the meanings intended to be conveyed to the user, with this being presented in the language determined by the locale.

Transforming meaning into the words of some language is known as language generation. The architecture of Fig. 1 makes language generation, or a limited form of it, integral to the architecture. This also fits in well with some current approaches to HCI, which suggests that the interactions with the user should be mediated by a knowledge model of the system, a model which is intended to capture the user's view of the software. This user conceptual model is itself locale sensitive and should be changed during localisation.

4.2 Language generation

The handling of error messages gives particular problems when they need to be composed from several parts. When the parts come together they may need to be modified in their grammatical form so that a grammatically acceptable text results. More generally word order may need to be changed and small atomic sentences combined, in order to produce a natural looking solution. In the Glossasoft (Hall and Hudson 1997) project two approaches were taken.

The first approach was the use of message templates, messages with slots into which you can substitute actual values depending on the context in which each message is generated. After substitution there will need to be some grammatical tidying up.

For example, instead of using four messages to announce that one of the four disk drives of the system is damaged, you could use the message template:

```
Disk <number of disk> is damaged
```

If the user is unfortunate enough to have two damaged disks, the application could also use this message template to announce that:

```
Disks 1, 2 are damaged
```

but would need to access a morphological generation routine to make the appropriate adaptations. Using such extended message templates is especially important for languages with many inflectional word forms, for example synthetic languages such as Finnish, as it avoids the need to maintain all the different word forms inside the message catalogues. Using extended message templates also improves the organization of message catalogues.

At VTT a trial was undertaken on the software product OsiCon, a risk analysis support tool, and the approach was demonstrated successfully for English and Finnish.

However, you still have to maintain different extended message templates for the different languages supported. So far we have been looking at "canned text" systems. Canned text systems, in which the repertoire of messages is fixed and immutable, are typical of the situation we are usually faced with in localisation and internationalization. All possible messages have been anticipated and stored. This is even the case with the message template approach discussed above.

tion. All possible messages have been anticipated and stored. This is even the case with the message template approach discussed above.

The alternative approach is to use a general language generation approach (eg. Allen 1987). Language generation is capable of creating a range of messages that could not have been anticipated, messages which are contingent upon user actions and an evolving knowledge base within the system.

Central to this whole process is a knowledge model of the application and its supporting software systems, about which messages are to be generated in the context of specific user actions and difficulties.

Some event occurs in the system which triggers the whole process. This could be a request for help, or some erroneous user action which requires a message to be given to the user. The system must in general know something about the user, because what it wants to say to the user will depend upon what the user knows. As a function of what the system assumes that the user knows, the system will select information to present to the user. The outcome of the selection would be a number of abstract linguistic elements, or "speech acts".

This message is still very abstract, and particular word choices (or phrase choices) must next be made. This choice will affect the focus of the message to be generated, which can influence the "tone" of the message. For example, there is a choice between active and passive voices "You typed the wrong character." and the "The wrong character was typed." dropping the agent in the second choice to remove the accusatory tone that would be unacceptable in many locales.

This leads to some knowledge structure, which indicates what text is to be output, and now it remains to generate the actual text. We can generate the text directly from the data structures, using a collection of mutually recursive routines which embody within them the syntax of the target language, or we can use a table-driven or rule driven approach.

It is only in this final stage that the specific features of the natural language being generated become important. Preceding these, the knowledge model itself may be largely culture independent, though it may make some cultural assumptions such as a specific conceptual model of the computer.

At Demokritos a small example for Hewlett Packard's VUE interface system was tried out, with several different levels of user expertise being selectable. This showed that the approach was very promising, but also very computer intensive. Clearly the template approach is to be preferred if possible.

5. Application to Decision Support Systems

Decision Support Systems for Sustainable Development may be very diverse, as can be seen in the various papers of this book. These will involve information from population distribution, agriculture and health, through to finance. They will be underpinned by technologies covering graphical information systems requiring the display of maps, database management systems with large repositories of information, and knowledge-based systems with methods of inferring facts from those actually stored, as well as the distribution of both processing and data. These systems will need to be used by people covering a wide range of technical experience and expertise, drawn from a wide cultural and linguistic group.

Decision support systems have the general architectural form shown in Figure 2. At this point we are focusing on the supply of information for decision making available in a number of servers across national boundaries and thus across locales, to a number of terminals or clients where decisions are made. To this architecture will need to be added the internationalization architectures discussed above. This intersection architecture will influence the general DSS architecture in the following ways:

- The application software in the terminals in countries 1 to N will be localized for the user communities involved – there may be several. Where the user interface involves the display of maps in geographical information systems, map presentation conventions will need localization.

- Place names will be important in such systems, and constitute a special case – typically these would be recorded in the native language, but approved transliterations into other scripts and phonetic systems would need to be used, as in the use of Pin Yin for Chinese.

If at all possible information resources will need to be kept in locale-neutral form, or in some standardized form that enables generation and presentation in local form. Numerical data is the simplest, and in many cases may involve no more than display routines, though units for systems of measurement need to be taken into account. Time and dates could be more difficult, but are tractable. Text may give real problems, unless it can be stereotyped in some way, perhaps using some narrow domain knowledge model with some simple form of language generation, or by using a controlled language and a simple translation system like translation memory.

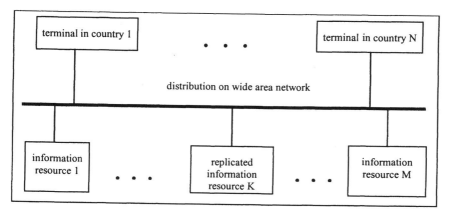

Figure 2. Outline architecture for Decision Support Systems.

Thus implicit in the approach must be the storage of data in some abstract form, with this mapped to the local terminal/client computers into a form that is acceptable to and understandable by the local users. This approach should also be applied to the local viewing software which should be internationalized to a core product which is locale-neutral, with local mappings to the locale of the user.

In developing a multi-lingual DSSs we will need to carefully characterize all the information that is being stored and shared in terms of its locale assumptions, and then develop locale-neutral forms. Where text is concerned we will need to work to understand it, removing as much of the locale sensitive components as possible. This should be done across all the range of DSSs for sustainable development, so that the localization processes can be used across the range of systems, saving costs and increasing the range of languages and locales that can be covered.

The process of decision making, and what constitutes appropriate information to use in decision making, can vary greatly between cultures. The articles in this book give illustrations of this variety, though there is an undercurrent of acceptance of or aspiration to western modernist rational decision making. Social values may take precedence over efficiency, and family may take precedence over nation. Different ways of making decisions will affect the way the local client systems are configured, and may deny any ability to make decisions between individuals who are not co-located.

To illustrate this let us contrast the USA with Japan. Hofstede (1991) characterizes cultures in four dimensions:

- power distance – should decisions be imposed or should people be consulted? (p27)

- individualism versus collectivism – should the interests of the individual prevail over the interests of the group? (p50)

- masculinity versus femininity – is toughness and assertiveness valued over tenderness, nurturing and modesty? (p83, 92)

- uncertainty avoidance – are uncertain or unknown situations felt threatening? (p113)

Hofstede made a large survey of IBM employees around the world, and scored countries in the range of (roughly) 0 to 100. The USA scores were respectively (40, 91, 62, 46), concluding that the culture is mildly consultative, highly individualistic, and mildly caring and accepts uncertainty. By contrast the Japan scores were respectively (54, 46, 91, 92), concluding that the culture is consultative, collectivist, highly assertive and tough, and dislikes uncertainty. This kind of characterization is very crude, but indicative. More detailed commentary suggests that while in the US decisions may be made by individuals for the advancement of the individual or some externalized corporation and accepting uncertainty, in Japan the decision would be made collectively and for the collective good to reduce uncertainty. Heaton 1998 reports how decision taking in computer aided design requires a move from single person workstations to a collective shared design surface in which all participants can see each other and sense the arrival of consensus. Shatzberg, Keeney and Gupta (1997) report on the failure of email in Japan due to this collectivism and the need for people to see each other in order to be able to reach consensus, deferring to the most senior person if appropriate.

The decision making tools described elsewhere in this volume may be easy to localize in the sense of Sections 2 to 5. However localizing them in this deeper sense of the cultural process of decision making is much harder.

6. Conclusions

We have seen that software for decision making can and should be localized so that it works in the language of the decision makers and respects local cultural conventions. Localization of language and simple cultural information like dates can easily be accomplished using today's technologies, but the sharing of information between diverse groups may well require more advanced methods of language generation.

However, the actual process of decision making may not be capable of computer support to the same degree. Decision making is deeply rooted in culture, and may not be capable of distribution mediated by technology. But that does not mean that localisation is not worth doing, so that information can be shared and made accessible. Accurate information is after all a prerequisite for sound decisions.

7. References

Allen, J. (1987). *Natural Language Understanding*, Benjamin Cummings Publishing Company, Inc: California

Apple Computer Inc. (1992). *Guide to Macintosh Software Localisation*, Addison-Wesley, 1992, ISBN 0-201-60856-1

Apple Computer Inc. (1992). *Human Interface Guidelines: The Apple Desktop Interface.* Addison Wesley, ISBN 0-201-17753-6

Danlos, L. (1987). *The linguistic basis of text generation.* Translated from the French (1985) by Dominique Debize and Colin Henderson. Cambridge: Cambridge University Press.

Duncan, D. E. (1998). *The Calendar.* London: Fourth Estate

Hall, P. and R. Hudson (1997). *Software without Frontiers*, Wiley 1997

Heaton, L. (1998). "Preserving Communication Context', in Ess and Sudweeks (Eds.) *Cultural Attitudes Towards Technology and Communication*, Proceedings of the CATaC 98, London, 163-186.

Hofstede, G. (1991) *Culture and Organisations. Intercultural Cooperation and its Importance for Survival. Software of the Mind.* McGraw-Hill.

INSTA (Inter-Nordic group on Information Technology Standardisation) (1992). *Nordic Cultural Requirements on Information Technology*, INSTA Technical Report STRI TS3 1992.

Jandt, F. E. (1995). *Intercultural Communication. An Introduction.* Sage.

Jones, S., C. Kennelly, et al. (1992) *The Digital Guide to Developing International User Information*, Digital Press, 1992.

Kano, N. (1995). *Developing International Software for Windows 95 and Windows NT.* Microsoft Press.

Keesing, R. M. (1975). *Cultural Anthropology. A Contemporary Perspective.* Holt Rinehart.

Kennelly, C. H. (1991), *The Digital Guide to Developing International Software*, Digital Press, 1991.

Luong, T. V, J. S. Lok, D. J. Taylor, K. Driscoll (1995). *INTERNATIONALISATION Developing Software for Global Markets*, Wiley 1995

Madell, T., C. Parsons and J. Abegg (1994). *Developing and Localizing International Software*, Hewlett-Packard Professional Books, 1994.

Microsoft (1993), *The GUI Guide*, Microsoft Press, 1993.

Pillai, D., and B. L. Swamikannu (1911). *Indian Chronology, A Practical Guide*, Madras.

Shatzberg, L, R. Keeney and V. K. Gupta (1998). 'Cultural and Managerial Comparisons: an Analysis of the Use of Email and WWW in Japan and the United States'. *Proceedings of the 1997 Information Resources Management Association International Conference*, Vancouver, Idea Group Publishing, 296-300

Taylor, D. (1992), *Global Software: Developing Applications for the International Market*, Springer-Verlag, 1992.

Unicode Consortium (1991/2). *The Unicode Standard. Worldwide Character Encoding.* Version 1.0, Volumes 1 and 2. Addison-Wesley 1990 and 1991.

Unicode Consortium (1997). http://www.unicode.org/.

Uren, Emmanuel; Robert Howard, Tiziana Peritonni (1993), *Software Internationalisation and Localisation: An Introduction*, VNR Computer Library, 1993, ISBN 0-442-01498-8

16 SOFTWARE INTEGRATION FOR ENVIRONMENTAL MANAGEMENT

Victor S. Chabanyuk
and Olexandr V. Obvintsev

1. Introduction

Environmental management always has been a complex activity, but over the last decade it has become even more complicated because we have begun to understand just how harmful our economic activities are to the environment. As a result, sustainable development now is recognized as the main objective of environmental management. To achieve it, special automated information systems to manage the environment sustainably have been developed in many countries, beginning in the early 1990s.

The execution of environmental management processes today requires the use of many computer tools (see Chapter 19). In software alone, there are spreadsheets; database systems to store the large volumes of data required for decision-making, together with data modeling methods; simulation modeling techniques such as system dynamics; mathematical programing software to carry out the calculations of decision theory; knowledge-based systems and expert systems to capture the rules of decision-making; neural network systems and case-based reasoning to help in unstructured problems; visualization systems, including map displays and virtual reality, as well as the more conventional charts and graphs; and group working software to help group decision-making. One set of tools stands out as being of special interest for sustainable development: geographic information systems (GISs). These include components of most of the tools listed above, and therefore are the major focus of this chapter.

Section 2 describes the activities of the Administration for Radiation Protection of the Population and Radioactive Waste Management (APPR) of the Ukrainian Ministry of Emergency Measures and Chornobyl Affairs (MECA) as well as the general architecture of the environmental management geoinformation system (GeoEMIS) – the radioecological geographic information system (RGIS). The main software elements of the RGIS are described and used as examples in subsequent sections. Hierarchical restrictions on software integration are described in Section 3. Desktop integration technologies, including compound documents management, structured storage and persistence services, uniform data transfer, as well as automation and scripting services are discussed in Section 4. Meta-level integration restrictions are presented in Section 5: EMIS software needs to be integrated to existing industrial standards like CORBA/OpenDoc and/or Microsoft DCOM/OLE. Section 6 presents the experience with the practical application of the ideas of CORBA/OpenDoc via DCOM/OLE, which are demonstrated through the example of the RadEco system. The last section provides conclusions and offers recommendations for solutions to software integration problems in developing countries on the Ukranian model.

2. Radioecological management at the Ministry of Emergency Measures and Chornobyl Affairs

This section describes the ecological measures carried out by the APPR of MECA (known as the Ministry for Chornobyl until 1996), which was set up in 1991 to address the complex of problems due to the 1986 disaster at the Chornobyl Nuclear Power Plant. Its aftermath saw large areas of 12 (out of 25) regions (oblasts) of Ukraine contaminated with radioactive pollution.

Each of the various problems caused by the Chornobyl disaster are managed by a MECA group, or administration. The APPR is responsible for two: protecting people from radioactive contamination and managing radioactive wastes. Its administrative functions can be defined as radioecological management.

The primary radioecological management tool authorized by the national program to mitigate the consequences of the Chornobyl disaster is financed by the Chornobyl Public Finance Fund. Two sections of the program will be considered here: "Information support for the radiation control system" and "Ecological cleanup of the environment."

First, let us look at the Ministry's radiation control system, which is carried out by its information support section.

2.1. The radioecological management process

2.1.1. Monitoring plan. The radiation control system consists of two monitoring networks: base and inspection. The base network consists of various departments that regularly monitor contamination of different environments at predetermined points. The inspection network is not fixed. It is determined by the current monitoring objectives, while the actual measurements are carried out by various organizations on a contract basis. It is the job of the responsible APPR official (research officer) to develop the environmental monitoring plan, either annually or as instructed by management. The officer is expected to

1. update or determine the research goal;

2. determine a research object; and

3. take into account financial, technical, time, and other research restrictions.

The information support for the radioecological monitoring process has two obec-
tives. The first is to inform — to gather and store data on the contamination of ecosystems and inform management and the population about contamination levels. The
second is to manage — to prepare information for decision-making on ecosystem
monitoring plans, distribution of compensation to those living in contaminated areas,
preparation of the regulatory and legislative documents for changes to the scale and
volume of the compensation, and countermeasures for abiotic ecosystems.

The research object is determined by which characteristics of an ecosystem should
be monitored for gamma-, beta-, and alpha- contaminants. To distinguish the various
characteristics, it is necessary to record the following base observations: 1) time (moment) of observation; 2) space (territory of observation); 3) environment of observation; 4) observer; 5) number of observations (collected samples); and 6) the set of
monitored entities from which samples are collected.

2.1.2. Monitoring and data collection. Monitoring usually means analysis of samples according to standard procedures on special devices at specialized radiological
laboratories. The gathered data are values of monitoring variables, which, taken together, provide a sketch of the main characteristics of the environment object. The
data are then sent to various national and regional departments. Every department that
collects data should ensure 1) input of values of parameters and variables (contaminants) into the local database (LDB), data set and/or the monitoring record (report); 2)
verification of data; 3) geocoding of the entered data; 4) conversion of values of the
variables and parameters into abstract digital form; 5) carrying out of routine monitoring functions (registration, data entry, deregistration); 6) selection and visualization, by various methods, of the contamination data and data on monitored objects; 7)
export and transfer (for example, by e-mail) of entered data into integrating components; and 8) preparation of accounts on the condition and amount of data entered.

Data on natural environment contamination obtained from various organizations
are entered into an integrated data bank or treated without integration.

2.1.3. Data processing. Data processing is carried out by input into local databases,
datasets, and/or an integrated data bank of functions of the following classes: operations with data, aggregation functions (sum, average, median), chart functions, cartographic functions, functions of 1-, 2-, and 2.5-dimensional visualization, and interpolation functions.

Depending on software, data can be categorized and grouped together as 1) operational functions with databases, 2) statistical functions, 3) graphic functions, 4) cartographic functions, or 5) combinations of the above.

The functions listed above are characterized by the fact that they do not alter data
(for example, queries) or generate new data from monitored variable values for only
one out of six parameters. We call these functions of zero and first generation.

In practice, it is necessary to apply two or more parametric functions; for example,
time-spatial functions (they are also known as prognosis models or models of carry).
However, experience shows that to apply such functions in administrative organizations such as MECA is problematic because of their size and complexity.

2.1.4. Interpretation. After the data are treated in any way, it is necessary to interpret
them with reference to the research goal in order to determine their usefulness, after
which the researcher can draw conclusions and prepare the material for decision-
making.

Interpretation of the data can be made through visual analysis of the condition of the contaminated natural environments and ecosystems; determination of the total number of contaminants within a given area; aggregation of data on the ultimate allowable concentration (UAC) for the different environments; matching of actual contamination data with UAC standards; assignment of the various areas in the appropriate classification; search for and revision of the information on the physical properties of different natural environments; construction of diagrams representing change in contaminant values; and charting of various thematic maps, in particular, maps of the area under study and maps of the actual values for each contaminant; preparation of pollution maps by interpolation methods.

2.1.5. Feedback. If it is impossible to obtain an answer to the research questions, the researcher should treat the data in some other way. This iterative process may lead to a set of complementary systems, each of which presents the data from a different perspective. Such complementary systems can give a better understanding of a problem than any single system. Sometimes, the system or systems obtained after the data has been processed is too complex to interpret usefully. In such cases, the systems should be simplified.

After the data has been processed and interpreted, supplementary data may be collected to increase their reliability. They also may be reconsidered on the basis of new data.

2.2. Radioecological GISs

To process the large volumes of radioecological data collected more efficiently, the APPR invited Intelligence Systems GEO Ltd. to develop the RGIS. The goal of the first RGIS phase — RGIS1 — is information support in radioecological management for the contaminated agricultural areas of Ukraine. The RGIS1 was developed and became operational in 1997. The second phase of the system (RGIS2), implemented in 1998/99, covers all abiotic environments. The RGIS architecture is shown in Figure 1.

The RGIS automates four environmental management processes: 1) planning, organization and collection of data on contamination, natural resources, and the economy; 2) integration, accumulation, and organization of data into databases; 3) manipulation, analysis, and processing of data; and 4) preparation of data to facilitate making and implementing decisions to protect particular ecosystems and optimize their use. Its predominant use is for the processing and presentation of spatially coordinated data and geoinformation methods. Because of this, it is known as a GeoEMIS.

Following the development of the RGIS1, it was concluded that the software requires two types of integration: 1) intertask — integration of separate tasks of an automated subsystem (inside the automated system), and 2) interprocess — integration of separate business processes (between separate automated systems).

The RGIS1 encompasses three technologies: 1) GeoRegister, 2) GeoBIS, and 3) RadEco. GeoRegister technology is intended to create desktop georegistration systems that record and organize the characteristics of spatial entities. Depending on the type of monitoring being done, environmental registration systems (GeoRegisterE) and registration systems for human activities (GeoRegisterA) can be created.

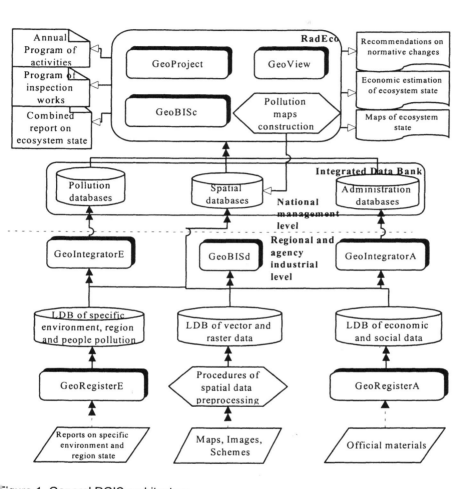

Figure 1. General RGIS architecture.

Besides GeoRegister*_*, georegistration technology includes 1) nonautomated construction procedures of base topographical digital maps from space images and scanned raster images of topographical maps and schemes, 2) preprocessing procedures of vector digital topographical maps, and 3) tuning software of georegistration systems (software and information) with GeoTuner.

Geodata base information system (GeoBIS) technology is intended to create and manage an integrated data bank, which is formed from local databases accumulated in registration systems or from files arriving from local suppliers. It consists of environmental data from the geointegration software GeoIntegratorE and data on human activities from GeoIntegratorA, as well as data from GeoBIS, which has desktop (GeoBISd) and corporate (GeoBISc) variants. GeoIntegrator performs the following tasks: coordination of local and integrated classifiers; geocodification of data; registration of processed data in the system journal; data verification; and conversion of input data formats into an integrated data bank format.

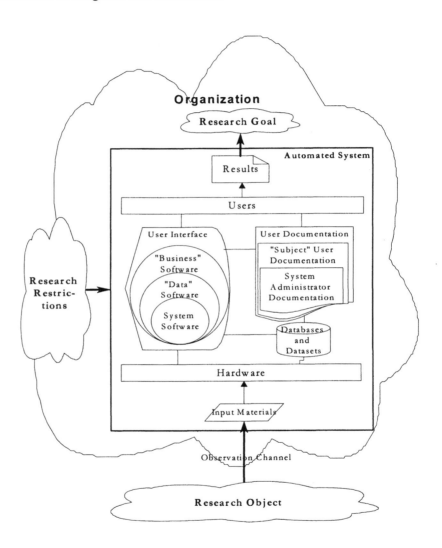

Figure 2. Architecture of a typical RGIS automated subsystem.

GeoBIS allows the RGIS to form, execute, and store typical and atypical queries; revise passport data; install a filter on query outcomes; represent query outcomes as diagrams and graphics of cumulative distribution functions; use thematic cartography and construct thematic maps; export query outcomes into spreadsheets and databases; form typical and atypical combined reports of the text, tables, graphics and maps; and print the reports.

RadEco (radio ecology) technology is used to automate the radioecological management processes carried out at the APPR. Like GeoBIS, it includes GeoProject and GeoView software and nonautomated procedures for statistical data processing and construction of contamination maps.

GeoProject performs the following tasks: formulation of the contents of a program of works; determination of the amount of financing required; nomination of the executors; management of agreements signed within the limits of the program of works; record-keeping and control of the implementation of MECA's Program of Works; and the preparation of reports tracking the implementation of the program.

GeoView accepts new maps and workspaces; creates the raster image of a map; finds and selects maps by meta-characteristics; creates, saves, and edits workspaces; creates and saves previous and current projects; generates the distribution kits of maps and workspaces; builds isoline maps, Voronoy diagrams, and square contamination maps; supports a meta-information database; orders procedures for the reception, registration, storage, search, and distribution of maps; and transmits information as well as analytical and historical materials to assist decision-making by senior Ministry officials by means of a ranked colour form.

Nonautomated workflow procedures for statistical data processing and the construction of contamination maps supplement tasks carried out with GeoBIS and Geo-View software. They consist of instructions on completing a procedure, the software and hardware used to carry out a task, and a control sample. Other software includes the geoinformation package Arc/Info and the statistical package S-PLUS.

3. Hierarchical restrictions of EMIS software integration

The problem of software integration has existed since computers first appeared; therefore, it is difficult to define the problem so as to satisfy everyone. However, it is possible to define some boundaries to the problem beyond which it is not practical to consider. In this section boundaries are determined by proceeding from information system (IS) hierarchies.

Integration is defined as the association of separate elements in a unit. Integration of automated systems is the association of several automated systems of different prescription into a uniform multifunctional automated system (Pershikov and Savinkov, 1991).

3.1. Hierarchy in current application architecture

It is difficult to imagine modern information technologies without a client/server architecture of applications, which were actively developed with the appearance of the first local area networks (LANs) and now dominate complex ISs. There were three "waves" in the development of client/server architecture. The third wave may be described as "three-tiered."

The three-tiered architectural approach to client/server solutions has steadily gained momentum as the leading model for corporate application development. This approach separates the various components of a client/server system into three tiers of services that must come together to create an application: user services, business services, and data services. These tiers do not correspond necessarily to physical locations on the network. Rather, they are conceptual layers to aid in the design of robust component-based applications.

The model that describes these services and their interaction is known as the services model. It is not tied to any particular product suite, technology, or vendor. Formally defined, a service is a collection of related features that respond to requests for specific activities and/or yield information on the basis of specifications of interface and behavior, and are accessed through a consistent interface that encapsulates its implementation. There are three categories of services. These have their own attributes, as Table 1 shows.

Table 1. Categories of services.

Service type	Service attributes
User services	Presentation of information and functionality, navigation, protection of user interface consistency and integrity
Business services	Shared business policies, generation of business information from data, protection of business integrity
Data services	Definition of data, storage and retrieval of persistent data, protection of data integrity

Each of these services has common features and provides common functions. The services are networked together and operate cooperatively to support one or more business processes. In this model, an application becomes a collection of user, business, and data services that meet the needs of the business process or processes it supports.

The selection of three layers of application came about because of the need to create large, complex ISs. To create such systems, it is necessary to divide them into smaller components that are easier to work with. This division is carried out by two methods: abstraction and encapsulation. The level-by-level architecture based on abstraction has several main characteristics: 1) precisely defined layers – these can be constructed one upon another so that each new layer ensures more complex abstraction in comparison with a lower layer; 2) formal and obvious interfaces between concrete layers that can invoke the services of a lower layer; the interfaces determine high-level operations that activate more detailed behavior which is executed on a lower layer; and 3) concealed and protected details inside each layer. The concealment of details is important both for abstraction and encapsulation.

Each of the three layers, or tiers, is in turn subdivided into three sublayers, which are located top-down: 1) navigation and control – the upper sublayer provides a way to search the services and their activation, and guarantees that they will grant their service; 2) active services – active objects in a layer (applications, support services to the business rules, and databases); 3) integration – the lowest sublayer provides services that allow many independent objects within the limits of one layer to work together as if they represent a composite object.

In summary, see Table 2, which shows the main objects of the layers and sublayers of the final table of a subsection.

Table 2. Three-tiered architecture of application and its sublayers.

Layer of application	Navigation and control	Active services	Integration
User services Business services	Windows Services co-ordinator	Documents Business rules	OLE Environment of services support
Data services	OLE DB. Query processor	DB	Transactions co-ordinator

3.2. Hierarchy of information system levels

In the previous section, the IS hierarchy was considered in a narrower sense; that is, relative only to software applications executed on the system hardware. However in the definition of the automated IS and its architecture shown in Figure 2, there are some objects that are not absolutely "computer" objects — users and documentation. EMIS users, as a rule, work in environment protection organizations, which have differing mandates determined by the organization, laws, and normative acts regulating environment protection activity.

In the literature on ISs in the broader sense, there is general agreement on the selection of the three levels of abstraction in each system (Iivari, 1989): 1) organizational, defining the organizational role and context of ISs; 2) conceptual/infological, defining "implementation independent" IS specifications; and 3) datalogical/technical, defining technical IS implementation.

Again, as in the previous section, the levels of abstraction or stratification descriptions conform to a hierarchical structure of concepts with four characteristics: 1) the hierarchical relationship is linear; 2) the levels describe the different characteristics of the same system; 3) the top levels contain, in some sense, a more abstract description of the system than the lower ones; and 4) the structure of concepts of the system contains definitions of the relations between levels.

Mylopoulos (1991) calls the indicated levels of abstraction "worlds": Usage World, Subject World, and System World. They are constructed by abstracting: the host organization; the universe of discourse; and the technology. Note that the top levels of the IS define the lower; for example, it is impossible to create a technological level that contradicts the organizational.

3.3. Subject hierarchy of environment management information systems

Restrictions are determined by the infological or subject level of the IS. In the RGIS, for example, there are five precisely selected hierarchical epistemological levels (Klir, 1985): 1) source system, 2) data system, 3) production system, 4) structured system, and 5) meta-system, which together create a subject hierarchy of the system (Figure 3).

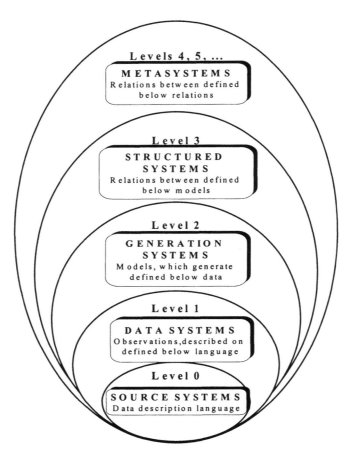

Figure 3. Hierarchy of the epistemological levels of systems (Klir, 1985) (simplified presentation).

Interaction with the external world is carried out through the source system for obtaining the data system, which is simulated by systems at higher levels. GeoRegister software is used to provide data for the RGIS. Local data systems, local zero and primary generation systems, the integrated (structured) data systems, and structured zero and primary generation production systems are implemented in desktop and/or corporate GeoBIS software. RadEco software technology, which includes GeoProject, GeoBIS, and GeoView is used to implement structured systems and meta-systems.

Generally speaking, the structured system represents by itself a set of source systems, data systems or generation systems, which have a joint parametric set. The systems which create the structured system are called its elements.

There are few reasons to represent the full system by a collection of its subsystems. For example, one of the subsystems is concerned with observation or measurement. As time is one of the parameters, it is frequently technically impossible or unreasonable to simultaneously observe (measure) all the variables that relate to the research goal. In such a case, it is possible to collect only partial data for the largest possible subset of variables. In other cases (especially at the beginning stages of development), the user of the system is compelled to use data assembled by different organizations or researchers and covering only a few of the variables necessary for the work.

Another reason for systems structuring is the complexity, which is connected to the need for visibility of the process and design of the considered system. Other reasons have to do with the availability of the limited set of suitable ready elements (units), the efficiency of implementation, and different problems of reliability, monitoring, and maintenance of the designed system.

In the word "meta-system" the Greek prefix meta has three meanings: 1) "meta X" can be something observed (that has a place) after X (X is the predecessor of meta X); 2) "meta X" can indicate that X has changed (a general term for change); or 3) "meta X" can denote something higher than X, something of a higher kind.

The meta-system X is a triple $MX=(W, X, r)$, where X is any set of systems whose parametric sets are subsets of W and r is a procedure of substitution of one system by another ($r: W \to X$).

The mechanisms of the structured system and meta-system are independent from each other's mechanisms of system integration and can be applied in any order.

Systems classes and subclasses are created with the help of the appropriate software, which was developed for the automation of corresponding business processes. Unfortunately, the software has a very different nature; therefore at present it is impossible to unify it. There remains the integration approach. However, general system hierarchy implies that the more complex the automated business process is, the more levels the corresponding information system should have. In our judgment, the management intelligence system (MINTS) class is much more complex than the management information system (MIS) class. However, before construction of a practically useful MINTS, it is necessary to construct a corresponding MIS. From practical experience with EMIS development in Ukraine, we know that the software integration options for the MIS class are restricted. In practice, it is very difficult to create MINTSs; as a result, they cannot be considered as subjects for integration at present, at least not until the severe mathematical formulations of EMISs of the MIS class have been carried out.

4. Intertask integration of the software

4.1. Object linking and embedding — technology for integration of desktop systems

The following object linking and embedding (OLE) technologies, which form a framework for compound documents, will be considered: 1) compound document management; 2) structured storage and persistence; 3) uniform data transfer; and 4) automation and scripting.

We will not consider OLE – COM because the layer of business services is based on it and it is described in subsection 5.2. In general, it is necessary to say that OLE consists of a set of interfaces that define the set of related functions. The interface determines the contract between components. OLE components ensure the protocol of the interface interaction that allows clients to acquire pointers — in execution time and in interfaces — that are supported by the components. OLE components are defined by classes that implement one or several interfaces and a class factory, which can produce components of this class.

4.1.1. Compound document management determines the rules of association that allow independently developed components to subdivide the window of the container. Using the container as mediator, the components should interact to create documents without visible seams. The containers activate components that are embedded into the document. The components map their data in a place assigned to them and thus interact with the user.

The use (or purpose) of the compound documents defines the interfaces between the container application and server components that it supervises. In this case, the server is a visual component that "serves" the container. The container/server interfaces define protocols for activation of servers and editing of their contents "in place" inside the container window. Container applications supervise the storage and the windows for imaging of data of the compound document. Each server component supervises its own data. It is possible to use the container/server relation for the visual integration of data of different formats. The server data are called embedded if they are saved inside the compound document of the container and linked if they are saved in the other container file.

4.1.2. Structured storage and persistence. The OLE structured storage system provides a file system within a file. The current implementation of this architecture is called compound files. These create the "file within a file" by introducing a layer of indirection on top of existing file systems. Compound files break a file into a collection of storages (or directories) and streams (or raw data values). This internal directory system can be used to organize the contents of a document. OLE allows components to control their own storage in the compound document. The directories describe the streams; the hierarchical structure makes it easy for OLE objects to navigate within the document. Compound files provide some rudimentary transaction facilities for committing information to storages or restoring their previous states.

OLE also provides a set of interfaces that allow a client to communicate with a persistent object. These interfaces define the capabilities of a persistent object. At one end, a persistent object may know nothing about structured storages; it only knows how to store and manipulate its state in a regular file. At the other end, an object knows how to navigate the storages of a compound file. In the middle are objects that only know how to manipulate a single stream. In OLE, the client creates an object and then hands it a persistent store that contains its state information. The object then initializes its state by reading the storage object. The client also can ask the object to write its state to storage.

4.1.3. Uniform data transfer. OLE provides a generalized intercomponent data transfer mechanism that can be used in a wide range of situations and across a variety of media. Data can be exchanged using protocols such as the clipboard, drag-and-drop, links, or compound documents. The exchanged data can be dragged and then pasted or dropped into the same document, a different document, or a different application. The actual data transfer can take place over shared memory or using storage files. Asynchronous notifications can be sent to a linked client when source data change. The OLE uniform data transfer provides a single interface for transferring data that works with multiple protocols.

4.1.4. Automation and scripting. Automation and scripting services allow automation clients (called controllers) to supervise server components. Automation clients use the help of a scripting language, or means, such as Visual Basic. The automation bases on the OLE dynamic invocation facilities are called dispatchable interfaces. They allow the client to call a method or to manipulate a property through the mechanism of late binding. Dispatching ID is transmitted to a method of call, which decides what method to call in run time.

Automation clients must discover at run time what interfaces an automation server provides. This includes the methods an interface supports and the types of parameters that are required by each method. It also includes the properties a component exposes to its clients. Clients can obtain all this information at run time from Type Libraries – the interface repository.

4.2. Example of OLE use: construction of a combined report from compound documents about the state of an environmental problem

To demonstrate how OLE technologies can be used to solve software integration problems in environmental applications, we give the example of the construction of a reference on the number and distribution of settlements in the 4th radioactive contamination zone throughout the regions (or oblasts) of Ukraine through the application of GeoBIS, a subsystem of RadEco.

After the contamination caused by the Chornobyl nuclear disaster, the polluted territory of Ukraine was divided into four zones. The population of each zone receives appropriate compensatory payments for residing there. The 4th zone of radioactive contamination — the zone of strengthened radioecological control — is an area with a density of ground contamination well above levels recorded before the disaster. Cesium isotope counts there range from 1.0 to 5.0 Ki/km^2, strontium counts from 0.005 to 0.01 Ki/km^2; thus the estimated effective equivalent dose (EED) of human exposure in the area, with allowance for factors such as the radionucleotide migration in plants, far exceeds the 0.5 mSv (0.05 Ber) per year levels recorded before the disaster.

Data on cesium 137 and strontium 90 ground contamination and the estimated EED of human exposure are saved in "contamination databases" which can be accessed by clicking on the map in Figure 4. Ultimate standards and administrative-territorial division data are saved in "administration databases" while topography is saved in the "spatial databases" of the RadEco system.

Before the application of GeoBIS, the user should prepare the template of the combined report on the state of this artificial ecosystem — settlements in the 4th radioactive contamination zone — by setting necessary titles, information on signatures, helpful information about the strengthened radioecological control zone, and so forth.

The number of settlements is obtained from the following generation function:

$$^x\!f = \sum \chi(1.0{<}v_1{<}5.0)\ \chi(0.005{<}v_2{<}0.01)\ \chi(0.5{<}v_3)\ \chi(w_2{\in}{}^x\!W_2$$

where: $x \in$ {list of the contaminated oblasts of Ukraine}; $v_1 \in$ [1.0; 5.0] denotes cesium 137 contamination; $v_2 \in$ [0.005; 0.01] denotes strontium 90 contamination; $v_3 > 0.5$ denotes the EED; and the parameter $w_2 \in {}^x\!W_2$ describes the contaminated areas in the oblasts of Ukraine.

The GeoBIS response to a query is presented in Table 3.

Table 3. A GeoBIS query result.

Oblast title(x)	Number of 4th zone settlements ($^x\!f$)
Kyiv	438
Zhitomir	363
...	...

The user does not need to finish this work with GeoBIS, but can insert the table into the template of the document by using a format of the word processor MS Word. As the tabular presentation of data is not always clear, the user may want to see this information in the form of a schedule or diagram. To achieve this, the GeoBIS application transmits the table to the MS Excel spreadsheet where a column diagram is created. The user can insert the diagram into the template of the document again by using MS Word. The resulting information now can be presented as a thematic map. The map also can be transferred into the necessary place in the template of the combined report on the state of the ecosystem.

In this simple example, almost all the OLE technologies are used. The compound document technology uses the container application MS Word, in which the combined report is modeled. MS Excel and MapInfo act, in turn, as server applications. For want of data transfer between applications, the services of uniform data transfer are used. The MapInfo GIS package executes the cartographic functions of the GeoBIS to construct thematic maps. For the example discussed above, a thematic map showing the settlements in the contaminated raions (municipal areas) and oblasts (regions) is shown in Figure 4.

5. Meta-system restrictions of EMIS software integration

5.1. The meta-system level of an EMIS

The class of automated systems is considered the automated meta-system. Naturally, the class of automated systems of the architecture in Figure 2 can have only one unique object. However, each developer of the system is interested in having as many instances (and consequently users) as possible. On the other hand, each developer must customize the system to the specific requirements of the user. This additional effort can be alleviated if the existing customizations can be reused.

The meta-level on which the concrete automated system is considered (Bergheim, *et. al.*, 1989) is called application, and the meta-level on which the automated meta-system is considered is called conceptual.

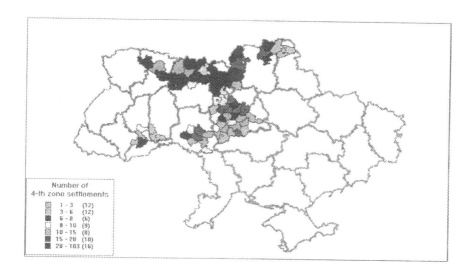

Figure 4. Number of 4th radioactive contamination zone settlements.

In order to understand the nature of meta-system objects better, let us consider Figure 2 as a "user's guide." At the conceptual meta-level there is a corresponding template of the document. Both developers and users will benefit if the template corresponds to the standard document. This means that at the conceptual meta-level it is best to use standard solutions.

In Bergheim, four hierarchical meta-system levels, which form the framework of IS science, are outlined:

1. ω-level (operational): the changes of state in application are described. An application is understood as a formal representation of some part of reality (for example, as the state of the database is changed over time).

2. α-level (application – meta-level of a ω-level): the application itself is described. One example is concrete GISs implemented on an Arc/Info GIS-package. Other example might be concrete topographical or geographical maps (in case of manual or non-automated GISs). The *state* of an α-level in a concrete moment is an instance in a ω-level.

3. β-level (conceptual [German *begriff*]) – meta-level of a α-level): the ways to receive instances of an α-level are circumscribed. In other words, the question is about the formalism used to obtain the specifications at an α-level (for example, the Arc/Info package).

4. γ-level (generic – meta-level of a β-level): the way to receive instances of a β-level are also circumscribed. In other words, at this level the question is how to build different formalisms. It focuses on the problem of what parts of reality it is possible to determine (for example, position-based [layered] approach to GISs, which is used in Arc/Info, or the feature-based [object-oriented] approach).

We cannot pursue here an indepth analysis of Bergheim, but we note that it is about the scientific substantiation of meta-system limitations on software integration. Over the last few years, the software industry has standardized many conceptual-level meta-objects. Two competing standards define the limits in software integration: CORBA/OpenDoc and DCOM/OLE. Both standards are based on the concept of the component.

A component (Orlafi *et. al.*, 1996) is a reusable, self-contained piece of software that is independent of any application. A minimalist component is a marketable entity, without full application, that can be used in unpredictable combinations and has a well-specified interface, an interoperable object, and an extended object. Components are *bona fide* objects in the sense that they support encapsulation, inheritance, and polymorphism.

5.2. Comparing CORBA/OpenDoc and DCOM/OLE as the technological basis of corporate systems integration

One of the main problems with the integration of corporate environmental management systems is the general architectural and structural principles of their construction. These principles have evolved from file/server applications, through client/server architecture, up to component systems. The distributed component systems have obvious advantages when compared with their predecessors. Primarily, they have flexibility, scalability, and the possibility of changing a system that is now at a lower level of components. Substitution of components frequently does not influence serviceability of the system as a whole, and modern component technologies allow such substitution even "on the fly," without disconnecting the whole system.

The integration of the component system would be simpler with an operation that used rather large block-components. The technological basis for the interaction of components that have a higher level than contemporary operating systems is not available, however. It is necessary to set standard rules for the creation and accommodation of components. We will consider two existing industrial standards, CORBA/OpenDoc and DCOM/OLE, that offer somewhat identical, but essentially different, approaches to the construction of component distributed systems.

5.2.1. The CORBA/OpenDoc model. A consortium of more than 500 companies, called the Object Management Group (OMG), developed the CORBA standard in 1989. The development of CORBA has differed from the development of other software standards because principles and specifications preceded concrete realizations. CORBA uses an object-oriented approach as a basis for its component model. The component consists of objects that have a defined interface and behavior. It allows both single and multiple inheritance (implementation) of objects.

The general structure of CORBA is indicated in Figure 5. The basis of CORBA is the component or object bus called the "object request broker" (ORB). The common services, frameworks, and vertical and horizontal facilities are connected to this bus. CORBA allows the system to create ordinary objects and thereafter to build into it transactions, locks, and storage by multiple inheritance of appropriate services.

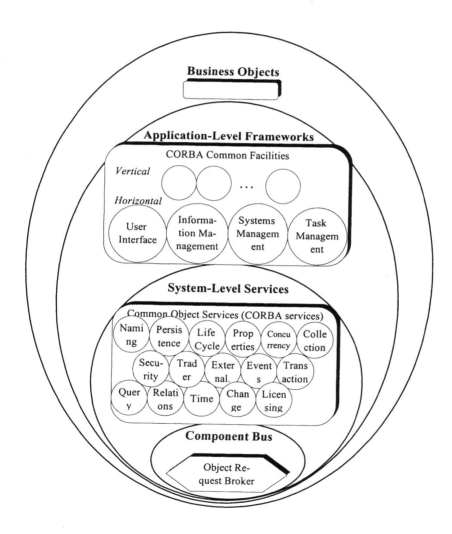

Figure 5. General CORBA architecture (Orlafi *et. al.*, 1996).

The common services are implemented by components with defined interfaces. At present, interfaces of common services are standardized by OMG. The frameworks are implemented by sets of components, integrated by their functionality. The horizontal and vertical facilities are made out as frameworks with the possibility of including additional business objects. The vertical facilities are developed in directions: electronic documenting, information supertrunks, production, distributed simulation, search and production of oil and gas, and accounting.

CORBA offers several scripts of component systems integration on the basis of common services (low level), horizontal facilities (average level), and vertical facilities (high level). Because of the need for integration on the basis of common services, general components have been developed and described, and certain behaviors have been assigned to them by multiple inheritance of common services. The components are compiled, registered in a repository, and assembled into the system.

Because of the need for integration on the basis of horizontal facilities, vertical business objects have been developed that use the interface of corresponding horizontal frameworks. These objects are compiled, registered in a repository, and assembled into the system.

Only the setting up of existing frameworks can solve the problem of integration on the basis of vertical facilities, however. When necessary, additional vertical objects can be developed which use the interface of horizontal and vertical frameworks.

5.2.2. The DCOM/OLE model. The main concepts connected with OLE technology were considered in the previous section. Now we will consider the basis of the OLE component object model (COM) and its extension to the distributed component object model (DCOM).

DCOM determines interfaces between components within the limits of one application or between applications. DCOM, as well as CORBA, separates the interface of a component from its realization and allows possibilities for dynamic definition and call of interfaces that represent an object. Objects and components can be distributed across the network. In this case the network address of a component is determined and fixed in the Windows registry with the help of configuring that ensures it can be invoked, if necessary.

The interfaces are subdivided into incoming and outgoing interfaces. The incoming interfaces are called from the outside components and the outgoing ones are called by the component itself.

DCOM does not support inheritance fully, because it is treated in the object-oriented approach. Instead, DCOM offers single-level multiple inheritance of interfaces. As a result, any object can inherit (implement) many interfaces but the inheriting of implementation is impossible. The partial substitution to inheritance of implementation is made by the technology of containment, where the component includes the services of other components and transfers them to the client through the uniform channel.

DCOM determines the interfaces of class factories and licensing. It also provides directory service for the Windows registry. The DCOM client calls for the service of a component by transmitting the unique class identifier (CLSID) and process identifier (PID), which unequivocally identify the service. IT also offers the service of "connectable objects," or raising and processing of events. The raising of events is implemented by outgoing interfaces and the processing of events by incoming ones. The recipient of events can directly register these events, which the system will react to, and ignore others.

DCOM requires each component to implement the IUnknown interface. This interface is used for the dynamic determination of the interfaces of a component and for unloading a component when it is no longer useful. As well, DCOM allows developers to describe their own interfaces, which expands the standard ones. Operations with their own interfaces do not differ from operations with standard DCOM interfaces.

The integration script for the component system from DCOM/OLE looks like this: first, describe own interfaces and implement components; next, add ready-to-use components; then, distribute components across the network and register them; finally, start the system.

5.2.3. The reasons for choosing DCOM/OLE as the technological basis of corporate systems integration. From the viewpoint of integrity and the prospects for its ideas, the CORBA/OpenDoc model looks better than DCOM/OLE. The limitations on inheritance cause nonsufficient structuring in DCOM/OLE. There are inherent limitations with the technical and software platform as well. In both models, the services of the lower and average levels are insufficient. What are the factors, therefore, that have caused us to choose DCOM/OLE?

In Ukraine, the predominant hardware platform is the IBM PC compatible personal computer, while the desktop operational system is Microsoft Windows. This dominance is even more pronounced than in the developed countries. At present, the manufacturers of the supporting software CORBA have not presented broad and convenient methods of development for a Windows platform. Microsoft, on the other hand, already has included DCOM in Windows'98. Furthermore, the inclusion of DCOM into the OS base, especially the main memory, is more economical.

Even more impressive is the availability of software products to support the model. First of all, there are the Microsoft Office products, which are the de facto standard in office applications. All manufacturers of GISs, even those not oriented to a Windows platform, have software ready to support DCOM/OLE.

DCOM/OLE also allows the integration of components already developed by the standard methods of registering components in the system register and by methods of configuring component distribution on the network.

Last, but not least, is the price factor. All composites of DCOM/OLE cost less than CORBA/OpenDoc.

5.3. The main principles of construction of an environmental management corporate system

5.3.1. General architecture. It is most important to construct corporate EMISs in a multicomponent architecture. As noted earlier, the most popular program architecture is a three-tiered architecture with a selection of user services, business services, and data services (VBE4, 1995), offered by Microsoft. In this model, the services are implemented by components.

The model allows for flexible distribution of the system across the network by using computing resources optimally. Our concern is that dividing the selection into only three levels of services is rather artificial. One should be able to select more than three levels when working with data operations. At the same time, to consider the system only from the viewpoint of data processing is essentially limiting and does not completely correspond to an actual situation. In any system, there are general system services that can be subdivided into levels, mapping a degree of abstraction of the used concepts. EMISs, as a rule, include services of interaction with maps that is not usually offered in three-tiered models. That is why it is better to speak about a multi-tiered distributed architecture rather than a three-tiered one.

5.3.2. Division of the system and components. Considering the multitiered distributed architecture of applications, it is possible to select separate projections of a structure of the system. If we consider a three-dimensional model, the vertical and horizontal divisions are selected. The vertical divisions subdivide the system into separate functions; for example, operations with data, processing of maps, and so forth. The horizontal divisions subdivide the system into sublayers, each of which takes its place in a hierarchy of layers and is responsible for fulfilling part of the general functions. The number of sublayers for each vertical division is not constant and depends on the complexity of the executed functions.

An example of horizontal division into sublayers is data processing. Different sublayers can be selected: user interface, meta-services, services of a logical store level, or services of a physical store level.

The user interface sublayer is responsible for displaying data, navigation, and changes of data by the user. The meta-services sublayer ensures data retrieval of meta-information and data transfer to a sublayer of a user interface. The sublayer of services of a logical store level ensures data retrieval from a physical store level and transfers it to a meta-services level. The sublayer of a physical store level receives information from physical stores (databases, directory structures, and so forth) and transmits it to the sublayer of a logical store level

Specific components can be included in different sublayers in different vertical divisions. Thus, meta-services can be considered as an intermediate sublayer from the point of view of exporting data.

5.3.3. Associations of components and their classification. The separate components are of interest from the point of view of system engineering in component architecture. But if one wants to integrate corporate systems, separate components are nevertheless not large enough blocks. From the point of view of integration, it is more expedient to consider associations of components, that is sets of integrated components. It is also possible to consider such associations of components for both vertical and horizontal divisions of the system.

If we consider associations of components for vertical division, it is possible to select components of data management or components of error handling. If we consider associations of components for horizontal division, it is possible to select components of sublayers of a user interface, components of intermediate sublayers, and system components.

Associations of components used to solve problems of a certain class are called frameworks. Frameworks provide interfaces sufficient for the construction of a system with a certain functionality. Frameworks also allow the association of new components to the system and put forward requests for such associations. The creation of high-powered and simultaneously flexible frameworks for the solution of different classes of problems is one of the main goals of corporate systems integration.

6. Interprocess software integration: the RadEco system

6.1. General architecture of RadEco

The general description and list of tasks that are solved by RadEco are outlined in Section 2. It was decided to conduct the system engineering in multicomponent architecture with the use of DCOM as a basis for component integration. The subsystems of RadEco are: Core – system functions that are uniform for the whole system; Admin – administration of users and resources; QueryBuilder – construction of the queries; ThemeMapBuilder – construction of thematic maps; MapBank – cartographic bank; DataReceive – data receiving; BackgroundTechInterface – interface with background technologies; ClassifierProcessing – management of the classifiers; Departmental – administrative functions. Each subsystem may in turn be subdivided into subsystems that then create a vertical division of the system.

6.2. Construction of an information base

The RadEco system integrates different types of complex information related to mitigating the consequences of the Chornobyl disaster. This information can be divided into databases, documents, and maps.

The most important part of the information system is stored in the databases. The data is input into textual documents such as the final reports. The maps are used for data visualization.

Both a relational and an object-oriented database management systems (DBMS) were considered for the system: ultimately we chose a relational DBMS. Certainly, an object-oriented DBMS would have been more useful from the point of view of storing complex data types like the maps, for example. However, the greater part of the basic data of the system arrives in the form of relational tables. The object-oriented possibilities of the relational DBMS are not standardized at present, and that certainly reduces potential possibilities of integration of the system for the new information environment.

The concept of information stores was used to compensate for the lack of a logical information base structure. Information stores are subdivided into SQL stores where database access is carried out by the SQL requests; hierarchical stores with the hierarchical directory structure; file stores; and register stores such as the Windows system register.

The functions of information processing manipulate logical stores, which in turn are mapped on physical stores: databases, directories, and files. This approach ensures a high level of both data program and platform independence.

An additional generalization of the information of the system is the construction of meta-declarations of the information as structured meta-information. Thus, there is a high-level description of the information of the system that facilitates user access and enables the system to include new information sets and link them with existing ones.

6.3. Components and their distribution across the network

A number of components and associations of components have been developed.

All components of the system use System Core interfaces, which are implemented by a system controller component. The interfaces of a system core are intended to be used for error handling, management of the system log, and localization of the system.

An association of interfaces and components of data processing has been developed for the implementation of data operations. Low-level data services: the DBKernel interfaces connect with different physical stores (DBMS, file, and others); the RDODBKernel component connects with databases through RDO (remote data objects); and DirDBKernel components connect with data organized in a file system and implement DBKernel interfaces.

Data services of a logical store level: the DataSource interface connects with different logical stores (DBMS, file, and others); and the DataController component connects with logical stores and implements the DataSource interface. Meta-data services: the MetaObject interface connects with meta-data objects; and the MetaServices component connects meta-data with a user interface and implements the MetaObject interface. To this association of components we can add the interface of request construction (QueryBuilder), which uses meta-services for data retrieval. The association of data access components is shown in Figure 6.

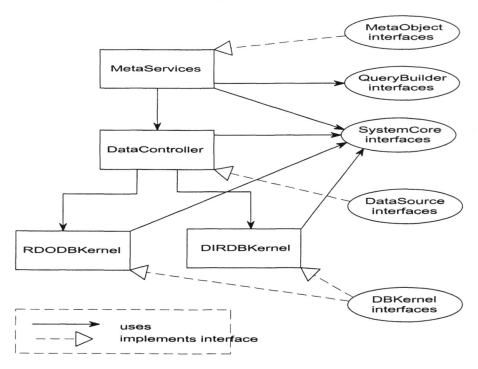

Figure 6. Components of data access.

Other developed component associations are intended to be used for processing cartographic information. The many different cartographic systems have their own facilities of operation with their own map formats on the DCOM/OLE platform. They are MapInfo, MapX, ArcView, MapObjects, and Excel Data Map. To process maps represented in different formats, RadEco uses a uniform map interface.

The association of interfaces and components, and its uses, is listed below: 1) The map interface is the common interface of external access to cartographic objects. 2) The DataMap component implements general classes for communication with maps. 3) The MapInfoMap component implements access to maps through MapInfo and implements the map interface. 4) The MapInfoMap control displays a map with the help of MapInfo. 5) The ToolbarMap control gives the general possibilities of user interface on operations with maps.

The accommodation of corporate system components in the network plays an important role in maintaining the efficiency of the system operation. Theoretically, DCOM allows for the possibility of any component accommodation in any network node. In practice, the accommodation of components in the network should satisfy certain limitations. The components connected to a user interface should be placed on client systems. The components of intermediate sublayers should be located on client or special application servers. The components, which directly ensure data access, should be placed as close as possible to a source of data.

RadEco can work on both a single computer and on a LAN while maintaining simultaneous access by many users. Components of low-level DB data access and map bank access are placed on the application server. All other components are placed on client systems, although some of the components can be distributed across the network and organized so that clients can have access to one copy of each component.

7. Conclusion

Approaches to program integration are circumscribed in Ukraine because of the need to develop an EMIS. Three of the latest trends in IS development set certain limitations on the development of an EMIS. They are

1. the three-tiered (user, business, and data layers) architecture of the client/server applications for ISs in a narrower sense;

2. the three-level (user, subject, and system levels) hierarchy of ISs in a broader sense; and

3. the four-level (operational, application, conceptual, and general meta-levels) IS meta-hierarchy.

Given these limitations and the experience of using the EMIS in Ukraine, it is possible to recommend the following.

On the general meta-level of EMIS, it is necessary to use some kind of object-oriented approach. On the conceptual meta-level, the principal aim is the construction of components. The computer industry has already created several important β-models for the EMIS. We have described some of them: 1) a three-tiered architecture for client/server applications; 2) the intertask software integration system OLE; and 3) the interprocess software integration systems, CORBA/OpenDoc and DCOM/OLE. For countries with a level of development similar to Ukraine's, we recommend Microsoft's DCOM/OLE model.

We would like to conclude with a few words on the possible evolution of the Geo-EMIS/RGIS. Most important is the structural evolution of the system as a whole. Our aim would be to steer the project in the direction of a framework for environmental management systems. The main idea would be to produce an "almost ready" system with interfaces to include new objects and components. The assembly of the system could be executed by development and completion of the system as a whole and by integration of separately developed components into a ready framework, without the necessity of changing other parts of the system. The candidates for inclusion are the diverse model blocks and components, components of decision-making support, and so forth. We trust that the long-term gains and advantages will justify our hopes of completing the project.

References

Bergheim, Geir, Erik Sandersen, and Arne Solvberg (1989). "A Taxonomy of Concepts for the Science of Information Systems," in Falkenberg E.D., Lindgreen P. (eds), *Information System Concepts: An In-depth Analysis* / Proc. of the IFIP TC8/WG 8.1 Working Conf. on Information System Concepts: An In-depth Analysis, Amsterdam *et al.*: Horth-Holland, 269-321.

Iivari, Juhari (1989). "Levels of abstraction as a conceptual framework for an information system," in Falkenberg E.D., P. Lindgreen (eds), *Information System Concepts: An In-depth Analysis* / Proc. of the IFIP TC8/WG 8.1 Working Conf. on Information System Concepts: An In-depth Analysis, Amsterdam et al.: Horth-Holland, 323-352.

Klir, George J. (1985). *Architecture of Systems Problem Solving*, NewYork et. al.: Plenum Press.

Mylopoulos, J. (1991). *Conceptual modelling and Telos*, DKBS-TR-91-3.

Orlafi, Robert, Dan Harkey, and Jeri Edwards (1996). *The Essential Distributed Objects Survival Guide*, New York et. al.: John Wiley and Sons, Inc.

Pershikov, V.I. and V.M. Savinkov (1991). *An explanatory dictionary on computer science*, Moscow: Finance and statistics.

VBE4 (1995). *Building Client/Server Applications with Visual Basic*, Microsoft Corporation.

17 DESIGN OF DECISION SUPPORT SYSTEMS AS FEDERATED INFORMATION SYSTEMS

David J. Abel, Kerry Taylor, Gavin Walker and Graham Williams

1. Introduction

The design of a decision support system (DSS) presents many challenges. While the term "DSS" has been applied to cover a wide range of systems supporting decision-making processes, a DSS is most typically defined as an information system used to support the solving of ill-defined problems (Simon, 1973). Broadly, ill-defined problems are characterized by the feature that both an understanding of the structure of the problem and the recognition of 'good' solutions are developed progressively as the investigation proceeds. Typically, problem-solving requires some creativity and is essentially open-ended in terms of the data needed and the processing to be performed. This means that the usual information systems design approaches that focused on identification of data, the manipulations needed, and workflow have limited applicability.

Rather, DSS development most frequently focuses on assembling a collection of tools such as a database, some models, and some visualization systems. The intent is that then the user can use the DSS as a workbench and choose tools relevant to each step in the investigation. Design, therefore, has a strong systems architecture orientation, where an 'architecture' is defined by the set of functions present, their assignment to components or subsystems, and the interconnections between subsystems.

Three types of components are present: *tools, connectors,* and *facilitators.* *Tools* include the modeling, presentation, and database components. Given the open-ended nature of the target problems for a DSS, the collection of tools is extensive and will grow over the life of the DSS. In many cases, an organization will wish to incorporate pre-existing software systems as components of a DSS because of the cost of developing special-purpose components and the organizational benefits of retaining established and trusted operational systems.

The remaining components are required essentially to overcome problems associated with assembling and using extensive collections of tools. The *connector* components serve to interconnect the tools. A simple example is a utility to take an output file from one modeling system to generate an input file for another. More widely, connectors overcome differences in commands and data between components, at times accommodating the subtle but significant differences in underlying data and process models. The role of connectors is especially important in integrating subsystems developed independently or sourced from different disciplines.

The *facilitator* components manage the complexity of extensive collections of tools, including cases where intelligent selection of tools must be made. An important example of a facilitator is a model management subsystem which provides model substitution and integration services. Essentially, facilitators encapsulate the functions of groups of DSS components.

This chapter considers first the defining characteristics of a DSS to establish the distinctive design requirements. It then reviews the classical statements of DSS architectures to clarify the tool, connector, and facilitator component types. Implementation strategies are discussed in terms of the loosely-coupled systems, modular and federated approaches. A DSS for water quality management in urban water systems, HYDRA, is presented as a case study to demonstrate design problems and a strategy. The chapter concludes by considering some issues in applying the design to DSSs intended for decision-makers rather than domain specialists.

2. DSS design strategies

2.1. Design issues

The term 'decision support system' is widely used with some divergence in its intended meaning (Armstrong and Densham, 1990). For the purposes of this study, we adopt the sharper definition of a DSS as a system supporting the solving of ill-defined problems. A looser definition would be that a DSS is an information system combining a range of technologies, such as database and modeling, for complex planning and management tasks.

An ill-defined problem is essentially one whose formulation changes in the course of the search for a solution. Change can occur in some or all of the structure of the physical process being investigated, the constraints on admissible solutions, the scope of the problem, and the objective of the investigation. Consider, for example, the problem of partitioning a forest region into zones for logging or preservation. An initial objective might be to maximize the income to government from logging licenses, provided that the areas withheld from logging are adequate for the preservation of the species of flora and community types currently in the region. As the investigation proceeds, this simple formulation of the problem might be extended in the following ways:

Structure of the problem. Roads must be built to harvested areas to haul out logs, and because the shape and size of a reserve is important to its continuing viability, planning must consider the routing of roads.

Constraints. An endangered animal species is found in the region. To guard against its extinction, the areas with suitable habitat must not be assigned for logging.

Scope of the problem. It is realized that the soil types and terrain in part of the region are conducive to erosion and consequently siltation in streams. Siltation in turn affects fisheries downstream so that the scope of the

Objectives. It is recognized that the benefit from logging is not restricted to the revenue from logging licenses but should take into account other economic activity such as infrastructure investment. Conversely, the impact on other industries and activities such as tourism should be included in assessing the net value of logging activity.

The likelihood of extensive change suggests that the usual structured approaches to analysis and design of information systems are largely inapplicable. Instead, the emphasis is on establishing a structure for the system that can accommodate change gracefully so that additional facilities can be included as the need for them becomes apparent. These facilities will include the data, information, and knowledge collections needed as background information for the user and as source data for prediction and optimization. Often, change requires extension to database schemas rather than introduction of new database management systems. More significant are changes to the models and modeling systems, where modeling encompasses statistical analysis, prediction, or estimation using simulation, optimization, and inferencing and other knowledge-based techniques. Here, expansion of the scope of the system or recognition of significant changes in problem structure can require new subsystems. For example, a DSS previously concerned with land-use zoning might be extended to consider water-quality issues by adding a hydrological model.

Similarly, as the workflow to carry through an investigation is unpredictable, an important design objective is to allow the user to carry through complex, multistep specifications and evaluations of alternatives flexibly and efficiently. This can be done, for example, by providing transparent and efficient transfer of data between the various components of the system.

An important design decision is whether to build the DSS by integrating existing software systems or by implementing all subsystems from scratch. Integrating existing systems is attractive primarily because it avoids the costs of redeveloping software for sophisticated and complex facilities. Additionally, establishing a predictive model for operational use requires usually a significant investment in acquiring data and calibrating the model. Organizationally, if a modeling system has become established, there will be a degree of confidence in it and adoption of an alternative will be resisted. On the other hand, the integration task itself can be costly and there are inevitable compromises in performance (through data transfer between the systems) and access to the full capabilities of the systems (Fedra, 1993).

2.1. Approaches to a DSS design

The important design goals of extensibility and catering to an unpredictable workflow suggest that an architectural approach is highly desirable. It is not surprising that the literature shows a strong emphasis on the architecture, both in identifying the functional modules required and how these modules should be organized and connected.

In the DSS literature, Bonczek *et al.* (1981) and Dos Santos and Holsapple (1989) identified four principal types of components for a DSS. These are: a language for formulation of requests between the user and the DSS; the output presentation system; a problem processor that embodies all the information-processing capability of the DSS; and the static store of data and knowledge about the problem domain for the DSS. The problem processor itself might be a collection of models of the physical processes and solvers—software systems applying a model to generate predicted data or an optimal solution to a problem. As the sophistication of DSSs grew and as new technologies were recognized as applicable, additional component types were proposed. For example, Chang *et al.* (1994) proposed a component for information and knowledge management and processing. For the purposes of this paper, we will refer to the database, processing, and user interface components collectively as tools.

Sprague and Carlson (1982) applied a closely similar classification of component types in a study of DSS architectures. Labeling the component types as dialogue, modeling, and database, they suggested four DSS structures: the Network, the Bridge, the Sandwich, and the Tower. These architectures apply differing approaches to connection and communication between components. In the Network DSS, for example, a component can be coupled through its component interface to other components of any type. In the Bridge, a single new component type (the bridge) provides a single point of access from the dialogue component to all models and databases.

The Sandwich assumes a single dialogue component accessing multiple modeling components, but with all modeling components using a single database. The Tower extends the Sandwich by accommodating multiple databases accessed through a single data extraction system. Some of the points of the comparative assessment of these architectures by Sprague and Carlson appear to relate to limitations of the technology of the early 1980s. A recurring theme, however, is the difficulty of designing, implementing, and maintaining connections between components that are independently designed.

The topic of coupling autonomous systems has received extensive attention in the field of federated databases and its many specializations (Sheth and Larson, 1990). The fundamental objective here is to enable interoperability between heterogeneous database systems. Heterogeneity has many dimensions, including the underlying data models, different representations of the same physical objects, and management responsibility. Sheth and Larson identify four specialist functions—the *filter*, the *transformer*, the *constructor,* and the *accessor*—as the minimal set needed to couple heterogeneous database systems.

Abel *et al.* (1994a) proposed that these four functions were applicable to the broader task of systems integration of dialogue (interface), modeling, and database systems, with some minor redefinition. In terms of use in a DSS, the four coupling functions (with some variation to the definitions of Sheth and Larson) can be defined as follows.

The *filter* function tests whether an operation is valid in terms of the supported functions of a component. For example, a filter for a solver using linear programing as its solution technique might check that the form of objective function and constraints is consistent with the linear programing model.

The *transformer* function translates an object from the underlying data model of one component to the data model of another. For example, a geographical information system (GIS) might represent a stream system as a connected set of lines, while a hydrological simulation system represents the streams as arcs of a network using a graph model. A representative transformation function would take the GIS representation of the streams and develop the equivalent network representation.

The *constructor* function translates between command languages and schemas for components. First, it can take a statement of an operation on an object, expressed in terms of the command language and objects in the schema of one component, and generate commands to another component for the equivalent operations on equivalent objects. For example, to maintain the representations of streams in the GIS and hydrological system in step, an update to the streams in the GIS would need to be propagated to the hydrological system. A *constructor* function would generate an update command on the network representation. Additionally, a *constructor* can assemble data provided by two components. For example, it might combine the location of a sampling point from the GIS and historical records on water quality from a database to generate a more complete record for the sampling point.

The *accessor* function accepts a command or a set of commands to a component for execution, and checks success or failure of the operation.

Abel *et al.* (1994b) considered distribution of the connector functions in a system architecture, particularly considering *network* and *star* configurations. The network configuration is characterized by direct component-to-component connections. In most cases, the network configuration requires many simple connector functions that are customized to the linked components. The star configuration centralizes the connector functions in a hub. The major advantage is extensibility for new components and resilience to change in the functions and implementation of components, but at the cost of greater design and implementation complexity to generalize the connector functions.

Implementation of a DSS, however, must also cater to the complexity of an extensive set of tools. Two forms of complexity are especially important. The first deals with extensive collections of data, particularly where the data are stored in a number of databases. Navigation of the databases to find pertinent data becomes difficult and might be a barrier to the full exploitation of the available data, information, and knowledge. This problem is likely to become especially acute as DSS design exploits the collections now published on the Internet (Abel *et al.* 1997).

The second form of complexity deals with choosing models and solvers from extensive collections. In this case, the difficulty is more likely to lie in the matching of a specific formulation of a problem, or a part of a problem, to a model and solver. Each model and solver has a certain limited domain of application. At one level, a solver is based on a certain representation of a type of problem (such as prediction or optimization) and a computational process. It is then restricted in the types of problems it can handle. For example, a hydrological simulation system might be based on equations of flow which provide accurate prediction only within certain ranges of flow rates. At another level, calibration of a model (estimating some parameters) might be performed within certain ranges for key variables. Use of the model outside those ranges might then lead to unreliable results.

We propose a new class of component in addition to the tools and connectors. We define facilitator components to be those providing services to accommodate complexity. Equivalently, a facilitator component acts as an agent of the user or a tool to locate functions or to choose between functions. Two forms of facilitators can be recognized in the literature. In DSSs, some forms of model management systems can be seen as facilitators. Specifically, model management systems able to perform model substitution, that is, selection of a model by the problem context, can be classed as facilitators. Systems acting as model generators (producing a model for input to a specific solver) and systems offering simple model integration (coupling models) are more accurately classed as *constructors*. In the database field, query optimization, with its services to determine an optimal plan of database operations, can be recognized as a facilitator function.

3.   The HYDRA system

3.1. Systems motivations

The federated systems approach is being explored in the HYDRA Project, a collaboration of CSIRO Land and Water and CSIRO Mathematical and Information Sciences (Abel *et al*. 1993). The strategic goal of the project is an enhanced ability to build DSSs by developing a methodology that can be applied reproducibly and reliably and to develop a core set of tools supporting the methodology. The research strategy is to apply federated systems concepts to integrate hydrological modeling systems, databases, and spatial information systems into a DSS of limited scope. The project is addressing the objectives incrementally. In the first year, the feasibility of equipping existing modeling systems with a maplike interface was investigated as a first experiment in coupling systems with significantly different underlying data and process models. In the second year, a federated systems approach to coupling subsystems was tried. Now, in the third year, the emphasis is on using this acquired experience to devise and test a design and implementation methodology that simplifies the development of the connector modules.

The research is concentrating on building an experimental DSS for water quality management in the Hawkesbury-Nepean catchment as the application focus. This region includes the western suburbs of Sydney, Australia and is subject to increasing urbanization. The planning issues for water quality center on understanding the effects of modified land use (replacement of horticultural and pastoral uses by residential use) and of assessing the effects of changes in the sewage treatment network to maintain and improve water quality. Changes in sewage treatment include commissioning new sewage treatment plants (with a choice of processing technologies), decommissioning existing plants, and using other mechanisms such as wetlands for treating water before discharging it into the rivers.

The Sydney Water Corporation, the responsible water agency for the region, has extensive experience in modeling the region. It uses several modeling systems (solvers), each with particular domains of application. For example, the SALMON-Q solver can model tidal flows in channels but not surface flows across land, while the HSPF solver can model flows across land and in channels but does not support tidal flow. The corporations's modeling policy is to divide the entire catchment into subcatchments and major rivers and to select the most appropriate solver for each subcatchment. Broadly, a subcatchment is a region draining to a single exit point. In part, this policy reflects the high cost of calibrating a model for a geographic region. This way, the entire catchment can be represented as a tree of subcatchment regions, each matching a model (Figure 1). In this context, a model describes the objects present in the subcatchment, the processes linking those objects, and the calibration parameters estimating lumped or unobservable physical properties.

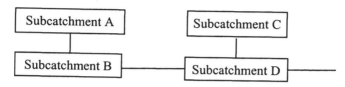

Figure 1. The catchment represented as a tree of regions.

The project to date has focused on the usability of the system and coupling the current hydrological modeling systems. Usability has been addressed principally by providing a graphical user interface (GUI) which organizes and presents the system facilities in ways familiar to the user. That is, a conscious design goal is to shield the user from the underlying modeling systems. One element of the strategy is to structure the GUI around specification of a scenario and examination of the outcomes of the scenario. A scenario comprises a set of changes in land use and sewage treatment plants and a limited set of external factors such as rainfall patterns.

Another element of the GUI is the use of visual formalisms which apply familiar presentations of data. The primary types of display are a map of the region and charts. These are working surfaces rather than simple outputs. For example, the map includes both the objects able to be manipulated and some navigational data to allow the user spatial orientation. A change in land use or a new sewage treatment plant is specified by editing the map with the object-oriented approach of the EDSS (Abel *et al.*, 1991; 1992). Under this paradigm, clicking on an object in the map will cause the object to present a menu of operations which can be performed on it. For example, as illustrated in Figure 2, clicking on a sewage treatment plant will generate a menu showing the current characteristics of the sewage treatment plant and allow the input variations by changing the values for various attributes. These attributes include, for example, the nitrogen levels in the discharge from the plant and the plant's maximum intake volume.

3.2. Connector and facilitator functions in HYDRA

The starting point for design of the current form of HYDRA was two solvers—HSPF and SALMON-Q, a GUI structured around a maplike presentation of the geographic region and chart displays of predicted data. A strong design intent was to simplify the introduction of further functions, both in additional data and in additional modeling systems. This has been supported by a strongly modular architecture and by the specification of generic interfaces between layers of the architecture

The connector functions in HYDRA provide the mapping between data and process models of the subsystems, manage data, and command transfer between subsystems. In general, there is not a one-to-one mapping between object types. For example, the GUI model includes a river object which is spatially a line. The matching object type in HSPF and SALMON-Q is a reach, essentially a part of a stream which is uniform in its hydrological characteristics. Reaches are not described spatially. Rather, in HSPF and SALMON-Q, reaches are defined as elements of a hydrological network model, with specialized nodes for point and diffuse sources, reach exits, and so on.

The primary facilitator function in HYDRA determines the models needed to be run to predict the outcomes for a scenario. HSPF and SALMON-Q can be computationally very expensive, with execution time for a subcatchment model in the order of hours. Where a scenario deals with a few subcatchments, it would be wasteful to rerun all subcatchment models. HYDRA, therefore, includes a re-evaluation minimization function which tests for subcatchments touched by a scenario. Strictly, the nature of the physical system implies that all subcatchments and rivers downstream from a modified subcatchment will be affected. The user must designate the subcatchments for which predicted data will be examined.

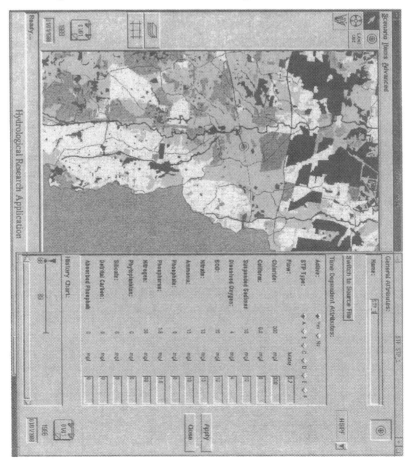

3.3. HYDRA architecture

In a top-level view, the HYDRA architecture (Figure 3) has two major subsystems: the specification subsystem consisting of the GUI and the system manager, and the prediction subsystem with the model manager and the solvers. This decomposition follows the concepts of the external data type service of Schek and Wolf (1993). Importantly, it confines the scope of interactions between the components of the system. In outline, scenario specification and evaluation of outcomes involve only the GUI and the system manager. The model manager and solvers are invoked only to predict the outcomes of a scenario once it has been defined fully. Also importantly, this approach segregates the data and process models used. The specification subsystem applies a single data and process model, designed to match the user's conceptual model of realworld objects such as rivers, sewage treatment plants, wetlands, and so on. Each solver has its own data and process model and operations within its solver driver, and combinations are restricted to that particular data and process model. Translation between these models, such as mapping between a sewage treatment plant in the specification subsystem and a point source in HSPF, is confined to the model manager.

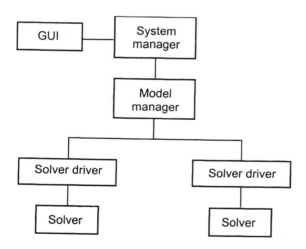

Figure 3. HYDRA overview architecture.

We introduce the subsystems first and then consider in more detail the model manager as the key site for connector and facilitator functions.

3.3.1. Subsystem overview. The GUI performs its usual role of managing the interaction with the user and the presentation of data. It views the system manager as its sole source of data, including predicted data. It is presently implemented with the Tcl/Tk tool, chiefly for ease of implementation.

The system manager performs three major roles:

- Crucially, it provides the GUI with equivalent services to a database system by maintaining the current scenario and providing navigational and predicted data. To the GUI it is accessed as if it is a database holding both prestored data (data manipulated during specification of the scenario) and data generated by running models. Its command language (used by the GUI) consists of select (fetch), append (create a new instance of an object), delete (destroy an object instance), and update (change values for the properties of an attribute). The command set also includes an evaluate operator to trigger materialization of predicted data.

- It provides *filter* operations by inspecting commands from the GUI and applying constraint checks on the validity of specifications for the elements of a scenario. To do this, it tentatively diagnoses the referenced object in a command by the model (the solver) which will ultimately deal with the object, through reference to a statement of the domain of application of each model. It then tests the constraints imposed by the model. For example, the HSPF models used by the Sydney Water Corporation do not include the full set of nutrients that can be modeled with SALMON-Q. If a sewage treatment plant has some of the excluded nutrients in its outflow and it is in a sub-catchment modeled by HSPF, the system manager will flag an error to the GUI.

- Finally, it provides *constructor* and *accessor* functions by initiating the materialization of predicted data (through the model manager) and by fetching predicted data.

Scenarios are maintained as sets of versions of objects that can be manipulated. When requesting materialization of predicted data, the system manager passes a full statement of the scenario to the model manager. It also informs the model manager of the catchments touched during specification of the scenario.

The model manager (Figure 4) performs *transformer*, *constructor* and *accessor* functions.

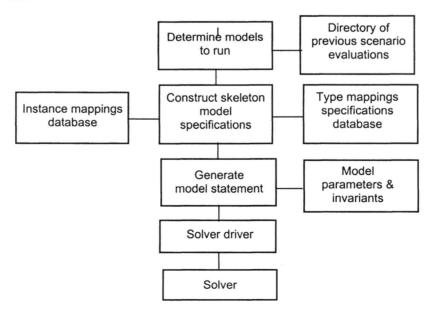

Figure 4. Model manager architecture.

The model manager determines the models to be run by the modeling systems and then generates commands to the solver drivers for execution. These commands are expressed as a generic set of operations (update, append, and delete) on the types of objects supported by the various solver drivers.

Each solver (modeling system) is coupled to the system manager and the model manager by a solver driver, which translates the commands from these subsystems into the native command set of its solver and executes the operation. The command set for a solver driver, in addition to those commands required by the model manager, includes the command select for retrieval of predicted data by the System Manager and for reporting information on the availability of predicted data from previous model runs.

3.3.2. The connector and facilitator functions.

The model manager combines the major *constructor*, *accessor* and *transformer* functions in HYDRA. The flow of control is as follows:

- The specification subsystem presents a scenario as a set of objects defined according to that subsystem's data model. It also provides a statement of the subcatchments touched directly by this scenario (that is, subcatchments with new or modified objects).

- The model manager determines the models which must be run to provide the requested predicted data. This centers on determining the subcatchments indirectly affected by the scenario. For example, in Figure 1, changes to sewage treatment plants in Subcatchment A will affect the water quality and quantity at the exit for Subcatchment A, which is the entry for Subcatchment B. Predicting data in Subcatchment D will then require running the models for Subcatchments A, B, and D. More interestingly, the scenario might include changes in Subcatchments A and B, but the changes to Subcatchment A might be identical to those of a previous scenario. It is then recognized that the predicted data from Subcatchment A can be re-used here.

- In order to run each model, the model manager constructs a command set corresponding to the objects in the relevant subcatchment. This involves principally mapping from the specifications subsystem schema to the schema of the selected solver. The model manager refers to two databases. The type mappings specifications database describes type-to-type translations. For example, it associates a specifications subsystem sewage treatment plant type with an HSPF point source type. It also stores the associations between objects in the solver database and the specifications subsystem. For example, the model manager maintains the knowledge that the sewage treatment plant called 'New plant' in the specification subsystem is referred to in the HSPF model for the South Creek Catchment as the point source 21 in reach 5.

- Finally, the model manager passes the command sets to the solver drivers. The solver drivers refer to a database of other data and information needed to construct a command set in the form required by its solver. This other data and information include calibration parameters and variables that cannot be varied within a scenario (such as soil types present and nutrient loads generated by different types of land use).

In review, this design supports the general objectives of the project primarily by separating the *transformer*, *accessor*, and *constructor* functions into distinct reducing the initial implementation costs and easing the introduction of additional solvers. Adoption of a common data model for all solvers (as data sources which have an append, select, update and delete command set) means that a uniform treatment of solvers in the model manager is possible. In turn, this enables a standardized treatment of mappings between data models.

The design has been realized in a modular software architecture that aims for maximum reusability of components, especially solver drivers and the model manager. This is enabled by a separation of the syntactic concerns of the precise specification of mappings and model commands from the semantic concerns that identify the relationships between concepts represented in different tools under varying problem-specific circumstances. These concerns align conveniently with the typical separation of skills between software engineering experts and domain experts such as hydrologists and then support a cooperative DSS development process.

For more information on implementation the reader is referred to Taylor *et al.* (1999).

4. Conclusions

This paper has shown that the design of DSSs has unusual and challenging requirements so that an emphasis on system architecture is highly desirable. A design methodology based on the explicit inclusion of connector and facilitator components has been presented and demonstrated through the HYDRA system. Previous and current experience suggests that the approach is effective, and the system is in operational use at Sydney Water. However, a comprehensive evaluation of the approach would require applying it to the development of a DSS in an application domain other than the hydrological one for which it was developed.

The HYDRA system shows that the use of connector and facilitator functions shields the user from the specific solvers used. Clearly, this is significant in making the modeling technology more accessible to disciplinary groups such as planners and managers who may not be experienced in hydrological modeling, for example. It offers them the opportunity to make better-informed decisions. The danger is that the models might be used inappropriately or that predicted data might be misinterpreted. The builder of a DSS then must assume the responsibilities now held by modelers. There appears to be a strong case for the incorporation of strong *filter* mechanisms into a DSS such as HYDRA on the basis of a formal description of the application domains for solvers and models. Devising these mechanisms might well be a challenge for both modelers and DSS designers.

The task of building a DSS has been approached as one of integrating existing systems. In this context, the focus on connecting those systems as components is easily defended. There is, however, the alternative of building a DSS from scratch. It is plausible that this will become more readily achievable as technological advances provide tools to compete with existing specialist systems such as GIS. For example, the spatial database capabilities of a GIS are now available through extended relational database systems (Abel, 1989) and maplike displays can be generated with a number of interface-building systems. A GIS is no longer quirement for extensibility will still be present. A connector/facilitator approach to design will then remain useful as a means of enforcing a firm separation of components to support change and extension.

Acknowledgements

The HYDRA Project is partly sponsored by the Sydney Water Corporation and the CSIRO Urban Water Systems Research Priorities Project. The current project team includes Richard Davis, Sue Cuddy, Trevor Farley, and Michael Reed from CSIRO Land and Water, as well as Amir Deen and S. Maheswaran from Sydney Water. Previous team members from CSIRO Mathematical and Information Sciences include Philip Kilby, Paul Morris, Donald Syme, Tai Tran, Michael Kearney, and Zhexue Huang, and from CSIRO Land and Water, Dan Zhou.

References

Abel, D. J. (1989). "SIRO-DBMS: A Database Tool-Kit for Geographical Information Systems," *International Journal of Geographical Information Systems* 3(2), 103-116.

Abel, D. J., R. G. Ackland, *et al.* (1991). "EDSS: A Prototype for the Next Generation of Spatial Information Systems," in *Proceedings of AURISA '91*, Wellington, New Zealand, 19-22 November 1991 (Wellington: Australasian Urban and Regional Information Systems Association), 73-84.

Abel, D. J., S. K. Yap, *et al.* (1992). "Support in Spatial Information Systems for Unstructured Problem-Solving," in *Proceedings: 5th International Symposium on Spatial Data Handling*, Charleston, South Carolina, USA, 3-7 August 1992, 434-443.

Abel, D. J., P. J. Kilby, *et al.* (1993). "The Spatial Water Modelling Program: a Case Study in Integrated Spatial Information Systems," in *Proceedings of the Twenty-second AURISA Conference*, Adelaide, Australia, November 1993, 406-411.

Abel, D. J., P. J. Kilby and J. R. Davis (1994a). "The Systems Integration Problem," *International Journal of Geographical Information Systems* 8(1), 1-12.

Abel, D.J., P. J. Kilby and M. A. Cameron, (1994b). "A Federated Systems Approach to Design of Spatial Decision Support Systems," in *Proceedings of Sixth International Symposium on Spatial Data Handling*, Edinburgh, UK, September 1994, 46-59.

Abel, D.J., V. J. Gaede K. L. Taylor and X. Zhou (1997). "SMART: Towards Spatial Internet Marketplaces," Technical Report 97/28 CSIRO Mathematical and Information Sciences, Canberra, ACT Australia.

Armstrong, M. P. and P. J. Densham (1990). "Database Organisation Strategies for Spatial Decision Support Systems," *International Journal of Geographical Information Systems* 4(1), 3-20.

Bonczek, R. H., C. W. Holsapple and A. B. Whinston (1981). *Foundations of Decision Support Systems*. Academic Press, New York, USA.

Chang, A. M., C. W. Holsapple and A. B. Whinston (1994). "A Hyperknowledge Framework of Decision Support Systems," *Information Processing and Management* 30, 473-498.

Dos Santos, B. L. and C. W. Holsapple (1989). "A Framework for Designing Adaptive DSS Interfaces," *Decision Support Systems* 5(1989), 1-11.

Fedra, K. (1993). "GIS and Environmental Modelling," in Goodchild, M., Parks, B. and Steyaert, L., (ed.), *Integrating GIS and Environmental Models*. Oxford University Press, NY, USA.

Schek, H.-J. and A. Wolf (1993). "From Extensible Databases to Interoperability between Multiple Databases and GIS Applications," in *Advances in Spatial Databases*, Lecture

Sheth, A. P. and J. A. Larson (1990). "Federated Database Systems for Managing Distrib-
    uted, Heterogeneous and Autonomous Databases", *ACM Computing Surveys* 22 (3),
    183-236.
Simon, H.A. (1973). "The Structure of Ill-defined Problems," *Artificial Intelligence* 4,
    181-201.
Sprague, R. H. and E. D. Carlson (1982). *Building Effective Decision Support Systems*.
    Prentice-Hall, New Jersey.
Taylor, K.L., G. Walker and D. J. Abel (1999). "A Framework for Model Integration for
    Spatial Decision Support Systems," *International Journal of Geographical Information
    Science*.

18 KNOWLEDGE DISCOVERY IN DATABASES AND DECISION SUPPORT

Anantha Mahadevan, Kumudini Ponnudurai,
Gregory E. Kersten and Roland Thomas

1. Introduction

Two types of decision support systems (DSS) are discussed in Chapter 2: model-oriented and data-oriented. The assumption underlying model-oriented support is that accurate models of the decision problem are available or can be constructed prior to decision making. This approach includes optimization, simulation, decision analytic and statistical models. Various statistical methodologies can be used both to construct models and to estimate their parameters. Traditionally, these methods have been applied to relatively small data sets; the resulting models were used to verify given hypotheses, identify relationships among variables, and formulate predictions and scenarios.

Statistical methods require significant expertise in their application and interpretation of results. These methods are difficult to use when there is little explicit knowledge about the issues and when the problems are ill-defined but described by large or very large amount of data. Therefore, these methods are normally used by analysts and researchers rather then directly by decision makers. While one may suspect that valuable knowledge is buried in the data, it was not until recently, that tools for knowledge extraction and presentation in an easily readable form have became available.

Three types of developments made data-oriented DSSs viable and effective. Artificial intelligence (AI) adopted many statistical methods and embedded them in systems that efficiently search for patterns, derive rules, classify data according to some higher level concepts, and construct models from very large databases (Briscoe and Caelli 1996; Kohavi 1998). AI researchers have also developed methods to deal with such databases, including neural networks, genetic algorithms, and rough sets (Aasheim and Solheim 1996; Bigus 1996; Pawlak, 1995). The second development involves technologies for the construction of multidimensional databases and data warehouses that provide raw materials used by AI systems (Inmon, 1996). Finally, data manipulation and visualization techniques allowed the user search and view databases from different perspectives and to display models in forms that could be understood by decision makers (Fayyad, 1996).

The process of deriving or extracting knowledge from data is known as *knowledge discovery in databases* (KDD). It comprises several distinct steps one of them being *data mining* which specifically deals with model construction. Data mining is used to make generalizations from a given set of data. *Generalization* involves the construction of a model from the data that can be used to describe the population or predict the behavior of its members. While generalization is also one of the key issues in statistics, there is a significant difference between the two. In statistics, the issue of generalization involves the definition of a population to which to generalize. This brings in issues of random sampling. Without careful attention to the sampling method one cannot guarantee statistical generalizability of models.

In data mining, there is no attempt to sample randomly from some identifiable population. In fact, during the first four steps of KDD, the database is viewed more as a population than as a sample. In the later steps, however, the results of the data mining and modeling processes are extended to units outside of the original database. One can regard this as akin to the generalization of mathematical models of processes that are typical of the physical and engineering sciences, where the generalization is non-statistical in nature. Alternatively, one can identify this process with the statistical notion of a finite population generated from a super-population (see, for example, Skinner et al., 1989, p. 14).

In the data mining context, the finite population under study (that is, the database), would be regarded as a random realization from some hypothetical super-population. As an example, this would be a natural conceptualization for a database consisting of scanned data records from stores in a particular chain, for a single month. A key assumption for the extrapolation from such a data mining exercise to future time periods would be that the processes studied, or discovered, will not change structurally over time (an implicit assumption in all economic forecasting). Either way, the generalization of the results relies on two key assumptions that underlie data mining methods:

- a model can be approximated with some relatively simple computational models, at a certain level of precision, and

- the data available in the target database is sufficiently representative of the processes of interest that the generalization is warranted.

It is clear that these assumptions are not very precise. This may be the reason why, in statistics, data mining (also known as data dredging) has traditionally been referred to as a sloppy exploratory analysis with no prior specification of hypotheses (Glymour, Madigan et al., 1997). Recently, however, data mining has become popular because it addresses the needs of business and other organizations in ways that other information systems do not address. Ease of use and intuitive interfaces coupled with often surprising and interesting results have led to a general, if sometimes uncritical, acceptance of data mining.

Several data mining methods based on rule induction are discussed and compared in Chapter 13. In this chapter we discuss the KDD process from the decision making and decision support perspectives. The benefits and shortcomings of data mining are presented with a focus on the potential contributions for decision support.

The KDD process is outlined in Section 2. To illustrate the model construction activity, we have used a popular and accessible decision tree method. In Section 3, the method is introduced and the resulting model discussed and compared with experts' knowledge and with another model obtained using a rule induction method. The results of the comparison give grounds for three experiments discussed in Section 4. A database with over 1,400 records has been used to formulate three descriptive models. Statistical methods are then used to evaluate these models and the results are presented. The potential of KDD for decision support in sustainable development is discussed in Section 4 together with a brief discussion of the future developments in this area.

The difficulties one may encounter in the effective use of the KDD and data mining software do not diminish its usefulness. There have already been many successful applications, some of which are presented in the Appendix.

2. Knowledge discovery

Knowledge discovery in databases (KDD) is the process of identifying significant patterns in data or deriving compact, abstract models from data. It is used to obtain nontrivial information, that is, information that requires search and inferencing rather than straightforward calculation of some predefined quantities (e.g., totals and averages). KDD has evolved from the intersection of machine learning, pattern recognition, statistics, data visualization and high performance computing. These areas of study have been integrated to address problems of interpretation of large databases at a higher-level than provided by management information systems, query facilities and on-line analytic processing.

The KDD process is both interactive and iterative. It includes issues of problem and scope definition, element selection, data standardization, model elicitation (i.e. data mining), and use of recurring patterns in models for developing strategies or procedures (Fayyad 1996; Brachman and Anand 1996). These steps are depicted in Figure 1.

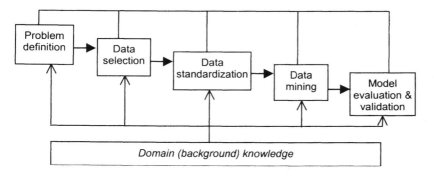

Figure 1. Knowledge discovery process.

Two important features of the KDD process are indicated in Figure 1. First, the process is iterative with possible loops between any two steps. Second, it relies heavily on the knowledge pertinent to the problem, that is, domain (background) knowledge plays a significant role in each step of the process. The process is used to extrapolate models from databases, while incorporating background knowledge throughout.

Domain experts, technology specialists, database documentation, and the end users can provide the domain knowledge necessary for this process. Knowledge about the issues under consideration is used to outline the initial problem and respective data elements. Knowledge is also required for the selection of data relevant to the problem and its standardization. Extrapolated models need to be verified, either through confirmatory statistics or background knowledge.

2.1 Problem definition

The first step in the KDD process is to obtain an understanding of the application domain, specify the expected outcomes of the process (user goals and expectations) and define the domain (background) knowledge that may be needed.

Decision support based on model-oriented DSS is restricted to variables and models present in the system. The task of problem definition is not affected by elements outside of the stipulated model. In the absence of a model the initial KDD step requires defining both the scope of the study and problem. This is performed in an iterative manner; the user undertakes subsequent stages several times before arriving at a satisfactory problem definition.

The prerequisite for both model- and data-oriented decision support is the identification of the decision problem. This includes specification of the decision maker's objectives and decision variables (see Chapter 2). DSSs are used to determine the variable values so that the objectives can be achieved.

The objectives of the decision maker guide the support process and provide goals for the KDD process. In other words, the KDD problem definition includes the specification of the goal of the process from the decision maker's point of view. Two types of goals can be distinguished: *verification* and *discovery* (Fayyad et al., 1996).

The verification goal requires prior specification of hypotheses by the decision maker; contrary to some observations, data mining is not necessarily "hypothesis free" (Glymor, 1997). The hypotheses one needs to formulate include the existence of the relationship between the goal (dependent) variable and other variables (independent) in the database, the type of relationship (for example, causal or functional) and the identification and exclusion of variables that may be disregarded in the verification process. From the decision support viewpoint the goal is to determine the appropriateness of existing DSS models in light of the existing data.

The discovery goal assumes that the process autonomously searches for relationships that define concepts and models. Discovery can be used to obtain descriptive models of an object or process or to formulate predictions of future behavior of some entity. Further, this step involves analyzing the problem to assess if it is appropriate for data mining techniques. If it is, then an assessment is made concerning the availability of data and the type of techniques to be employed. Discovery, for the purpose of decision support, is oriented toward the search for new models that can be used in a DSS.

These two goals are not mutually exclusive; verification may lead to discovery and discovery requires verification. Their role is to obtain the initial focus for the study whose ultimate goal is to enhance DSSs and provide decision makers with relevant information and knowledge.

2.2 Variable and data selection, and preprocessing

The process of selecting appropriate elements (variables and samples) from the source database (assumed to be the population) relates to restricting the scope of decision support. In this step of the KDD process the variables and the data on which data mining is to be performed are chosen.

Domain knowledge plays a significant role in this step. Dealing with large data sets makes it relatively easy to obtain variables that are highly correlated or appear to be strong predictors of dependent variables. These variables, however, may not be casually related, for example, the observed correlation may be due to chance, or it may be a spurious correlation due to the concomitant variation with some undetermined causal variable.

As an example of another non-causal relationship, consider the problem of plant disease diagnosis. The database may contain variables measuring the symptoms, diseases and treatments of the observed plants. Assume further that a treatment is applied only after a diagnosis has been made. Treatments are highly correlated with diagnoses but, obviously, the causal effect is "from diagnosis to treatment" and not vice versa. If the treatment variable is not removed from the dataset then symptoms need not be considered in order to determine the disease. It may be thus sufficient to select treatments to "predict" a disease. Obviously such a model is of little use to a decision maker who wants to be able to diagnose diseases and only after that decide on a treatment.

Removal of confounding variables and variables that describe implications of goal (dependent) variables is an important component of data selection. It is no less important, however, to retain variables that appear to be insignificant, or that might seem to have little in common with the decision support tasks and goal variables. While this may lead to a discovery of accidental relationships or spurious correlations, it may also provide interesting and previously unknown results.

One of the objectives of the KDD process is to discover new knowledge and not necessarily to reinforce the decision maker's conviction. Data mining methods allow large amounts of data to be processed to identify variables that otherwise might have been ignored. This is one of the reasons for the feedback loops indicated in Figure 1. Preprocessing and data visualization help the decision maker remove variables that should not be considered and keep those that may have some prescriptive or descriptive impact. This is part of the initial exploratory analysis used to provide a preliminary understanding of data.

We mentioned earlier that in KDD the database is treated as a population. Data selection may involve sampling from the database to obtain a sample on which data mining can be performed. Models are generated (estimated, in statistical parlance) from the sample, known as the training set. Their validity is subsequently verified using the remaining (holdout) records from the database.

Data selected for further processing may be incomplete and require cleaning and standardization. Decisions must be made about what to do with missing data fields. The task of element standardization is influenced by the choice of data mining methods. Further, data transformation may also be required. Processing of transformed data may lead to models with more explanatory power than processing the source data (John, 1997).

2.3 Data mining

The data mining step involves the selection and application of methods to construct models from data. The methods are used to formulate summaries and to identify significant structures and relationships present in data. Data mining is used to develop models that are understandable and that can be intuitively explained. Statistical significance and the validity of model generalization beyond the database are not considered in this step.

Some authors consider query operations, on-line analytic processing, visualization and statistics as data-mining methods (Simoudis, 1996). Though these information processing methods are used in data mining, they do not meet the main objective of data mining which is exploratory model construction.

Three types of models are distinguished in decision science: normative, descriptive, and prescriptive (Bell et al., 1988). *Normative models* provide guidelines for right actions and involve the formulation of principles of alternative evaluation and choice, proposed as rules that decision makers ought to follow. Since KDD is involved with historical data, it may provide information on how decisions were made and what their implications were rather than provide information about principles.

Descriptive models are used to represent actual decision processes and the results of decision implementation. They are formulated on the basis of the observation of decision makers' activities and the analysis of their decisions. One method of constructing such models is through the formulation of hypotheses, and their verification with statistical methods. Data mining, a data-driven approach, can also be used to construct descriptive models.

Prescriptive models focus on providing decision makers with support to make decisions more effectively and to think more constructively. They are used to facilitate the analysis and assessment of information, and to manage decision problem complexity. Prescriptive models can be obtained through the comparison of the expected and achieved outcomes, for example, from the comparison of plant treatments and their results. Data mining methods can be, we believe, used for this purpose.

There are other types of models that are generally not considered in decision science because they do not directly involve the generation of alternatives, their comparison and selection. These are *predictive models* used to determine the future behavior of some entity. These models are important for decision support because they provide information about processes and entities that are outside of the decision maker's control. They are also used to determine parameter values for descriptive and

prescriptive models. Data mining has often been used for the construction of predictive models (Fayyad, et al. 1996).

In summary, we distinguish three data mining operations useful for decision support:

1. Predictive modeling - used to generate a model that can be used to determine the
   future behavior of some entity.

2. Descriptive modeling - used to generate models that provide high-level summary
   information and that explain relationships between variables in the database.

3. Prescriptive modeling - used to construct models for the determination of decision variable values required to meet objectives specified by the decision maker.

The three types of models can be constructed by the application of one or more
data mining tasks. The tasks considered by data mining include:

1. predicting the class to which an object belongs;

2. predicting the dependent variable value given values of independent variables;

3. formulating and describing clusters of similar objects ;

4. describing a group of objects;

5. finding and describing relationships and associations among variables;

6. identifying deviations and changes in a distribution or behavior of objects;

7. identifying variables that control the values of other variables.

The first two tasks can be used for the development of predictive models, tasks three,
four and five for the development of descriptive models and tasks six and seven for
construction of prescriptive models. Note, however, that, some tasks may serve more
than one function. For example, tasks used for description can also be used to specify
elements of predictive and prescriptive models. That is, the assignment of tasks to
models is indicative rather than exclusive.

Data mining methods have been developed to perform specific tasks. These methods include association rules, cluster analysis, Bayesian classification, decision trees,
rough sets, neural networks, and genetic algorithms (Kohavi, Becker et al. 1998;
Pawlak, 1995; Srikant and Agrawal 1998; Weiss 1998).

2.4 Model evaluation and interpretation

Effective use of the KDD process requires an appraisal and confirmation of the
extracted models. This may be conducted in the form of traditional model testing and
through confirmation obtained from the related literature and from domain experts
(Glymour et al. 1997). Measures of "interestingness", or measures of intuitive explanation, are often used to sift significant patterns or rules from other output. These
measures include strength of rules (for example, number of rule precedents), statistical
indicators (for example, goodness of fit) or simplicity values (Fayyad 1996; Mienko
et al., 1996). Data visualization is also used to facilitate the evaluation of the significance of extracted information and models.

To use KDD results effectively in decision support, interesting and valid models need to be incorporated into existing knowledge. This may be difficult due to structural differences between the proposed models and existing knowledge. The latter may also be imprecise. The representation embodied in models may differ from the experts' knowledge, requiring the resolution of syntactical conflicts. The meaning of terms and concepts in the database, and those obtained from data mining, may conflict with those of the experts, requiring the resolution of semantical conflicts. Direct involvement of domain experts and decision makers is, therefore, essential in this step. Their verification and interpretations are necessary to convert models into knowledge and to incorporate them in DSSs.

The activities conducted in this step often lead to the repetition of earlier steps of the KDD process.

3. Mining data with decision tree methods

Decision tree methods allow the determination of relationships between sets of variables. They are used to select independent variables to partition data recursively. Classification methods are used for categorical dependent variables and regression methods are used for continuous dependent variables (Breiman, Friedman et al. 1984).

Several approaches to decision tree construction have been proposed including classification and regression trees (CART), chi-square automatic interaction detection (CHAID), and C4.5 (Briscoe, 1996). These approaches differ in the significance measurements used to select independent variables. The CART technique partitions data into two mutually exclusive subsets such that at least one of the resulting subsets has lower dispersion than the previous set of data. CHAID uses chi-square tests to calculate the association between the dependent variable and a chosen descriptive variable. The C4.5 method partitions data into mutually exclusive subsets such that each of the resulting subsets has lower entropy (dispersion) than the previous set of data.

Decision tree methods produce output that is easy to read and interpret. It can be represented as trees or rules; the latter format is especially useful for the development of knowledge-based DSSs. These methods have been implemented in many data mining software packages, for example AnswerTree (SPSS, 1998) and MineSet (SGI, 1998). They can be used for the construction of predictive, normative and descriptive models.

We are interested in KDD and data mining inasmuch they can be used for decision support. Our objective is not to provide an exhaustive analysis of a particular method, nor to present a complex application. Rather, we attempt to outline the opportunities of the application of KDD to obtain models for decision support. We will also discuss some remedies for the difficulties that are inevitably encountered. For this purpose we have selected a popular and easy to understand data mining method and applied it to a small database.

In this section we use the CHAID method to formulate a predictive model for a soybean disease problem; in Section 4 the same method is used to formulate three descriptive models.

3.1 Soybean disease problem

One dimension of expertise is the ability to assess the relative importance and informativeness of symptoms present in a diseased plant and to identify the disease from the symptoms alone. The difficulty in diagnosing some soybean diseases, for example brown spot, alternari spot and frog eye leaf spot, is that they often have similar symptoms. This, according to Mahoney (1996), causes misdiagnosis of diseases even by domain experts. A method that identifies clusters of symptoms for particular diseases may provide experts with additional information and enhance their knowledge.

Michalski (1980) notes that available diagnostic information regarding soybean pathology surpasses by far what a single expert can encompass. This is due to similarity of symptoms and differences in local conditions that may introduce deviations between symptoms for any given disease. Further, symptoms for a particular disease may change slightly over time. The exploration of historical data with methods for identifying deviations and changes may reveal new trends and patterns.

A classification method that can cluster symptoms and diseases and a model that can then be used to predict a disease given symptoms should be useful for decision makers with no deep knowledge of soybean diseases. Knowledge obtained from the KDD process may be used for explanatory and training purposes. Also domain experts and trained pathologists may use data mining tools to validate and possibly extend their knowledge.

A generally accessible small soybean database is available from the MLRepository (1999). The data set consists of 307 cases, with a dependent variable of 19 classes (values). Each class represents a soybean disease and each case in the data set is diagnosed with only one type of soybean disease. Thirty-five descriptors of plant and environmental factors were recorded for a diseased soybean plant. The symptoms were clearly observable conditions obtained with no sophisticated mechanical assistance (Michalski, 1980).

3.2 Decision tree

To illustrate the application of data mining for the soybean disease problem we use the CHAID method implemented in the AnswerTree software (SPSS, 1998). Decision trees generated for all disease classes are too large to be depicted and discussed. For the purpose of this discussion we have selected six of the nineteen classes (values of the dependent variable) for which there are 200 cases in the database.

The CHAID method compares the association between the dependent variable and the independent variable using Pearson's chi-square test. It requires the specification of a minimum significance level value. Independent variables for which significance is below the minimum value are disregarded.

The second parameter that needs be specified is the minimum significance level for merging branches. The CHAID method allows the user to group values (classes) of an independent significant variable, to form groups of values (categories). The procedure adds values to a group until the minimum significance level is achieved. The last required parameter is the minimum number of cases to be considered. If, at a node, the minimum number is reached the procedure terminates and no further testing is done at this node.

The minimum significance levels for independent variables and for merging branches were both set at 0.05. The minimum number of cases was set at 10.

The decision tree for the soybean problem is given in Figure 2. For illustrative purposes we have truncated the tree to cover the brown-spot disease.

The dependent variable is Diagnosis with six possible values listed in each node. The root (top-level) node comprises all 200 records of the database. The distribution of all records according to the identified diseases, that is values of the dependent variable, is given (for example, there are 40 occurrences of the value *photophthora-rot*).

At each node below the root level a significant predictive variable (calculated using Chi-square tests) is selected. Records are collected for specific value(s) of this predictive variable. For example, the variable Int_Discolor (internal discoloration in the stem) is the first predictor of diagnosis (Chi-square = 200 on five degrees of freedom) and the node for Int_Discolor = 0 is presented below the root level (see Figure 2). Predictive variables are recursively applied to the data, thus the application of a variable on a node is dependent on the variable(s) applied on preceding node(s). This condition is termed the "recursive" nature of decision trees (Weiss and Indurkhya 1998). Hence, variable Plant_Growth depends on Int_Discolor.

Branches indicate the alternative descriptions obtained from one or more values of a predictive variable in a node. In Figure 2 a branch is depicted at the second level for two values of the variable Plant_Growth.

The size of the tree is controlled by the user. If desired, the tree can be constructed so that it represents most records in the training set. This would increase the applicability of the model to those records, but reduces the chance that the model would generalize to the population. This situation is referred to as "over-fitting" the model to the training data. Analysis can be performed using only one dependent variable at a time.

The accuracy of classification of the model was 86.5%. For this model accuracy rate was calculated using the same observations that were used to build the model. When the model is tested for accuracy, the most probable outcome of a terminal node is assigned the observation that meets specified conditions. This accuracy rate may not be a reasonable indication of the performance of the model in classifying new cases, as the model was tested on the same cases that were used to build it.

3.3 Rule-based model

Decision trees can be represented with rules. Each path leading from the terminal node to the root node is a rule. For each rule the likelihood of observing a particular class of the dependent variable is calculated (it is a percentage of all the observations at a terminal node).

Four rules corresponding to the four numbered paths in the tree depicted in Figure 2 are given in Table 1.

Only rule 2 presented in Table 1 implies a single disease. The remaining three rules are not discriminatory because their conclusions indicate two or three possible diseases with a different degree of prevalence. This is the consequence of different diseases having similar symptoms.

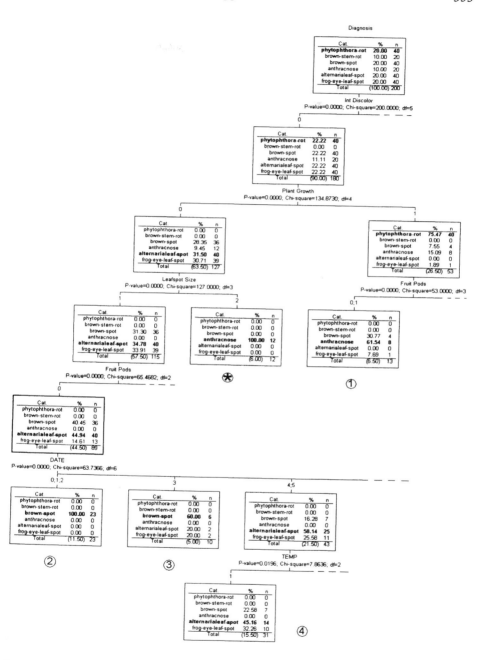

Figure 2. Soybean disease decision tree

We have also obtained rules that have little value. Consider the rule obtained from the path of which the terminal node is indicated in Figure 2 with an asterisk:

"If there is no internal discoloration in the stem, the plant growth is normal and the leaf spot size descriptor is not applicable, then the disease is anthracnose (100%)."

This rule is misleading because it suggests that regular (healthy) conditions indicate anthracnose disease.

The example of a misleading rule indicates the need for a preliminary analysis of observed models even before experts and decision makers became involved in the model evaluation and validation step of the KDD process. This analysis may also involve removal of rules with many precedents if there are other rules that contain only a subset of these precedents and have the same consequents. Rules with many precedents are highly specialized and considered "less interesting".

Table 1. Rule set generated with the AnswerTree CHAID method

Rule	Symptoms	Disease (%)
1	There is no internal discoloration in the stem AND the plant growth is abnormal AND the fruit pods are normal or diseased	Anthracnose (61.54%) Brown spot (30.77%)
2	There is no internal discoloration in the stem AND the plant growth is normal AND leaf spot size is above 1/8 inch AND the fruit pods are normal AND disease occurred in April, May or June	Brown-spot (100%)
3	There is no internal discoloration in the stem AND the plant growth is normal AND leaf spot size is above 1/8 inch AND the fruit pods are normal AND disease occurred in July	Brown-spot (60%) Alternaria leaf-spot (20%) Frog-eye-leaf-spot (20%)
4	There is no internal discoloration in the stem AND the plant growth is normal AND leaf spot size is above 1/8 inch AND the fruit pods are normal AND disease occurred in August or September AND temperature is normal	Brown-spot (22.6%) Alternaria leaf-spot (45.2%) Frog-eye leaf spot (32.2%).

To analyze the model and assess its potential usefulness one can repeat CHAID for different values of the method parameters and compare the results. A model is robust if, for the given dataset, changes in parameter values do not lead to significant changes in the model. Such a model is of greater value for decision makers because it is less sensitive to occasional records or outliers in the database.

Models are often verified using holdout data. The initial model is constructed from a sample drawn from the data base and assessed through the analysis of the remaining data. The two datasets are known, in data mining, as training and verification data sets respectively. The soybean dataset is small and therefore we used the entire set to construct the decision tree and rules. The use of training and verification data, and the estimation of the performance of the rules are discussed in Chapter 13.

3.4 Model comparison

Models obtained from data mining need be evaluated and verified (see Section 2.4). There are several possibilities, including the use of a different data mining method, comparison of the models with other models available in the literature, and the assessment of the model by the domain experts.

To compare rules obtained from the CHAID method we selected rules generated with AQ11, an inductive program developed by Michalski (1980) and applied to a different soybean database with 290 records. The observations were gathered from questionnaires completed by plant pathologists. The predictor and dependent variables available in these two data sets were identical. Rules that define brown spot disease obtained from the three sources are given in Table 2.

Table 2. Expert knowledge, and AQ11 and AnswerTree results for brown spot disease.

Expert knowledge	Inductive program AQ11	AnswerTree CHAID
Leaf condition is abnormal	Leaf condition is abnormal	
Leaf spot halos are present		
Leaf spot water soaked margins are absent		
Leaf spot size is above 1/8 inch	Leaf spot size is above 1/8 inch	Leaf spot size is above 1/8 inch
Disease occurred in May, August or September		Disease occurred in May, August or September
Precipitation is above normal	Precipitation is above normal	
	Crop is repeated in the same field for more than a year	
	Not the whole field is infected with the disease	
	Leaf is not malformed and roots are normal	
	No yellow leaf spot halos	
	No leaf spot water-soaked margins	
		No internal discoloration in the stem
		Fruit pod is normal

The three rules given in Table 2 contain some common precedents. There is one precedent present in all rules. Several precedents are present in two but not all rules. The rule generated with AQ11 is most specialized and has seven precedents. CHAID generated the rule with the least number of precedents (4), thus one may assume that this rule is most general. Interestingly, the expert's knowledge does not contain negative precedents that indicate that a particular symptom is absent for the brown spot disease. Both programs generated such precedents.

No rule from one source contradicts any other rule and none is a generalization of another. The question is if one should use all the rules or amend the expert's knowledge with precedents that appear in the generated rules but that are absent in the

statement of experts. We believe that the answer to this question should be left to domain experts.

4. Model verification

In the preceding section we compared two models obtained from the application of
data mining methods with expert knowledge. While such a comparison is useful, additional and detailed analysis is often required especially if the decision problem is
complex or domain knowledge is not available. The data mining literature suggests
model verification by means of sub-samples of the database (Kohavi, 1995). In the k
fold cross-validation method, the database is partitioned into k, $(k \geq 2)$ subsets. The
model is constructed in k iterations, where in each iteration a different subset is used
for calculating the accuracy of the model constructed using the remaining $k - 1$ subsets. Another approach found in the literature is to sample the database k times and
average the results (Weiss and Indurkhya, 1998).

In this section other approaches to model verification are presented. For this purpose we use the database containing over 1,400 transcripts of negotiations conduced
via the INSPIRE system and two questionnaires filled out by the users. The system
and negotiations are discussed in more detail in Chapter 11.

For the purpose of model verification three models are constructed and analyzed.
As in Section 4, the CHAID AnswerTree method (SPSS, 1998) is used with a minimum significance level of 5% for considering significant predictors and for merging
branches. The decision trees are constructed using all available data.

4.1 Model confirmation

INSPIRE negotiations are anonymous and bilateral. Each user negotiates with their
counterpart for up to three weeks. Upon completion of the negotiation users are requested to fill in a post-negotiation questionnaire. Several questions pertain to the
user's perception of their counterpart. One of the questions asks about the user's willingness to work in future with their counterpart. We selected this variable (denoted
workwopp) as the dependent variable. Forty-four variables were considered as possible independent variables. In this model we seek an explanation rather than prediction; thus the model is descriptive and not predictive.

The decision tree generated with the AnswerTree CHAID method (SPSS, 1998) is
depicted in Figure 3. The following four variables have a relationship with the dependent variable (variable name in parantheses):
1.   Whether the counterpart was honest or deceptive (opphones).

2.   Whether the counterpart was cooperative or selfish (oppcoop).

3.   How much control the user had during negotiations (control).

4.   Whether an agreement was reached (agr).

There have been no similar experiments prior to the negotiations conducted via the
INSPIRE system. The users are allowed to preserve their anonymity, they can be from

any country and region, and they can use several decision support techniques. We do not have access to any expert knowledge that may be used to verify the above model. Although cross-validation may possibly improve the results, the problem of further verification remains.

To verify the model, logistic regression with the four descriptive variables and one dependent variable was conducted. An overall likelihood ratio test of the logistic regression model yields a likelihood ratio statistic of 200.0 on four degrees of freedom, with a significance level, p, less than 0.0001. This provides evidence that the four variables included in the model make a strong contribution to describing the user's willingness to work with his/her opponent in future. A standard goodness-of-fit test indicates that the model fits the data well ($p = 0.19$). In other words, there is no reason to suspect that important explanatory variables have been omitted (Agresti, 1996).

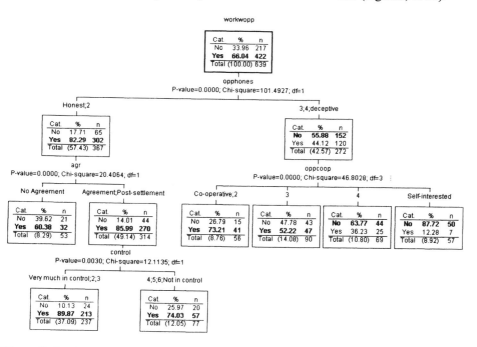

Figure 3. Decision tree for the users' desire to work with counterpart.

Logistic regression provides additional information. Each variable in the specified model has the following interpretation (under the assumption that the values of other variables are constant):

1. opphones: The odds of a user agreeing to work with their counterpart are 3 times greater for those who believe that their counterpart is honest, than for those who believe that their counterpart is deceptive.

2. oppcoop: The odds of a user agreeing to work with their counterpart is twice as great for those who believe that their counterpart is cooperative, than for those who believe that their counterpart is selfish.

3. control: The odds of a user agreeing to work with their counterpart are 7 times
    greater for those who believe that they are in control during negotiations, than for
    those who believe that are not in control during negotiations.

4. agr: The odds of a user agreeing to work with their counterpart are 0.27 times
    greater for those who reach an agreement, than for those who do not reach an
    agreement.

4.2 The case of two models

Negotiators are often satisfied with inefficient (not Pareto-optimal) compromises. The
second example considers the dependent variable opt (whether the final agreement
was efficient). There were twenty-two variables initially considered as possible predictive variables.

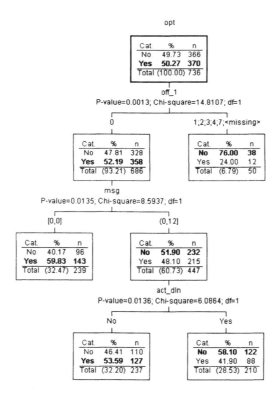

Figure 4. Decision tree for the agreement efficiency problem.

From the decision tree model depicted in Figure 4, it follows that the following
three variables may be significant predictors for the dependent variable:
1. Number of offers conveyed by user on the second-last day of negotiations (off_1).

2. Number of messages sent by user to their counterpart (msg).

3. Whether there is activity forty hours prior to the deadline (act_dln).

We conducted, as in Section 5.1, a logistic regression. In this model, however, only two variables (off_1 and act_dln) are significant, at the 5% significance level. Variable msg is not significant at this level.

The msg variable is discrete, $0 \leq$ msg $< +\infty$. The decision tree (see Figure 4) indicates that of the possible values in the variable, only two categories have significant association with the dependent variable (opt).

A second iteration of logistic regression was conducted, with the msg variable recorded into two categories: 1) no messages were conveyed by the user during negotiations, and 2) at least one message was conveyed by the user during negotiations. The descriptive variables with significant logistic parameters are off_1, act_dln, and msgbin (dichotomized msg variable).

An overall likelihood ratio test of the logistic regression model yields a likelihood ratio statistic of 24.6 with three degrees of freedom, ($p < 0.0001$). This provides evidence that the three variables included in the model make a strong contribution to explaining the probability of reaching an efficient agreement. A standard goodness-of-fit test indicates that the model fits the data well ($p = 0.17$). In other words, there is no reason to suspect that important explanatory variables have been omitted.

The interpretation of the logistic regression for each significant predictor is (keeping other variables constant):

1. off_1: An additional offer sent on the second-last day of negotiations reduces the odds of reaching an efficient agreement by .5 times.

2. act_dln: The odds of reaching an efficient agreement are 0.25 times lower when there is activity forty-eight hours prior to the deadline.

3. msgbin: The odds of reaching an efficient agreement are .54 times higher when no messages are sent by the user.

The presence of significant categories in msg might not have been considered unless there was prior knowledge that combining categories would increase the contribution that the variable makes to estimating the probability of success. The use of CHAID, therefore, indicates the possibility of increasing explanatory power when only significant categories are considered.

4.3 Interactions

NSPIRE users have access to data visualization and decision support tools. One of them is the history graph that depicts the flow of offers and counteroffers between a user and their counterpart (an example is depicted in Chapter 11, Figure 6).

In the third example we study the contributing factors to the use of the negotiation history graph. The dependent variable is graphuse (Whether the user viewed the history graph at any time during negotiations). There were twenty-three variables that might be related with the dependent variable. The decision tree for this problem is given in Figure 5.

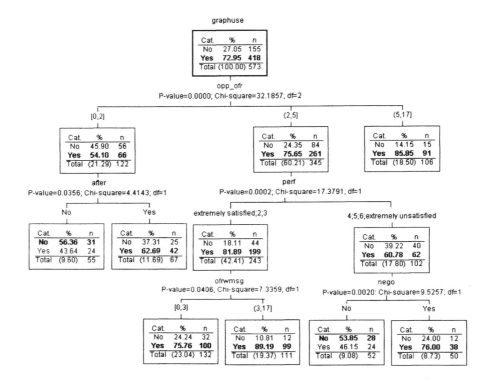

Figure 5. Decision tree representation of history graph problem.

From the initial set of variables the following five are related to the dependent variable:

1. Number of offers received from counterpart during negotiations (ofr).

2. Number of offers conveyed that included a message (ofrwmsg).

3. Whether the user would increase use of Internet after using INSPIRE (after).

4. Whether the user would use a similar system (to INSPIRE) in real life (nego).

5. The user's perception of their performance during negotiations (perf).

From the application of logistic regression we find that all descriptors, except for the variable after, are significant at the 5% significance level. In the previous example (Section 5.2), the msg variable was found to be significant if dichotomized. In considering the after variable, there is no further refinement possible because it is a binary variable. This raises concern with respect to the choice of this variable in CHAID.

Decision tree methods are used to construct models that represent interactive relationships among descriptive variables. Logistic regression measures the significance of the contribution a descriptor makes towards estimating a probability — interactions among variables are not automatically included in a logistic regression model, though they can be separately coded and included as additional descriptive variables.

Interactions can be efficiently explored with loglinear modeling (Agresti, 1996). We observed interactions between the dependent variable and the 3-way interaction between after*nego*ofrwmsg, and between the dependent variable and perf (a*b denotes interaction between variables a and b). Each m-way interaction ($m \geq 2$) implies the existence of lower order interactions. The variable perf has no interaction with any other variables, and so, it has a direct interaction with graphuse.

These initial results indicate a 3-way interaction between variables after, nego, and ofrwmsg. Such a high-order interaction is difficult to interpret. An approach to this problem is to divide the dataset according to a variable that appears in an m-way interaction. In this example, the dataset is divided into two subsets using the binary variable after. Loglinear analysis is used to construct further models on each of the constructed subsets.

For the subset of data where all observations take a value of after = 0, it is found that the dependent variable has an interaction with the generating class nego*ofrwmsg and a direct interaction with variable perf. For the subset of data where all observations take on the value of after = 1, it is found that graphuse has direct interaction with variables ofrwmsg and nego. This step reveals that division of the initial data set has reduced the complexity of the existing interactions (from third-order to second-order, in one subset). A two-way interaction still represents complex interaction, and, therefore, further division is suggested. Correspondingly, the subset consisting of observations with values of after = 0 is partitioned using the values in variable nego. The subset with after = 1 is held constant since there is no higher order interactions for that data.

From the original set of data, there are now three subsets (after = 0 and nego = 0; after = 0 and nego = 1; and after = 1). For the subset of observations where after = 0 and nego = 0, it is found that the dependent variable has an interaction with variable perf alone. For the subset of observations where after = 0 and nego = 1, it is found that the dependent variable has an interaction with variable ofrwmsg alone.

Conducting logistic regression on each subset of data (using only corresponding descriptors) yields significant contributions of the descriptors to the estimation of success: the user viewed the history graph at least once during negotiations. These models (and corresponding p-values at alpha of 5%) are presented in Table 8. This sequential division of the database to yield simple logistic regression is interesting in that it can be regarded as a secondary CHAID-type process. Presumably, some optimal compromise between the database division and simple model representation exist. This general issue is of importance for the use of KDD but it is beyond the scope of this chapter.

Table 8. Logistic regression models for each of three subsets of data

Conditions of after and nego	Descriptors
after = 0 AND nego = 0	perf (p = 0.0121)
after = 0 AND nego = 1	ofrwmsg (p = 0.0121)
after = 1	ofrwmsg (p = 0.0001), nego (p = 0.0224)

5. Discussion

In the previous sections we presented several potential difficulties with the use of the KDD for model construction and validation. The four examples are based on relatively small databases with problems characterized by less than fifty variables. Large and very large databases and problems with thousands of potentially relevant variables will only multiply the difficulties of model validation and interpretation.

The problems of applying KDD methods to real-life situations need to be addressed. At present, this requires direct involvement of experts and the use of statistical methods. This does not mean, however, that KDD is not useful and that it provides no new insights. It also does not imply that the role of statistical methods in model formulation and verification remains the same irrespective of the use of data mining methods. Statistical methods, as we have attempted to show in this chapter, complement and enhance data mining methods. The latter allow for easy and efficient model construction, while the former are used to assess, verify and interpret the results.

The situation with KDD systems is similar to the earlier generations of DSSs that could not be used directly by decision makers. Instead, they were used by analysts and MS/OR specialists. Hence, one strategy for the effective utilization of KDD software is by back-office analysts who have an understanding of the domain as well as the software and methods. These experts may provide decision makers with interesting and relevant models or embed them in the DSSs that the decision makers use directly. Another strategy requires a significant extension of the capabilities of the existing software for KDD so that it can provide integrated support for all steps of the KDD process, including model validation.

Systems for KDD are considered intelligent (Limb and Meggs 1994). This intelligence, however, is limited to the use of AI mechanisms in the data mining software. At present the effort is primarily on data mining with other KDD steps only partially supported. There is little integration between the support provided during the different steps of the process.

Knowledge discovery from databases is a novel approach for model construction and validation. It requires dedicated software, powerful computing and communication facilities and access to databases. Many developing countries face shortages of computers and software, however, efforts of their governments and international organizations help to alleviate these problems. In many chapters of this book large complex DSSs have been developed and implemented in developing countries. Several examples of GISs, that have requirements similar to KDD, are discussed in chapters 3 and 5. Voluminous data for both developed and developing countries is collected from satelites and by other means (NASA, for example, collects data at the rate of 50 gigabytes per hour). Investments in communication technologies will allow for efficient access to data and to KDD methods.

Application of model-based DSSs in developing countries have been criticized because they often do not account for local indigenous knowledge and ignore local practices. Their incorporation into systems can be achieved through direct studies; an example of this approach is presented in chapters 10 and 13. It is also possible, however, to extract information about local procedures and traditions from historical data describing communities and regions.

Annex. KDD and sustainable development

1. Forest management

The analysis of ecosystems, including studies of biodiversity, forest density, wildlife population and water pollution, is aided through images collected via remote sensing. Traditionally, statistical methods, such as linear discriminant analysis, were used to analyze images. The use of these methods was cumbersome because of the very large amount of data. Flinkman et al. (1998) applied the rough set method (Pawlak, 1991) to analyze interactions between land uses, vegetation types, forest density, and other biotic, abiotic and anthropogenic conditions in the Siberian forest. The objective of their study was to identify key attributes for the formulation of sustainable forest management policies.

Precise and robust classification of land cover is required to conduct environmental assessment and to formulate plans and management policies. Friedl and Brodley (1997) used three decision tree methods (univariate, multivariate and hybrid methods) to obtain models for land cover classification. The models were compared with models obtained from two statistical methods (maximum likelihood and linear discriminant analysis) and used cross-validation for model comparison. The study was conducted on a global database, a database for North America, and one describing a forest in California. According to the authors, the decision tree models have accuracy comparable to statistical models. The advantages of the data mining tools include their good performance in the presence of disparate and missing data, and the ease of model construction given the complexity of the data.

2. Precision agriculture

Farming requires efficient management of resources (such as water and fertilizer) that needs to account for changes in the environment. Dong (1998) used remotely-sensed data to construct descriptive models of agricultural practices. The association rules analysis was used for this purpose. The author states that harvests from different fields can be mapped to each season and varying yields analyzed according to seasons and locations to determine the relationship between seasonality and location on crop yield. Farm managers and extension workers may use this information to allocate resources in order to maximize yield.

3. Oil wells

Oil drilling has significant impact on the surrounding eco-system. Selection of the type of mud where drilling is undertaken has, in turn, impact on drilling efficiency and costs, bore-hole instability and cleaning, and the cost of oil extraction. Project managers in an oil field in Oso, Nigeria encountered problems (for example, high incidents of stuck pipe and filter cake build up) and cost overruns when their decisions to select mud types were based on experience, the available resources and personal preferences. To address these problems they used data mining techniques to choose between alternative mud types (Dear III, 1995).

A structured approach based on a decision tree method was devised in order to determine the relationships between the attributes of mud type, drilling (including costs) and eco-system. The analysis revealed that mud types that were recommended by experience often led to higher costs, while types proposed with data mining and previously not considered, allowed lowering of the long-term costs (Dear III, 1995).

4 Chemical pollutants

Concentration of chemicals in the environment is regulated to minimize their impact on the ecosystem. This requires frequent assessment of chemical levels. Ranking systems, based on subsets of aggregated factors, have been introduced for this purpose. Aggregation, however, often does not include interactions among these factors which are required to determine acceptable levels of exposure. According to Eisenberg (1998) some of the ranking systems are highly complex and cannot be understood by decision makers and, therefore, are considered as "black boxes". They provide estimates of chemical exposure but not information about the uncertainties and sensitivities associated with these estimates.

Eisenberg (1998) developed a methodology based on the CART decision tree method for the assessment of chemical exposure levels. It is used to identify properties of chemicals, conditions of the environment, and the relationships among chemical properties that are most important for classifying an area according to its exposure level. The CART method was used to derive classification rules of chemicals and identify factors affecting chemical exposure levels. These rules are also used for the analysis of changes in factors and corresponding changes in the classification (chemical exposure levels).

Acknowledgements

We thank David Cray for his comments and suggestions. This work has been supported by the Social Science and Humanities Research Council of Canada and the Natural Sciences and Engineering Research Council of Canada.

References

Aasheim, O. T. and H. G. Solheim (1996). "Rough Sets as a Framework for Data Mining", Norwegian University of Science and Technology: 148.

Agresti, A. (1996). *An Introduction to Categorical Data Analysis*, New York: Wiley.

Becker, B., R. Kohavi, et al. (1997). "Visualizing the Simple Bayesian Classifier". *KDD 1997 Workshop on Issues in the Integration of Data Mining and Data Visualization*.

Bell, D. E. H. Raiffa and A. Tversky (1988). *Decision Making. Descriptive, Normative and Prescriptive Interactions*, Cambridge, MA: Cambridge Univ. Press.

Bigus, J. P. (1996). *Data Mining with Neural Networks*, McGraw-Hill.

Brachman, R. and T. Anand (1996). "The Process of Knowledge Discovery in Databases: A Human-centered Approach", in U. Fayyad et al., *Advances in Knowledge Discovery and Data Mining*, Menlo Park, CA: AAAI Press, 37-58.

Breiman, L., J. H. Friedman, et al. (1984). *Classification and Regression Trees*, Belmont, Wadsworth International Group.

Briscoe, G. and T. Caelli (1996). *A Compendium of Machine Learning, Vol.1.* Norwood, Ablex.

Dear III, S. F., R. D. Beasley, et al. (1995). "Use of a Decison Tree to Select the Mud System for the Oso Field, Nigeria." *Journal of Petroleum Technology*, 47(10), 909-912.

Dong, J. (1998). "Mining Association Rules from Imagery Data", Computer Science Dep. North Dakota State University, Fargo, ND.

Eisenberg, J. N. S. and T. E. McKone (1998). "Decision Tree Methods for the Classification of Chemical Pollutants: Incorporation of Across-Chemical Variability and Within-chemical Uncertainty." *Environmental Science and Technology*, 32(21), 3396-3404.

Fayyad, U. M. (1996). "Data Mining and Knowledge Discovery: Making Sense Out of Data" *IEEE Expert*, 11(5), 20-25.

Fayyad, U. M., G. Piatetsky-Shapiro, et al. (1996). "From Data Mining to Knowledge discovery: An Overview", U. M. Fayyad, G. Piatetsky-Shapiro et al. (Eds.), *Advances in Knowledge Discovery and Data Mining*, Menlo Park, MIT Press.

Flinkman, M., W. Michalowski et al. (1998). "Identification of biodiversity and Other Forest Attributes for Sustainable Forest Management: Siberian Forest Case Study", IR-98-106, International Institute for Applied System Analysis, Laxenburg, Austria.

Friedl, M. A. and C. E. Broadly (1997). "Decision Tree Classification of Land Cover from Remotely Sensed Data", *Remote Sensing of Environment*, 62, 399-409.

Glymour, C., D. Madigan, et al. (1997). "Statistical Themes and Lessons for Data Mining" *Data Mining and Knowledge Discovery*, 1(1): 11-28.

Inmon, W. (1996). *Building the Data Warehouse*, New York: Wiley.

John, G. H. (1997). "Enhancements to the Data Mining Process", Computer Science Department, School of Engineering, Stanford University: 194pp.

Kohavi, R. (1995). "A study of cross-validation and bootstrap for accuracy estimation and model selection". *International Joint Conference on Artificial Intelligence (IJCAI)*.

Kohavi, R. (1998). "Crossing the chasm: From academic machine learning to commercial data mining". Mountain View, SGI, http://reality.sgi.com/ronnyk/chasm.pdf.

Kohavi, R., B. Becker, et al. (1998). "Improving simple bayes". Mountain View, Data Mining and Visualization group, Silicon Graphics, Inc.

Limb, P. R. and G. J. Meggs (1994). "Data Mining - tools and techniques" *BT Technology* 12(4): 32-41.

Mahoney, J. J. (1996). "Combining Symbolic and Connectionist Learning Methods to Refine Certainty-Factor Rule-Base", Department of Computer Sciences, The University of Texas. [on-line] http://www.cs.utexas.edu/users/ml/abstracts.html

Michalski, R. S. and R. L. Chilausky (1980). "Knowledge acquisition by encoding expert rules versus computer induction from examples: a case study involving soybean pathology." *International Journal of Man-Machine Studies*, 12, 63-87.

Mienko, R, et al. (1996). "Discovery-oriented Induction of Decision Rules", Cahier du LAMSADE, No. 141, Paris.

MLRepository (1999). Machine Learning Database Repository, University of California-Irvine, http://www.ics.uci.edu/~mlearn/MLRepository.html.

Pawlak, Z., I. Grzymala-Busse, et al. (1995). "Rough Sets", *Communications of the ICM*, 38(11), 89-95.

Sasisekharan, R., V. Seshadari, et al. (1996). "Data Mining and Forecasting in Large-scale Telecommunication Networks", *IEEE Expert*, 11(1), 37-43.

SGI (1998). "MineSet 2.5 tutorial", Silicon Graphics Inc.

Simoudis, E. (1996). "Reality Check for data Mining" *IEEE Expert* 11(5), 26-33.

Skinner, C. J., D. Holt and T. M. F. Smith (1989). *Analysis of Complex Surveyes*, Chichester" Wiley.

SPSS (1998). "SPSS in Data Mining". Chicago, SPSS Inc., http://www.spss.com.

Srikant, R. and R. Agrawal (1998). "Mining quantitative association rules in large relational tables", IBM Almaden Research Center.

Tukey, J. W. (1977). *Exploratory Data Analysis*, Reading, MA: Addison Wesley.

Weiss, S. M. and N. Indurkhya (1998). *Predictive Data Mining: A Practical Guide*, San Francisco, Morgan Kaufman.

IV DECISION SUPPORT SYSTEMS FOR SUSTAINABLE DEVELOPMENT

19 EXPERIENCE AND POTENTIAL
Patrick A.V. Hall, *et. al.*[1]

1. Introduction

This chapter has arisen from a workshop held at the United Nation University International Institute for Software Technology in Macau in 1996, and sponsored by the International Development Research Centre, Canada. The workshop was organized in response to the Agenda 21 report from the Rio Earth Summit, and particularly to its Chapter 40 on Information for Decision Making. The paper has been written for policy makers concerned with the development and decisions about development.

The process of development requires the making of decisions, selecting from among several possible alternative development paths the line of action that will return the most perceived benefits. Development should be sustainable and decisions directing its path should be made with sustainability in mind. In this context, the definition of sustainable development provided in the Brundtland Commission (1987) report, namely "development that meets the needs of the present without compromising the ability of future generations to meet their own needs", offers both a guiding principle and an objective for the decision process.

It is important that development decisions are made well and that they use the best information, methods and tools available. For development to be sustainable we need to make decisions that do not have long-term negative impacts, and to assess both

[1] This chapter has been authored by the participants of the IDRC and UNU/IIST workshop in Macau: D.J. Abel, B. L. Adams, C. Audroing, S. Bhatnagar, D. Björner, B.V. Bossche, R. L. Bowerman, V. S. Chabanyuk, C. Shijing, S. Chokkakula, J. R. Eastman, A. F. Dos Santos, E. D. Gamboa, S. Gameda, V. L. Gracia, G. B. Hall, P. A.V. Hall, H. He, T. B. Ho, G. E. Kersten, S. Koussoube, J. G. Krishnayya, M. K. Luhandjula, Z. Mikolajuk, M. F. Mulvey, R. Noussi, S. Pokharel, S. Poulard, A. Rafea, M. Rais, M. Ramachandran, M. Rao, H. G. Smith, V. Tikunov, W. Tracz, J. P. Weti, R. Woirhaye, J. K. Wu, A. G-O Yeh, H. Ykhanbai, H. Zhu. The editors of this volume revised the chapter and added references to its chapters and other published papers.

impacts and benefits prior to undertaking implementation actions. Because predictions of impacts, especially in the long term, can ever only be approximate, we can only make select development paths of limited duration. Decisions need to be revisited and revised as their consequences are revealed in practice.

We formulate a basis for the study and application of decision support systems (DSS) for the purpose of sustainable development. Our objective is to provide a broad position on issues related to DSS and their implementation. We provide researchers and practitioners with a perspective on current issues and future needs. We establish a basis for discussion of the unique requirements of developing countries and their sustainable development.

We characterize, in Section 2, the essential features of sustainability that will be significant in our consideration of DSS. We start by a short review of the evolution of the concept of sustainable development from Malthus through to Brundtland. This is a particularly western view of the history, and we also argue that sustainable practices have been widespread across communities and through the ages. We focus on the need for equity and the use of indigenous knowledge, and emphasize that it is essential that the decision making processes themselves should also be sustainable.

In Section 3, we review the process of decision making and its underpinning theory, noting recent trends away from rational decision theory to decision support theory and the need to be able to accommodate uncertainty and risk in decision making. The importance of people in the decision making process is emphasized.

The roles of computer-based decision support technologies are described in Section 4, working through the process from information gathering and storage, through knowledge bases, the support for decision making, to visualization and the support for group decision making. We then discuss the need for solutions which integrate many tools, with existing knowledge-based and geographic information systems giving partial solutions, and conclude by outlining an agenda for developing a domain-specific architecture for decision support systems for sustainable development (DSSfSD).

Section 5 considers the wider issues of deploying DSS in developing countries. Decisions require high quality data, trained personnel, up-to-date computer systems, and a continued commitment to maintaining the data and computer systems. The conclusions are given in Section 6 with a list of recommendations arising from this chapter and the experience of the authors.

2. The need for development that is sustainable

"Development" is assumed to be desirable, but what it means exactly is problematic. It is generally assumed to align with "progress", to mean improvements in living standards, health and welfare, and the achievement of other goals agreed to by the community concerned. What is considered to be development depends upon those intended to benefit from that development, and no universal definition is possible. The nearest we might come would be the World Bank's "Reducing poverty is the fundamental objective of economic development" (World Development Report 1990).

Development has always been with us, as societies have evolved and adapted to their environment, migrated to new environments, learned to use the resources available and discovered new resources that they could use. These developmental economic activities are inevitably associated with the consumption of natural resources. This consumption then raises questions about sustainability. Again, this concern has always been with us, and societies have always responded to the concern by selecting appropriate forms of economic activity. But as we move to a global society aware of the disparities around the world, the issue of sustainable development has taken on a new force.

2.1 The need for sustainability

Publication of the Brundtland Report in 1987 (WECD, 1987) has led to world-wide interest in the concept of sustainable development. However, this concept is by no means new.

Concern for the sustainability of life on our planet can be traced back two thousand years to Greek culture. Here it appeared first in the Greek vision of 'Ge' or 'Gaia' as the Goddess of the Earth, the mother figure of natural replenishment. Guided by the concept of sustainability, the Greeks practiced a system whereby local governors were rewarded or punished according to the appearance of their land.

More recently, major concern with the limited productivity of land and natural resources appeared with the publication of Malthus' essay on population in 1789 and Ricardo's Principles of Political Economy and Taxation in 1817. These thinkers worried that economic growth might be constrained by population growth and limited available resources.

Then came the Industrial Revolution and colonial expansion. Towards the end of the last century and the beginning of this century, an optimistic view arose that the prosperity of Western economies could continue unabated. Natural resources were no longer regarded as posing severe restrictions on economic growth as new technologies for making better use of resources and new resources were discovered. However, the fragility of our economic growth was soon revealed by the depression of the 1920s and 1930s, and more recently by the world oil crisis and economic recession of the 1970s. Neo-Malthusians began to have doubts about unlimited growth, stressing once again the importance of natural resources in setting limits to economic growth.

In April 1968, the Club of Rome which consisted of a group of thirty individuals including scientists, educators, economists, humanists, and industrialists gathered to discuss the present and future predicament of mankind published the book "The Limits to Growth" (Meadows et al, 1972). The book predicted that the limits of growth of the earth would be reached some time within the next one hundred years if the present trend of growth remained unchanged. The predicament came from the exponential growth in global population, resource depletion and industrial pollution, in the context of our finite resources.

In 1992, 20 years after the controversial "The Limits to Growth" was published, the same authors published a successor book, "Beyond the Limits", which re-examined the situation of the earth (Meadows et. al., 1992). With new evidences from global data, the book shows that there is still an exponential growth in global population, economic growth, resource consumption and pollution emissions. In 1971 they concluded that the physical limits to human use of materials and energy were just a few decades ahead. In 1991, after re-running the computer model with new compiled data and analyzing the lately developed pattern, they realized that in spite of the world's improvement policies, many resources and pollution had grown beyond sustainable limits. The earth may be approaching its limit faster than what we would have thought.

The concept of sustainability is evident, albeit implicitly, in several related concepts. For example, the notion of a carrying capacity is defined in wildlife ecology and management as "the maximum number of animals of a given species and quality that can, in a given ecosystem, survive through the least favorable conditions occurring within a stated time period" and in fisheries management as "the maximum biomass of fish that various water bodies can support".

Most recent interpretations of sustainable development are modified derivatives of the concepts of the limits of growth and carrying capacity. They not only stress the importance of resource availability in limiting economic growth but also draw attention to the need to develop methods that facilitate growth in harmony with the environment, emphasizing the potential complementarity between growth and environmental improvement. Hence the Brundtland definition of sustainable development as "development that meets the needs of the present without compromising the ability of future generations to meet their own needs". The IUCN definition as "a process of social and economic betterment that satisfies the needs and values of all interest groups, while maintaining future options and conserving natural resources and diversity" (IUCN, 1980, p. 2).

Population is a key factor to be considered in the implementation of sustainable development. Sustainable development can only be pursued if population size and growth are in harmony with the changing productive potential of the ecosystem. An example is the ECCO (Enhancement of Population Carrying Capacity Options) computer model which has tried to identify the trade-offs between population growth and standard of living, and between intensification of agriculture and soil conservation (Loening, 1991).

2.2 The need for equity

The above discussion focuses to a large extent on economic development. However, development must also help overcome the lack of equity between rich and poor, developed and underdeveloped, north and south. The concept of equity in the use of Earth's resources cannot be overlooked. Equitable resource use requires that everyone gets a fair share of the planet's resource base and that this principle holds equally within a country, between countries, between genders and from generation to generation. The intergenerational transmission of equity in resource use clearly is key to a sustainable future and this underscores the importance of the temporal dimension in the Brundtland Commission's definition of sustainable development.

There is also a need for equity in the consideration of expertise deployed in development. While the expertise that various communities around the world have may be very different, none is necessarily superior to the other. Indigenous peoples have lived in harmony and stability with their environment for generations, and this knowledge must be incorporated into the decision making process. For example, a network for the sharing of indigenous knowledge has been established in India (Gupta, 1995). Methods of development must be appropriate to the people and the environment in which they are applied.

Decisions that are made, and the rationale underpinning those decisions, must be acceptable to the people concerned. Decisions cannot be made by proxy by outside agencies, they must be made by the people in the communities themselves. It is this involvement of people at all levels that is absolutely critical in making development sustainable.

2.3 Implementing sustainable development

Although the term sustainable development is now widely used, there is, in general, no widely accepted operational framework through which to practice sustainability. Sustainable development does not mean no development. It means improving methods for resources management in the context of an increasing demand for resources. Sustainable development must facilitate economic development while fostering environmental protection.

In order to systematize the concept of sustainable development it is useful to utilize Barbier's (1987) model of interaction between three complementary systems - (a) the biological (and other natural resources) system; (b) the economic system; and (c) the social system. For these three systems, the goals of sustainable development may be expressed respectively as maintenance of genetic diversity, resilience, and biological productivity; satisfaction of basic needs (reduction of poverty), equity-enhancement, increasing useful goods and services; and ensuring cultural diversity, institutional sustainability, social justice, and participation.

Within this framework there is scope for different communities to seek different balances between these systems. Some interests may place high value on obtaining high environmental quality, while others may prefer to have improved living standards. Income, education, social structure and ideology are factors that determine the definition of sustainable development in a community. However, no community lives in isolation and the environmental impact of one community can affect everybody. Some rules need to be designed to guide people's behavior for sustainable development:

1. A given renewable resource cannot be used at a rate that is greater than its reproductive rate, otherwise complete depletion would occur.

2. Strict controls on the use of non-renewable resources are necessary to prevent their early depletion. Substitutes and new technologies are helpful in reducing the use of scarce resources.

3. The amount of pollution emissions cannot exceed the assimilative capacity of the environment. Abatement measures should be taken to reduce the influences of pollution.

4. We need biodiversity in the ecosystem because there may be unknown genes of high value to be found in some species. Species extinction can also introduce imbalance in the ecosystem. Some species can improve the human living

environment by generating soil, regulating fresh water supplies, decomposing waste and cleaning the ocean.

Different countries have different perceptions and thus different approaches to sustainable development. It is not possible to design universal measurements and indicators of sustainable development because of the different weights that are given to different components by different communities. What we are concerned with is enabling communities to make their own decisions about sustainable development, using the theoretical frameworks and tools that they themselves believe to be appropriate and can continue to use. Not only must the methods of economic activity and resources utilization be sustainable, but the decision methods too must be sustainable.

3. Decision making and DSS

Development involves making decisions as to the choice of a desired path to follow. Decision theory is, in and of itself, a highly complex field. Here, we take a very broad view of decision making as any situation where a decision taker has a choice between alternatives. In the simplest case there may be only one alternative and the decision is to take this or not. However, in reality there are usually numerous competing options or alternatives available in any course of action and thus the decision is correspondingly more complex. In contexts where decision making involves action it is important to evaluate also the implementation and results of decisions. When a deliberate course of action is laid out and subsequently implemented this constitutes planning. Decision making and planning may be at the group or individual level and in the former case reconciliation of different value systems is likely to be required. This may involve negotiation or trade-offs before a course of action that is acceptable to all groups is agreed upon.

3.1 Rational decision making

At the basic level, decision making involves a few simple stages Simon (1960): intelligence, design and choice. Simon's decision model is discussed in Chapter 2. In addition to the three stages identified by Simon, a post-decision stage of monitoring and evaluation should be included to follow up the outcome of a decision.

In decision science we identify a number of decision variables, numerical values or other values which can be represented by numbers, which will form the basis for our decision. In order to make choices we need to do three things. Firstly the values must be standardized. This involves scaling all the various values so that their numerical ranges are comparable and that they have a common interpretation, perhaps as a probability. Then, the standardized values are aggregated by combining various elements to form the basis for a judgment, as in factor analysis, multiple criteria analysis and so on. Finally the aggregated value or values are thresholded to produce a binary result or decision.

These rational decision methods are widely applied, and supported by suitable mathematical models:

- seeking to maximize accessibility, the location of primary health care services in the Central Valley of Costa Rica was decided using a location-allocation model (REDATAM, 1996);

- land use in the rapidly developing areas of Dongguan in south east China, adjacent to Hong Kong was modeled using an equity function (see Chapter 5);

- incentive strategies in rural development were formulated with a two-level multi-criteria model (see Chapter 7);

- decisions about sources of energy in Nepal were modeled using multiple objective programming (see Chapter 9); and

- recovery actions from the Chernobyl nuclear disaster in the Ukraine used modeling of radioactive contamination dispersion (see Chapter 17).

The above processes are grounded in statistics and mathematics, and implicitly assume that the various values required can be obtained and are accurate. There is an assumption that the relationships between the variables are known. Another assumption lies in the existence of a basis for decision through a utility function or functions involved in the aggregation step, and that these utility values have an agreed interpretation. These utility values are encapsulated within the value system of decision takers. The processes postulate that an optimal solution exists and is meaningful. In the development context, all of these assumptions are questionable. Two simple examples are:

- the absence of hydrological data to guide well location in Cameroon; and

- the importance of powerful people in influencing decisions in rural India (Bhatnagar, 1991).

3.2 Uncertainty and Risk

Data is never precise; its acquisition may lead to estimations and error; survey populations may be very small. This "fuzziness" needs to be taken into account, and a number of theoretical approaches are available to us from rational decision theory. The values may be given a probabilistic interpretation, or may be taken as deriving from the fuzzy sets of Zadeh. Then again, the Dempster-Schaffer belief theory might be used, and there are other approaches that could be taken. The particular approach to uncertainty will depend upon the needs of the decision maker and the information that he/she is using. This can itself lead to uncertainty in any calculations concerning the outcome of the decision making process. Therefore any decision has an attendant risk, and this risk should be calculated to make it explicit. Only then should it be used in decision making.

Examples addressing uncertainty and risk presented to the workshop were:

- normal data from surveys carried out by local administrators was found to be up to 30% inaccurate, attributed to lack of stake in the results of the survey by the administrator concerned (Bhatnagar et al., 1994);

- variables used in planning in a variety of sectors within Goa were treated as fuzzy variables (Krishnayya,1995);

- predictions of floods from rises in the sea level in Vietnam have to be treated probabilistically (Eastman et al.,1996); and

- water supply uncertainty in Zaire was treated using fuzzy sets.

Often decisions cannot be made on the basis of the theory outlined above. The theory assumes that you can be objective in measurement and in calculation, and this may not be possible. It assumes that formulae and equations can be solved, and this may not be possible either. This realization has led to Decision Aid Theory (Roy 1993). It may not be possible to find the best solution, and instead we should aim for a feasible solution, one that is good or at least satisfactory. We should aim to satisfy rather than to optimize.

3.3 Human-based decisions

Where we are unable to model the decision process, and as we have argued above this is usually the case for development, we must rely on humans to make the judgments necessary. People can weigh a number of alternatives and arrive at decisions, recognizing a good outcome when they see it, even if they are unable to articulate the reasons for the choice sufficiently precisely for it to be automatable. Where the people involved are drawn from the communities for whom the decisions are being made, they embody the value systems that are important for making an appropriate and sustainable decision. Where groups of people are involved, the possibly conflicting needs and values must be reconciled using appropriate processes.

In making sustainable development decisions in British Columbia, Canada, there were the conflicting interests of Salmon Fisheries, Forestry, Oil, and indigenous peoples, with significantly different value systems, which needed to be balanced (Kenney et al., 1990).

Where sustainability objectives need to be pursued with sparse and incomplete data/information, these limitations can partially be bridged by tapping into extensive experiential local knowledge (see Chapters 10 and 11). For conditions where quantitative cause effect relationships are absent or scarce, particularly across bio-physical and socio-economic domains, qualitative local knowledge can be essential to fill the gaps and correct inaccuracies. Local knowledge encompasses implicitly aggregated facts and information, incorporates uncertainty, and draws from experience, resulting in intuitive, general relationships or correlations between different factors affecting sustainability. The depth of such knowledge can be considerable (see Chapter 10). Capturing the knowledge of agricultural extension workers in Egypt, and later planning to use this knowledge base to train farmers and new extension workers is an example of this (see Chapter 13).

4. DSS - support for decision making

In order to render complex decision making manageable and reasonable in terms of the underlying goal of sustainable development, the decision making process as described above can be supported systematically by a variety of computer-based software tools. Kersten and Meister (1995) have undertaken a comprehensive survey of the tools available, concentrating on geographical and demographic information systems, and the more basic database and knowledge based systems (KBS).

We will describe the tools required in terms of their general type, focussing on the stage in the decision process being supported, from information gathering through storage to exploring alternatives to helping people make the decision.

4.1 Information collection and management

Decision making requires information, and this needs to be collected. One important source of social data is the governmental census. Other governmental and non-governmental sources of things like opinion polls, natural resources inventories and commercial registers may be useful. These will need to be extracted and transferred from their current databases to the decision maker's database. An important concern here may be the preservation of confidentiality, typically achieved through the aggregation of names to make the data anonymous. For a particular study it may be possible to obtain the raw data, but this process should be restricted, and must guarantee that the aggregation that is part of the decision making process will protect confidentiality. Data interchange formats and any special converters produced need to be discussed and accepted by the research community (REDATAM+, 1996).

Existing data would typically need to be supplemented by surveys focussed on the needs of the decision problem at hand. Computer aids, particularly the use of the internet, may help here, though with communications infrastructures in their current undeveloped state this may not be possible in developing countries.

Another important source of information that is now readily available is remote sensing from satellites. Here, information providers may make data available to local communities to help in their decision making. This data would typically feed into geographic information systems (GIS).

An example of the way remote sensing data might be distributed can be seen in India. Remote sensing data from satellites is collected centrally, then distributed to states who further distribute this data to districts; in parallel with this is a second line of distribution via the regions to the projects (Rao et al., 1994).

While the storage of most data can be achieved using standard database products, the storage of geographic data usually requires special methods. Geographic data can be stored as (1) "vectors" in which the geographical area is represented by a number of point, line and polygon objects whose coordinates are stored, and as (2) "rasters" in which the geographical area is divided into many small uniformly sized rectangles or pixels". Over these can be laid further networks of roads, rivers, boundaries and so on, as well as the locations of towns and similar features. Features can be divided along thematic lines, "layering" the area according to certain criteria and displaying thematic maps to highlight the different layers. These two representations are

equivalent and can be converted one to the other, but the representations favo
different calculations. Examples of the vector approach are Themaps (Krishnayya
1995) and *win*R+ (REDATAM+, 1996). IDRISI is an example of the raster approacl
(Eastman, 1992).

Large amounts of raw data from different sources and in different formats must b
verified and often converted into formats suitable for other components of a DSS
This exemplifies the well known problem of interoperability and data interchange
One solution suggested is to use Federated Database Systems (see Chapter 18).

Data integrity is critical for computer-based modeling and KBS. A computerize
DSS should provide facilities for verification of information integrity, and fo
discovery of discrepancies in received information. Statistical methods and rule-base
systems provide some tools for the analysis and preprocessing of data used fo
generation and evaluation of alternative decisions.

4.2 Modeling and rational decision support

It is important to explore the consequences of particular courses of action. To do thi
we need to build models and facilities. Models enable manipulation an
experimentation with variables representing characteristics of real systems within
predefined time scale. Long-term effects of suggested decisions can be analyzed in
short computer session.

The most common modeling tool is the spreadsheet, but equally important here ai
simulation modeling techniques. A complex DSS requires a collection of models. Th
software architecture should include facilities for model repository, selection c
appropriate models and composition of subsets of models to solve complex problems

The many methods used in rational decision making, such as multi-criteria analysi
and linear programming, are all supported by computers using mature softwar
packages. The uncertainty prevalent in development decision problems makes
necessary that these packages can handle uncertainty and risk.

These have become standard capabilities of off-the-shelf software, and no othe
special features are required for decision making in developing countries.

4.3 Visualization and the human interface

When people need to participate in the decision making process, they find it helpful t
have pictures and diagrams to help them visualize the situation about which they mu
make a decision. Routine facilities for these are the so-called business graphics of pi
and bar charts and graphs which show the relationships between numerical data. Als
of great usefulness is the display of a network of dependencies between parts of th
problem and their influence on the solution.

Of more recent origin is the ability to display the rich and complex maps i
multiple colors with a fine level of detail. These maps can display the spati
relationships between the elements of interest, and the geographical distribution
those elements through theme maps under user control. Humans have very powerf
spatial reasoning capabilities, and the display of geographic data can tap into th
reasoning power.

A range of visualization methods are possible to help those involved in the decision making process:

• simple diagrams helped users understand the financial planning methods proposed for them in the Philippines (see Chapter 6);

• simple graphs helped explain the trade-offs in making decisions in Canada;

• theme maps helped decision making in Chinese disaster planning (Yufeng et al., 1995); and

• anamorphic maps of the world show countries' gross national product, and population represented by size on the map.

As with all computing systems, it is important that the systems are easy to use by the persons concerned. It is important that technical expertise not be necessary. General usability analysis is applicable here.

Of particular importance in DSS is the ability of decision makers to work in their own natural language. While personal computers (PC) and Unix systems support the translation of software to other languages, some deeper cultural issues will need to be addressed which are not taken into account through these basic approaches. One example of a cultural dimension to be addressed is the choice of colors in theme maps, where for example, red signifies danger in Europe but joy in China (see Chapter 16).

4 Group decision making

Often decision making is a group process through which various stakeholders need to reach agreement. One important approach is the visualization method, where the representation of alternative choices is available to decision makers. Tools for group working and workflow may also be important and particularly so when the stakeholders involved are geographically dispersed and communication networks need to be exploited.

Simple computer conferencing methods may suffice, but more structured systems which enable the development of a debate by widely separated people over a period of time may also help. Systems are commercially available using "groupware" to support collaborative working, the holding of meetings and so on. The recent rapid growth of the Internet will be important in making this support widely useful.

There needs to be trial application within decision making for development.

5 Knowledge capture and representation

It is important to be able to capture local knowledge about a decision problem. One important way of doing this is through KBS and expert systems.

KBS generally contain a knowledge base and a problem solving or inference method. "Expert systems" is sometimes used as a synonym for KBS. The first generation of expert systems represented knowledge as "production rules" of the form if this situation is found in the data then undertake this action or add to the data in this way"; the inference engine applied these rules using "forward and/or backwar chaining" to control the sequence in which rules were applied. Since the beginning c the 1980s commercial products (shells) have been produced to support this approach Later on, frames and objects have also been included in the commercial products as second method for knowledge representation. Inference methods for frames an objects are supported by "methods" attached to them, and "inheritance". Althoug successful applications have been implemented using rules, the number of rules i these applications is small (25 to 50 rules).

In the mid 1980s, some scientists noted the disadvantages of using production rule and a new trend has appeared emphasizing inference at the "knowledge level" c human problem solving. This new wave of KBS was called second generation exper systems. In the USA, the Generic Task (GT) methodology and Role Limiting Metho have appeared, while the Knowledge Analysis and Design Structuring methodolog (KADS) has appeared in Europe.

The main idea of these second generation expert systems was to characterize th system as a task that performs a specific function such as diagnosis, planning scheduling and others. The task structure consists of one or more sub-tasks, and/or c primitive problem solving methods (PSM). In the GT methodology the granularity c the PSM is coarse whereas in the KADS methodology the granularity is fine. Eac PSM uses its appropriate knowledge representation scheme. Unfortunately, fev commercial products have appeared to implement second generation expert system: though some tools have appeared to help in using the KADS methodology in th design phase.

Capturing the rules of an expert system can itself require great expertise, and promising technology here is that of rule induction - the system 'learns' the decisio rules from examples of correct decisions (see Chapter 14). Another learning metho that is promising is Neural Networks.

Case Based Reasoning (CBR) is a promising approach, if a record of successfi decisions is available, and if the context of the decision can be characterized by number of attributes which then can be used to assess the similarity or otherwise of new situation to past cases. An example of the application of CBR is th determination of the equivalence of educational qualifications (see Chapter 15).

If possible the KBS being used should provide advanced facilities (e.g. generatin explanations on how and why particular conclusions have been drawn). Built-i uncertainty factors should be available to allow analysis using incomplete an unreliable data in decision-making procedures and evaluation of the probability c results. Collaboration between KBS could be useful in solving some problems in way similar to decision making by an interdisciplinary group of experts.

The application of KBS to capture expertise in sustainable development falls int two main categories: where the KBS is the core of the DSS, and where the KBS is a auxiliary to some other system. The first category is where there is human expertise i an area related to sustainable development such as crop management, pollutio handling and so on (see Chapter 13).

The second category is where there is a need for expertise in handling the resul from a certain model and/or software package in order to let the results of that syste be comprehensible to the decision makers. Examples are the output from sophisticated simulation model for economic growth, or the reports generated from large databases.

4.6 DSS Integration

It is agreed that a key capability of DSS must be the interoperation of tools obtained from different sources. We must be able to choose the appropriate tool for a particular job and transfer information into it and out of it as we explore the alternative decisions available to us. This transfer of information is difficult at present, though there is a move towards more open systems - data interchange is already well established in manufacturing and publishing, and standards for GIS are being developed.

There is an emerging trend of data standardization for GIS in developed countries. The National Committee for Digital Cartographic Data Standards (NCDCDS) and the Federal Interagency Coordination Committee on Digital Cartography (FICCDC) try to establish standards to ensure compatibility among digital spatial data gathered by different agencies (Digital Cartographic Data Standards Task Force, 1988). A similar effort is also being made in the United Kingdom and Canada. Attempts have been made in the People's Republic of China to arrive at a national standard of geographic co-ordinate system for GIS, classification system for resources and environmental information, and the delineation of boundaries of administrative, natural, and drainage area regions (see Chapter 5).

An alternative to this openness mediated by standards is to have a set of facilities already integrated from a single supplier. Many program development systems aimed at a particular class of problems may also have integrated with them many of the other facilities that we have listed above. Two of these are important for us.

Off-the-shelf KBS may include not only mechanisms for knowledge representation and inference, as described in section 4.5, but may also have all of the visualization and user interface development facilities that we described in section 4.3 (see Chapter 5). Further, it is possible to use rule based systems in a more general manner, i.e. as a programming tool that does not require the level of technical expertise that conventional programming requires, as we see in the effective use of KBS in China (see Chapter 5).

Off-the-shelf GIS are capable of integrating geographical data with other data from various sources to provide the information necessary for effective decision making in planning sustainable development. Typically a GIS serves both as a tool box and a database. As a tool box, a GIS allows planners to perform spatial analysis using its geoprocessing or cartographic modeling functions such as data retrieval, topological map overlay and network analysis. Of all the geoprocessing functions, map overlay is probably the most useful tool for planning and decision making - there is a long tradition of using map overlays in land suitability analysis. Decision makers can also extract data from the database of GIS and input it into different modeling and analysis programs together with data from other database or specially conducted surveys. It has been used in information retrieval, development control, mapping, site selection, land use planning, land suitability analysis, and programming and monitoring. GIS can be seen as one form of spatial DSS. GIS have been applied in many decision situations, from land usage in West Africa (Eastman, 1992) to tourism in the Cayman Islands (REDATAM, 1996).

4.7 A reference architecture for decision support systems for sustainable development

The current state of technology means that systems developed in different places do not connect together well. We believe that the time is ripe now to move on to the next generation of technology to support decision making. We need to define a high level reference architecture derived from approaches being developed in Federated Database Systems (Tracz, 1993).

A set of DSS "product lines" should be identified (eg. land, water, energy healthcare, etc.) where each one has its users and stakeholders clearly identified AND the interoperability issues have been addressed (eg. land and water models can be analyzed together). Each product line should be configurable, extendible and designed to "reuse" a common set of capabilities shared by other product lines. We will need to identify the major software components and the interfaces through which they will exchange data. We must support the integration of software tools from wherever they are available. To achieve this we would need to undertake the following process based upon the experience of NASA in their work on domain-specific architectures (Tracz, 1993).

4.7.1 Analysis phase
1. Create a list of DSS domains.

2. Rank the domains according to impact.

3. Gather several (dozen?) scenarios that reveal how DSS will/can be used.

4. Establish a common vocabulary describing DSS.

5. Create a thesaurus, if necessary.

6. Create a list of "capabilities" DSS should/could provide.

7. Classify these capabilities as being "common", "required", "optional" or "alternative" across DSS.

8. List the design and implementation constraints on current and anticipated hardware platforms, operating systems, databases to be supported, etc. This includes interoperability with legacy systems and the data formats they would impose.

9. Review the scenarios, attributed capability list and design/implementation constraint list with the various stakeholders (eg. end users, developers, funders, etc.).

10. Update this information (called a domain model in some circles) and iterate again, adding more detail until some form of consensus has been achieved.

4.7.2 Design phase

1. Develop a "layered", "configurable" architecture that provides the identified capabilities while satisfying certain design and implementation constraints. The layering will allow common functionality/capabilities to be used across product lines as well as provide for a common technology insertion point.

2. Specify/Modify/Adopt data standards for use by the system.

3. Design a configuration mechanism, whereby the architecture can be specialized to meet end user application-specific requirements.

4. Evaluate the architecture for non-functional requirements such as concurrency, reliability, fault tolerance, security, performance, throughput, interoperability, configurability, extendibility, scalability, etc. (One technique is to identify a set of scenarios that address each of these requirements.)

5. Define interfaces of components that make up the architecture.

6. Define the communication protocols used between components.

7. Publicize the architecture/data standards.

4.7.3 Implementation phase

1. Fund the implementation of the configuration mechanism (or modify an existing one).

2. Prioritize the set of components.

3. Incrementally fund the development of the components according to the prioritized list. Hopefully, the marketplace will respond and develop "plug compatible" components.

4. Pick one or two DSS domains and use the existing artifacts to "deliver" a DSS.

5. Verify implementation and calibrate.

Ideally, we should patent the architecture and offer a no-cost license to maintain control over it. As a minimum, we should copyright the interfaces. In addition, the approach needs to be disseminated through a series of workshops, tutorials, videos, and so on.

5. DSS for sustainable development

5.1 Lack of quality data

The lack of available data is one of the major hindrances in the use of DSSfSD. Data is vital. In developed countries, most data needed is readily available thus making the establishment of a DSS relatively easy, but data is not so readily available in developing countries.

The most readily available data is that from remote sensing, but this data is mainly limited to land cover information from which a very limited amount of information can be extracted. Base maps are often lacking or are outdated, compiled by different agencies with different accuracy and map scales, and geocoding systems making them difficult to be integrated into the system. Nevertheless much useful planning can be undertaken, as for example in China where satellite remote sensing data has been used in planning for disasters and changes in land use at the national level (Chen et al., 1994) and even at the local level (Wu, 1995).

Socio-economic data is generally lacking and is often limited mainly to census data, though this can be very useful. The capturing of socio-economic data requires field surveys which are expensive and time consuming.

However, the main obstacle still lies in government's recognition of the need for statistical information for planning and its willingness to mobilize resources in collecting data.

It is not only the availability of data that is a problem, but the quality too. In India it was found that locally collected data could be up to 30% in error (Bhatnagar et al., 1994). The solution seemed to lie in making the collectors of the data also the beneficiaries, so that they had a stake in the quality of the data collected.

The currency of data is very important in decision making and there is a need for institutional arrangements to determine, coordinate and monitor the frequency of data updating, and verifying the quality of the data collected.

The centrality of data to the adoption of DSS and the high costs and lead time needed to acquire data make it highly desirable to ensure that data is seen as a national asset serving multiple purposes.

A first step towards acquiring this asset is coordination. Early and relatively cheap measures would include a national register of available data to forestall repeated and duplicated acquisition. It might also be possible to encourage projects to extend their activities to acquire, at low incremental cost, additional data highly likely to be used by other projects. A second step would center on encouraging the adoption of standards for the content and representation of data, to provide a formal guarantee that the data will be applicable to other projects.

5.2 Need for Education and Training

The current practice of decision making in developing countries has not advanced much in comparison to the tools available to help. The skills of planners and the planning system itself may not be ready to utilize the data and functions available. Planners may not yet be aware of the benefits and potential applications of technology. Little effort has been spent on transforming data into information for making decisions. Consequently decision making in the interests of a few dominant stakeholders results.

There is a general shortage of human resources even in developed countries. This shortage is more severe in developing countries both in absolute numbers and in relative terms, where training can be made difficult due to a lack of expertise and a shortage of funds in universities that do not lead in DSS education and research. Very often, it is government agencies that buy and use the latest systems through funding from international agencies

Training programs are needed for five major groups of users - policy makers, decision takers, programmers, technicians and educators. Policy makers should be made aware of the uses and limitations of DSS. Decision takers in the field should have a general understanding of data, models, and relational data structures, and the use of DSS functions in different stages of urban and regional planning processes. A higher level of technological competence is needed for the training of programmers. They must acquire skills to manage the system and to develop application modules to meet local needs. Technicians need to be trained in data collection and entry, particularly the technical process involved and the likely types of errors encountered. Educators should be kept informed of the latest developments in DSS. Universities and higher educational institutions should invest more in DSS training and research in order to develop local expertise.

In the Philippines we have found (see Chapter 6) that developing human resources is a slow process, requiring many years during which relevant data is gathered and DSS are developed to fit local needs.

It will not necessarily be possible to transport training programs from developed countries, since decision making processes may be different. This has been discussed above in sections 3 and 4, and arises partly from the need to handle uncertainty and risk, but also can be attributed to cultural differences - how decisions are made and agreements reached may be very different.

5.3 Leadership and organization

The strong influence of leadership and organizational setting on the effective use and introduction of computers is very well documented. A few key individuals interested in computers become instrumental in the initial acquisition of equipment and guide its applications. The function of leadership is to set clear goals and objectives, to win acceptance among information system users for such goals and objectives, and to provide commitment to achieve project goals and tasks. Another critical function of leadership is coordination of the different departments sharing the information system.

Prior computer experience can also be critical, both in ensuring awareness of the computer's potential and for the infrastructure to support their use. DSS projects are very often initiated by international assistance agencies and there is a general failure to take account of the organizational setting and personal motivations of those involved. There is evidence of large investments having been made to acquire technology, but there is less evidence that the systems are functioning satisfactorily and contributing to national development efforts. Moreover, problems due to maintenance costs and in the transfer of expertise arise when the international assistance ceases.

5.4 Software development

Software for large scale systems is purchased mainly from developed countries. It is expensive and consumes much foreign currency which is often in short supply. There is a general lack of locally developed software. Attempts have been made to use low cost commercial software to perform DSS tasks, most often using combinations of commercial CAD (Computer Aided Design) packages such as AutoCAD with commercial database packages such as Microsoft Access. These systems, although limited, can make decision support available to departments and agencies with little funding. However, these low cost software systems still need to be purchased from developed countries.

There have been quite a number of software developments in developing countries. However, these developments are fragmented and most involve one to two researchers. Developing countries do not have the human resources and institutional setup to develop and maintain software like the commercial packages available in developed countries (Krishnayya, 1995).

There may be a need for different researchers in a country or region to pool their human and other resources to develop a package that can have good documentation, manuals and support, similar to commercial packages in developed countries. Networks need to be established both within the developing world and with the developed world. Already there are initiatives under way to do this in some regions (AfricaGIS'95, 1995), and this workshop has lead to further transnational networking.

Usability, and particularly the natural language of the interface, is a barrier to the adoption of technology. Most imported programs and manuals are written in English, but most users, decision makers especially, have a limited understanding of English. User-friendly application programs, which hide the technology from users by providing instructions or pull-down menus written in local languages, need to be developed so that local planners and decision makers can use DSS (see Chapter 16).

5.5 Maintenance

Most of the DSS hardware and software used currently is imported from developed countries. It often takes a long time to repair a piece of hardware, particularly when the necessary components are not readily available locally. Equally, it is difficult to consult software companies when problems arise. Also, most of the service and expertise is concentrated mainly in large cities, especially primary cities, making hardware and software maintenance more problematic for sites located elsewhere. Systems must be available on low-end platforms like PCs and must be fully serviceable in-country, as would be the case for locally produced software. Large countries with a substantial requirement for DSS and GIS systems should be encouraged to develop suitable software locally.

Funding to acquire the system is mainly available through central government funding or international assistance, but little is available to maintain the system. Very often, the system cannot be in full operation because one or two terminals and peripherals are out of order and the agency responsible does not have the funds to repair them. More serious is the fact that there may not be funding and institutional arrangements to update the data once it is created. Decision making requires up-to-date information: the system will be rendered useless without it. The development of DSS should be considered as a continuous process and not just a one-off project. The sustainability of DSS themselves is important.

6. Conclusions and secommendations

The workshop has established a shared understanding of the current state of development and application of DSSfSD. The papers written for the workshop will be available through the IDRC library, and a book will be produced documenting a comprehensive range of case studies of decision making for sustainable development and the methods and tools that were used.

This has left us with a very strong foundation from which to move forward. We recommend that financial and organizational support be found for the following actions:

6.1 DSSfSD practitioner community building

The network established at this workshop should be strengthened and enlarged through:

- the establishment of an Internet list service through which experience can be shared, using for example the established devices of FAQs (frequently asked questions), newsletters, and WWW home pages;

- the establishment of focussed interest groups within DSSfSD (eg. groups looking at the application of expert systems and GIS);

- the establishment of a journal on DSSfSD;

- the arrangement of a follow-up workshop or conference, possibly coupled with some other event like the forthcoming CARI conference in Africa; and

- the sharing of this communication in languages other than English, ideally in all official languages of the UN.

It is through this community and the communication channels established that results of the other actions proposed be disseminated.

6.2 Database of existing DSSfSD projects

Many studies of DSSfSD have been undertaken, but most of these have been in research laboratories. A database of theoretical and operational DSSfSD in developed and developing countries is needed to find out how DSSfSD are developed in the research laboratory and how they are actually used in the real world environment. It should record:

- theoretical and actual use of DSSfSD;

- planning and implementation stages covered;

- sectors within the developing country;

- type of decision addressed;

- software, model, and data used; and

- organizational structure for using DSSfSD.

This database will help us understand the current state of the art and practice, and help identify areas for further development.

6.3 Focussed DSSfSD projects

Leverage can be taken from the sharing of experiences and results. Suitable projects could be in the following areas:
- land management;

- tourism;

- planning;

- disaster reduction; and

In all these areas it is important to secure that local level solutions are used to development problems. This should build upon groups of existing projects, such as regional planning in the Philippines and India, agriculture in Egypt and Thailand, and the Indian honey-bee network.

6.4 DSSfSD software development and distribution

It is important that appropriate software is readily available from whatever sources are suitable. Work should be initiated to:
- define reference architecture for DSS which identifies major components and their interfaces;

- develop a workbench for DSS including KBS, GIS and modelling;

- identify and distribute free software for this architecture and workbench;

- ensure that DSS are appropriately multi-lingual and multi-cultural; and

- promote international standards and processes for DSS and GIS.

6.5 Training and awareness raising

Substantial programs for training and education in decision making and the use of tools in this process need to be made available to development planners at all levels. In order to bring this about we recommend projects to:
- identify and make available free training and educational materials;

- develop further materials as necessary;

- assess the need for follow-up awareness raising events and organize these as needed;

- establish sharing networks for planners and decision makers in the area of sustainable development; and

- compile reports of case studies as an aid to this.

The provision of free software will be important in facilitating this.

6.6 Advancing the foundation for DSSfSD

To support the improvement and enhancement of decision making in sustainable development, studies of the underpinning foundations need to be carried out. Projects should be initiated to:

- characterize the decision making process formally;

- establish new approaches to decision making appropriate to the different environments in which these decisions need to be made; and

- undertake social studies of the decision making process in different regions of the world.

7. References

Ahmad, Y.J., S.E. Serafy, and E. Lutz (Eds.) (1989). *Environmental Accounting for Sustainable Development*, Washington, D.C: The World Bank.

AfricaGIS'95 (1995), Inventaire des Applications SIG en Afrique, Published by Da Vinci Consulting s.a, Chaumont-Gistoux, Belgium

Barbier, E.B. (1987). "The Concept of Sustainable Economic Development", *Environmental Conservation*, 14(2), 101-110.

Batty, Michael. (1990). "Information Systems for Planning in Developing Countries, Information Systems and Technology for Urban and Regional Planning in Developing Countries: A Review of UNCRD's Research Project", Vol. 2, Nagoya, Japan: United Nations Centre for Regional Development.

Bhatnagar S.C. (1991), "Impacting Rural Development through IT: Need to Move beyond Technology" in Goyal M.L. (ed.), *Information Technology for Every Day Life*, Tata, McGraw-Hill.

Bhatnagar S.C., Rajan R., Das K. (1994), *A Manual for DLPDSS*, Ahmedabad, Indian Institute of Management

Chen Yufeng, He Jianbang (1994), "Applying Integrated Technology of Remote Sensing and Geographic Information System in the Monitoring and Evaluation of Major Natural Disasters", *Journal of natural Disasters*, April 1994

Dasmann, R. F. (1964). *Wildlife Management*, New York: John Wiley & Sons.

Digital Cartographic Data Standards Task Force (1988). "Special Issue on the Proposed Standard for Digital Cartographic Data", *The American Cartographer*, 15(1), 9-140.

Eastman J.R. (1992), *IDRISI: A Grid Based Geographic Analysis System*. Version 4.0, Worcester, MA: Clark University

Eastman .R., Jiang H. (1996), "Fuzzy Measures in Multi-Criteria Evaluation", in *Proceedings of the Second International Symposium on Spatial Accuracy Assessment in Natural Resources and Environmental Studies*, Fort Collins, Colorado

Gupta A.K. (1995), "Knowledge Centre: Building Upon What people Know", Conference on Hunger and Poverty, IFAD, Brussels.

IUCN (1980), World Conservation Strategy, International Union for the Conservation of Nature, Gland, Switzerland.

Kenney R.L., Von Winterfeldt D., Eppel T. (1990), "Eliciting Public values for Complex Policy Decisions", *Management Science*, 36(9), 1011-1030.

Kersten G. and D. Meister (1995). "DSSfESD Hardware and Software Catalogue", IDRC Report, Ottawa.

Krishnayya J.G. (1995), Regional Planning with THEMAPS Using Vector Analysis and Raster Analysis in a Case Study, Pune: Systems Research Institute

Loening, U.E. (1991). "Introductory Comments: The Challenge for The Future", in A.J. Gilbert and L.C. Braat (eds.), *Modelling for Population and Sustainable Development*, London: Routledge, 11-17.

Meadows, D.H., D.L. Meadows, J. Randers, and W. W. Behrens III (1972). *The Limits to Growth*, New York: Universe Books.

Meadows, D.H., D.L. Meadows and J. Randers (1992). *Beyond the Limits*, London: Earthscan Publication Limited.

NCE (1993). *Choosing a Sustainable Future*, National Commission on the Environment, Washington, D.C.: Island Press.

O'Riordan, T. (1993). "The Politics of Sustainability", in R. Kerry Turner (Ed.), *Sustainable Environmental Economics and Management: Principles and Practice*, New York: Belhaven Press, 37-69.

Rao M., Jayaraman V., Chandreasekhar M.G., (1994), "Organizing Spatial Information Systems around a GIS Core", Indian Space Research Organization (ISRO) Special Publication - ISRO-NNRMS-SP-70-94

REDATAM+ Project (1996), United Nations Economic Commission for Latin America and the Carribbean (Chile) and the University of Waterloo (Canada).

Redclift, M. (1987). *Sustainable Development: Exploring the Contradictions*, London: Methuen.

Repetto, R. (1986). *World Enough and Time*, New Haven, Conn.: Yale University Press.

Simon, H.A. (1960). *The New Science of Management Decision*, New York: Harper and Row.

United Nations (1994). Earth Summit CD-ROM.

Tracz W., Coglianese L., Young P., (1993), "A Domain-Specific Software Architecture Engineering Process Outline", *ACM Software Engineering Notes*, April 1993, 18(2), 40-49.

World Bank (1990). *World Development Report*, Oxford: Oxford University Press.

WCED (1987). *Our Common Future*, World Commission on Environment and Development, ford: Oxford University Press.

Wu Qingzhou (1995), "Lessons and Experience of Urban Flood Disasters of South China in 1994", in *Journal of natural Disasters*, (4) suppl., 128-131.

Yeh A.G.O., Li X. (1997), "An Integrated Remote Sensing and GIS Approach in the Monitoring and Evaluation of Rapid Urban Growth for Sustainable Development in the Pearl River Delta, China", *International Planning Studies*, 2(2), 193-210.

20 DSS APPLICATION AREAS
Gregory E. Kersten and Gordon Lo

1. Introduction

The concept of computer-based decision support was born in the early 1970s and is generally attributed to two articles. The first, written by Little (1970), introduced the notion of decision calculus as a "model-based set of procedures for processing data and judgements to assist a manager in his decision making." The second article, written by Scot Morton (1971), introduced support systems for managerial decision making.

The objective of this chapter is to provide references to various implementations of DSSs that were published in the 1990s. In the following nine sections we mention a number of DSS applications to environmental decision making and assessment, water resource management, agriculture, forestry, manufacturing, medicine, business and organizational support, and infrastructure. The list of references is not exhaustive but it demonstrates both scope and interest in practical research in the decision support systems field. It gives MIS managers, developers and decision makers a rich source of information about the specific domains where different DSS technologies have been used in both developing and developed countries.

1.1 General books and articles

Since the late 1980s DSS has been widely studied and taught in business, engineering, information systems and other university courses. A number of books providing a comprehensive introduction to the DSS area has been published. Recently published books include those by Dhar and Stein (1997), Holsapple and Whinston (1996), Marakas (1999), Mallach (1994), Sauter (1997), Sprague and Watson (1997), Turban and Aronson (1998).

DSSs have been applied in a variety of problems. A good survey of the application literature can be found in Eom, Lee et al. (1998), Santhanam and Elam (1998) and

Liang and Hung (1997). Surveys of negotiation and group decision support systems have also been conducted (Aiken *et al.*, 1993; Vogel and Nunamaker, 1990).

1.2 Web resources

Numerous World Wide Web (Web) sites provide DSS related resources. Several Web sites (Arnott, 1999; Demarest, 1999; Kersten, 1999; Power, 1999; Power and Quek, 1999) provide general information related to the design, development, evaluation, and implementation of DSSs. A list of DSS software publishers can be found in these sites and also in the *Data Warehousing Information Center* which is available on the Web (LGI, 1999). Quek (1999) maintains a site which contains a listing of Decision Support Systems Courses conducted by many institutions and universities across the world.

Demonstration or scaled down versions of DSSs can be downloaded from the Web. Many software publishers place demonstration versions of their products on the Web. DSSs for specific domains and types of problems can be obtained directly from the Web. For example, there are DSSs developed for agricultural purposes (Agriculture Canada, 1999; FARAD, 1999). Several water resources management DSSs can be downloaded from the Internet (CDSS, 1999; IIASA, 1999; Silvert, 1999). The Environmental Programs group of the North Carolina Supercomputing Center (NCSC, 1999) has developed an environmental DSS Web site that provides definitions and software available for downloading.

2. Environmental decision making

Environmental DSSs have been developed to assess the impact of utilization of natural resources and to evaluate the impact of agricultural and industrial activities on the environment. Guariso and Werthner (1989) discuss environmental decision support systems and provide architecture for such systems. Several applications of multiple criteria decision making in environmental management can be found in Paruccini (1994). References to many systems (including environmental DSSs and GISs) which were developed to solve environmental problems in Argentina, Chile, Columbia, Egypt and Poland are available on the Web site ERDAS (1999). Sandia National Laboratories (1999) developed an environmental DSS that performs risk assessment, based on potential risk to human health and the environment and on the cost of alternatives. Shaw et al (1998) describes several DSSs used for natural resource management in Australia.

Zhu et al (1998) discuss a knowledge-based spatial decision support system for effective environmental management. Gough and Ward (1996) provide a DSS framework that was applied for the management of Lake Ellesmere in New Zealand. Hipel et al (1998) developed GMCR used, among others, to model and analyze international environmental management disputes involving governments in both Canada and the U.S. The cross-disciplinary nature of environmental DSSs explains why relevant materials can be found in journals such as *Computers in Industry, IEEE Transactions on*

Systems, Man and Cybernetics, Journal of Environmental Management and *Journal of Water Resources Planning and Management.*

Hokkanen and Salminen (1994) explore the use of an integrated DSS that assist decision makers to identify optimal schemes for treatment, storage, transport and solid waste disposal. Several DSSs which use multiobjective methods and GIS technology have been developed for this purpose (see Hokkanen and Salminen, 1994; Pinter, Fels et al, 1995; Walker and Johnson, 1996; Swetnam, Mountford et al., 1998; Subramaniam and Kerpedjiev, 1998).

Paigee et al (1998) developed a DSS to assist risk managers to evaluate landfill cover designs for mixed waste disposal sites at Los Alamos. Moon et al. (1998) discuss LANDS (Land Analysis and Decision Support) System, which manages and integrates the data and models required to meet a broad spectrum of land management and planning needs.

The USDA-ARS in Tucson, Arizona has developed DSS to select land conservation management system in Mexico; its implementation revealed problems which are common to many applications of decision support technology in developing countries (Hernandez et al., 1998). Issues of environmental planning are also discussed by Barnikow *et al.* (1992) and Kainuma *et al.* (1990).

Several DSSs for the evaluation of land management practices and their effects can be found in Yakowitz et al. (1992, 1993, 1998a, 1998b) and Robotham (1998). DSSs for land management in developing countries are discussed by Mira Da Silva et al. (1998); Matthew and Peasley (1998); Jones et al. (1998) and Zhang et al. (1998).

3. Environmental impact assessment

An important topic within the scope of EDSS is environmental impact assessment. Several authors have described systems that can be used for this specific purpose (Barnikow et al., 1992; Kainuma et al., 1990; Kampke et al., 1993; Winsemius and Hahn, 1992; Wadsworth and Brown, 1995; Yakowitz et al, 1998). Muth and Lee (1986) present a model that can be used to assess the impact of natural resource exploitation. Ecozone II, a decision support system (DSS) was designed to facilitate EIA's in the sectors of agriculture, agro-industries and aquaculture in less developed countries (Howells, 1998).

GISs are often linked with forecasting models for environmental prediction (Hokkanen and Salminen, 1994; Specht and Owls, 1995). Chiueh et al. (1997) present the benefits of applying spatial decision support systems to assess soil contamination problems.

4. Water resource management

Water resource management is another facet of environmental decision making where DSSs have been applied. Water quality management decision support has been discussed by Berkemer et al. (1993), Camara et al. (1990), Lovejoy et al. (1997), and Xiang (1993). DSSs to support water reservoir decision analysis have been developed by Grobler and Rossouw (1991), and Simonovic (1992). Further, tools to support water delivery maintenance and planning have been presented by Pingry et al. (1991), Sutherland and Lambourne (1991), Thrall and Elshaw-Thrall (1990). DSSs for planning irrigation systems have been developed by Bandyopadhyay and Datta (1990); Tyagi et al. (1993) and Wilmes et al. (1990).

DSSs for strategic planning of water resources in Jordan have been developed by Alshemmeri (1997) and, in the Middle Nile Delta, by Abdel-Dayem et al. (1998). Riverside Technology (1999) designs, develops, and implements water resources management systems that are used in the Yellow River and Huai River in China, and the Nile River in Egypt.

A relatively simple ground water DSS was developed to assist in identifying salt water vulnerable areas and in developing management policies to prevent salt water intrusion in central Kansas (Sophocleous and Ma, 1998). DSS is also used for conjunctive management of surface water and ground water under prior appropriation; (Fredericks, Labadie et al., 1998; Sophocleous, Koelliker et al., 1999).

The system called WATERSHEDSS (WATER, Soil, and Hydro-Environmental DSS) was designed to help watershed managers and land treatment personnel to identify their water quality problems, and to select appropriate management practices. (Osmond, Gannon et al., 1997). Anderssen, Mooney et al. (1996) present a model of hydrodynamic behavior of water resources to construct wetlands in order to meet environmental regulations.

5. Agriculture

DSSs focusing on the overall improvement of agricultural production have been described by Goodrich (1998); Jacucci (1996), Gonzalez-Andujar et al. (1993), Power (1993) and Wagner and Kuhlmann (1991). Gameda and Dumanski (1998) developed Soilcrop system, which provides biophysical and socioeconomic criteria for determining the sustainability of cropping practices in different agroregions.

The comprehensive resource planning system (CROPS) is a multiobjective scheduling system that uses heuristics and constraint satisfaction to find acceptable farm-level plans in Virginia (Stone et al., 1998). DSSs have also been developed for management of agricultural operations and production control (Sorensen, 1998; Vickner and Hoag, 1998). These systems are playing an important role in agricultural management in developing countries such as Egypt, Indonesia and Mexico.

Shtienberg et al. (1990) and Yost and Li (1998) discuss several systems in the area of the disease identification and the specification of treatments.

Knowledge-based computerized DSSs have been designed to solve complex problems in soil acidity, phosphorus deficiency and nitrogen deficiency (Yost and Li, 1998; Kovacs et al., 1998). Jones et al (1998) developed a DSS to identify cropping and tillage options for profit maximization for a wide range of soil erosion control levels.

A DSS that aids decision making in agricultural and water resource management in the Eastern Nile Delta of Egypt was developed by Abu-Zeid (1998). Datta (1995) designed an integrated DSS for generating alternative water allocation and agricultural production scenarios for a semi-arid region. A geographical DSS was developed for the Ministry of Agriculture in Dominican Republic (Grabski and Mendez, 1998).

Amien (1998) developed an expert system that assesses the suitability of specific soil and climate conditions for supporting proper agricultural systems, and the selection of crops in Indonesia.

6. Forestry

Several DSSs have been used to assist decision making in forestry and natural resources (Marathe et al., 1991; Payandeh and Basham, 1993). Covington et al. (1988) developed a DSS called TEAMS that can be used as "a tactical planning system to aid forest managers in developing site specific treatment schedules". Reforestation problems have been addressed by Johnston et al. (1993). Ecosystem management, with the aid of DSS, has been proposed as a solution to many problems facing forestry today (Rauscher, 1999; Twery et al., 1998; Dewhurst et al., 1995). Financial matters related to forest management have been considered by Meyer (1992) and Payandeh and Basham (1993). The identification and treatment of diseases and pests have been analyzed by Power (1988).

Manley and Threadgill (1991) developed a DSS that imbeds a linear programming model to evaluate New Zealand's forests. Aggarwal et al. (1992) discuss methods of providing decision support for planning appropriate lumber harvests. Wood (1998) discusses the case of the Menminee Tribe's reservation timberlands. Forest managers determined that one third of the tribal forest could be converted to more productive and valuable tree species. Working together with Northern Arizona University they developed a DSS to devise a schedule for implementing the conversion Wood (1998). Wybo (1998) introduces a DSS dedicated to forest fire prevention and fighting.

Naesset (1997) discusses the use of GISs to search for sensitive areas that should be devoted to careful timber management practices. Tecle et al (1998) developed a multi-objective and/or multi-person decision support system for analyzing multi-resource forest management problems.

Forest resource managers in two U.S. Forest Service Ranger Districts use INFORMS-R8, a DSS to support common district planning activities (Williams and Holtfrerich, 1998). Ross and Hannam (1998) discuss three DSSs (TEAS, LOGSPERT9, HABASYS) used to manage legal, ecological, and social information required to make decisions about the conservation of forests on Protected Land in Australia.

7. Manufacturing

Deciding on appropriate investment in industrial/manufacturing ventures can be difficult. Park et al. (1990) and Rios (1993) provide some instances of the use of this technology. The analysis and planning of the manufacturing processes, including planning of materials requirements and manufacturing resources, can be assisted with specialized DSSs such as those suggested by Ozdamar, Bozyel et al. (1998), Migliarese and Paolucci (1993), Srihari and Cala (1992) and Suresh (1990).

Production related DSSs have been discussed by Agarwal and Tanniru (1992), Garza et al. (1992), Kleijnen (1993), Martin et al. (1993) and Migliarese and Paolucci

(1993). Numerically controlled machine tools are successfully used in manufacturing. For example, IDSSFlex is a DSS to analyze and evaluate flexible manufacturing systems (FMS) design alternatives (Borenstein, 1998).

8. Medicine

DSSs to support hospital services planning have been developed by many researchers (Beech and Fitzsimons, 1990; Sharkey et al., 1993; Kadas, 1995; Stodolak and Carr, 1992), likely in response to the importance and cost of these services. However, this is not the only type of medical decision making that has been supported using a DSS. Health-care systems are discussed by Datta and Bandyopadhyay (1993) and Doukidis and Forster (1990).

According to Hagland (1998) the number of physicians who use clinical information systems, including decision support tools at the point of care is constantly increasing. DSS applications allow the collection and manipulation of financial and clinical data on a variety of levels, including patient, procedure and physicians (DeLuca and Cagan, 1996). Keegan (1995) notes that the U.S. healthcare decision-support market is growing faster than the market for all other healthcare software.

Morgan (1996) illustrates the use of an intelligent DSS to solve information overload problems faced by health authorities. Leibovici (1997) developed a problem-orientated DSS to improve empirical antibiotic treatment. The development of proper nutritional balance for diets has been considered by Bandyopadhyay and Datta (1990), and Zwietering et al. (1992).

Tropical disease diagnosis has been investigated by Doukidis and Forster (1990). Mishra and Dandapat (1993) propose a system for EMG diagnosis. Further, patient simulation, using a negotiation metaphor, has been investigated by Kersten et al. (1993) and Kersten and Szpakowicz (1993). DSSs related to pharmaceutics was developed by Green and Krieger (1992) and Islei et al. (1991).

9. Business and organizational support

There are numerous examples of DSS applications in support of common organizational decision making activities. They include administration (Edwards, 1992; Johnson, 1996; Mohanty and Deshmukh, 1997); assessment of risk in international investments (Tessmer et al., 1993); corporate crises management (Mak, Mallard et al, 1999), banking network planning (Coats, 1990); credit decisions (Coffman and Brooks, 1992; Levary and Renfro, 1991; Fuglseth and Gronhaug, 1997); portfolio management (Kira et al., 1990); investment strategies (Huynh and Lassez, 1990); marketing (Arinze and Banerjee, 1992; Coffman and Brooks, 1992; Green and Krieger, 1992; Mak H-Y. and T. Buim 1996; Sisodia, 1992; Bruggen, Ahn and Ezawa, 1997; Smidts et al., 1998; Ghose and Nazareth, 1994); capital budgeting (Moribayashi and Wu, 1990); operations management (Proudlove, Vadera et al., 1998); scheduling (Ecker, Gupta et al., 1997; Djukanovic, Babic et al., 1996); strategic management (Martinsons and Davison, 1999); business reengineering (Barua and Whinston, 1998). Further, DSSs have been shown to improve the decision maker's capabilities by improving the visualization of the pertinent financial data (Lawton, 1993).

10. Infrastructure

Support for decision making related to trains and railroads has been discussed by Hoffman (1993), Hanif, S. and Arief, S (1998), Tanzi and Guiol (1998) and Missikoff (1998). Bielli (1992) developed a DSS for urban traffic management. A DSS for a district in the Commonwealth of Virginia was developed through the application of System Dynamics concepts (Garza, Drew et al., 1998). A sea navigation DSS has been presented by Grabowski and Sanborn (1992).

Several DSSs for truck routing and maintenance have been suggested by Bradley (1993), Lysgaard (1992), Ott (1992) and Shannon and Minch (1992). A DSS designed for use with construction projects has been described by Crosslin (1991). Project management DSSs include those described by Arinze and Partovi (1992), Courtney and Paradice (1993), Hastak, Halpin et al. (1996), Liberatore and Stylianou (1993), Archer and Ghasemzadeh, (1998) and Stewart (1991).

References

Abdel-Dayem, S., S. Abedel-Gawad and K. Abu-Zeid (1998). "Water Management Scenario Simulation for Decision Support in Multiobjective Planning", in S. A. El-Swaify and D. S. Yakowitz (Ed.), *Multiple Objective Decision Making for Land, Water, and Environmental Management*, Lewis Publishers, 629-639.

Abu-Zeid, K.M. (1998). "A Multicriteria Decision Support System for Evaluating Cropping Pattern Strategies in Egypt", in: S. A. El-Swaify and D. S. Yakowitz (Ed.), *Multiple Objective Decision Making for Land, Water, and Environmental Management*, Lewis Publishers, 105-120.

Adamopoulos, G. (1994). "Production Management in the Textile-Industry Using the YFADI Decision Support System", *Computer & Chemical Engineering*, 18, 577- 583.

Aggarwal, A. K., R. R. Vemuganti and W. Fetner (1992). "A Model-based Decision Support System for Scheduling Lumber Drying Operations", *Production and Operations Management*, 1(3), 320-328.

Agarwal, R. and M. Tanniru (1992). "A Structured Methodology for Developing Production Systems" *Decision Support Systems*, 8(6), 483-499.

Agriculture Canada (Apr 1999). "Demonstration DSS/ES for Manure Management", http://ozone.crle.uoguelph.ca/manure/default.htm.

Ahn, J. and K. Ezawa (1997). "Decision-Support For Real-Time Telemarketing Operations Through Bayesian Network Learning", *Decision Support Systems,* 21(1), 17-27.

Aiken, M., J. Krosp and J. Johnson (1993). "A Survey of Group Decision Support System Use", *SIGOIS Bulletin*, 14(1), 43-46.

Alshemmeri, T. (1997). "Computer-Aided Decision-Support System For Water Strategic-Planning In Jordan", *European Journal Of Operational Research*, 102(3), 455-472

Amien, I. (1998). "An Agroecological Approach to Sustainable Agriculture", in: S. A. El-Swaify and D. S. Yakowitz (Ed.), *Multiple Objective Decision Making for Land, Water, and Environmental Management*, Lewis Publishers, 465-480.

Anderssen, B., J. Monney, et al. (1996). "Decision Support for the Design of Constructed Wetlands", *Applied Mathematical Modelling*, 20(1), 93-100.

Archer, N. and F. Ghasemzadeh (1998). "A Decision-Support System For Project Portfolio Selection", *International Journal Of Technology Management,* 16(1-3), 105-114.

Arinze, B. and S. Banerjee (1992). "A Framework for Effective Data Collection, Usage and Maintenance of DSS", *Information & Management*, 22(5), 257-268.

Arnott, D. (Apr 1999). "The Monash Decision Support Systems Laboratory", *Monash University*, http://www.sims.monash.edu.au/dsslab/.

Badri, M. A. (1999). "A simulation model for multi-product inventory control management", *Simulation,* 72(1), 20-32.

Baird, I. A., P. C. Catling, et al. (1994). "Fire Planning for Wildlife Management - A Decision Support System for NADGEE-NATURE-RESERVE, Australia", *International Journal of Wildland Fire,* 4(2), 107-121.

Ballou, D. P. and G. K. Tayi (1996). "A Decision Aid for the Selection and Scheduling of Software Maintence Project", *IEEE Transactions on Systems , Man, and Cybernetics,* 26(2), 203-221.

Bandyopadhyay, R. and S. Datta (1990). "Applications of OR in Developing Economies. Some Indian Experiences", *European Journal of Operational Research*, 49(2), 188-199.

Barnikow, A.-M., U. Behrendt, et al. (1992). "DICTUM: Decision Support System for Analysis and Synthesis of Large-scale Industrial Systems", *Computers in Industry*, 18(2), 135-153.

Barua, A. and A. Whinston (1998). "Decision-Support For Managing Organizational Design Dynamics", *Decision Support Systems,* 22(1), 45-58.

Beech, R. and B. A. Fitzsimons (1990). "Application of a Decision Support System for Planning Services within Hospitals", *Journal of the Operational Research Society*, 41(12), 1089-1094.

Behan, R. W. (1994). "Multiresource Management and Planning with EZ-Impact", *Journal of Forestry,* 92(2), 32-36.

Berkemer, R., M. Makowski and D. Watkins (1993). "A Prototype of a Decision Support System for River Basin Water Quality Management in Central and Eastern Europe", Research Report, IIASA, Laxenburg, Austria.

Berner, E. S., G. D. Webster, et al. (1994). "Performance of 4 Computer-based Diagnostic Systems", *New England Journal of Medicine,* 330(25), 1792-1796.

Beulens, A. J. M. and J. A. van Nunen (1988). "The Use of Expert System Technology in DSS" Decision Support Systems 4(4), 421-431.

Bielli, M. (1992). "A DSS Approach to Urban Traffic Management", *European Journal of Operational Research*, 61(1, 2), 106-113.

Bonczek, H., C. W. Holsapple and A. Whinston (1980). "Evolving Roles of Models in Decision Support Systems", *Decision Sciences*, 11(2), 337-356.

Borenstein, D. (1998). "IDSSFlex: an intelligence DSS for the design and evaluation of flexible manufacturing systems", *Journal of the Operational Research Society,* 48(11), 734-744.

Bradley, P. (1993). "The Quiet Revolution in Trucking Services", *Purchasing,* 114(3), 36-38

Bruggen, G. H. v., A. Smidts, et al. (1998). "Improving Decision Making by Means of a Marketing Decision Support System", *Management Science,* 44(5), 645-658.

Camara, A. S., M. Cardoso da Silva, et al. (1990). "Decision Support System for Estuarine Water-quality Management", *Journal of Water Resources Planning and Management,* 116(3), 417-432.

Caporaletti, L. E. and S. E. Dorsey (1994). "A Decision-Support System for In-Sample Simultaneous Equation Systems Forecasting using Artificial Neural Systems", *Decision Support Systems,* 11(5), 481-495.

CDSS (Apr 1999)."Colorado's Decision Support Systems", http://crdss.state.co.us/.

Chiueh, P. -T., S. -L. Lo, et al. (1997). "Prototype SDSS for Using Probability Analysis in Soil Contamination", *Journal of Environmental Engineering,* 123(5), 514-519.

Coffman, J. Y. and R. Brooks (1992). "Interactive Decision Systems for Improved Profitability", *Credit World*, 80(3), 30-33.

Covington, W. W., D. B. Wood, et al. (1988). "TEAMS: A Decision Support System for Multi-resource Management", *Journal of Forestry*, 86(8), 25-33.

Coats, P. K. (1990). "Combining an Expert System with Simulation to Enhance Planning for Banking Networks", *Simulation*, 54(6), 253-264.

Courtney, J. F. and D. B. Paradice (1993). "Studies in Managerial Problem Formulation Systems", *Decision Support Systems*, 9(4), 413-423.

Crosslin, R. L. (1991). "Decision-support Methodology for Planning and Evaluating Public-private Partnerships", *Journal of Urban Planning and Development*, 117(1), 15-31.

Datta, S. (1995). "A Decision Support System for Micro-Watershed Management in India", *Journal of Operational Research Society*, 46(5), 592-603.

Datta, S. and R. Bandyopadhyay (1993). "An Application of OR in Micro-level Planning in India", *Computers and Operations Research*, 20(2), 121-132.

DeLuca, J.M. and R.E. Cagan (1996). *The CEO's Guide to Health Care Information Systems*, American Hospital Publishing.

Demarest, M. (Apr 1999). "Technology And Policy In Decision Support Systems", DP Applications Inc, http://www.dpapplications.com/library/tpdss.html.

Dewhurst, S.M., W.W. Covington, et al. (1995). "Developing a model for Adaptive Ecosystem Management: Goshawk Management on Arizona's Kaibab Plateau", *Journal of Forestry*, 93(12), 35-43.

Dhar V. and R. Stein (1997). *Intelligent Decision Support Methods. The science of Knowledge Work*, Upper Saddle River, NJ: Prentice Hall.

Djukanovic, M., B. Babic, et al. (1996). "Fuzzy Linear Programming Based Optimal Fuel Scheduling Incorporating Blending/Transloading Facilities", *IEEE Transactions on Power Systems*, 11(2), 1017-1023.

Doukidis, G. I. and D. Forster (1990). "Potential for Computer-aided Diagnosis of Tropical Diseases in Developing Countries. An Expert System Case Study", *European Journal of Operational Research*, 49(2), 271-278.

Edwards, J. S. (1992). "Expert Systems in Management and Administration - Are They Really Different from Decision Support Systems?", *European Journal of Operational Research*, 61(1, 2), 114-121.

Eom, S. B., S. M. Lee, et al. (1998). "A survey of decision support system applications (1988-1994)", *Journal of the Operational Research Society*, 49(2), 109-120.

ERDAS (Apr 1999). "ERDAS, Inc. – Before You Buy", http://www.erdas.com/before/casestudies/index.html.

FARAD (Apr 1999)."Food Animal Residue Avoidance Databank", http://www.farad.org/.

Finlay, P. and C. Martin (1988). "The State of Decision Support Systems: A Review", *Omega*, 17(6), 525-531.

Fredericks, J. W., J. W. Labadie, et al. (1998). "Decision Support Systems for Conjunctive Stream-Aquifer Management", *Journal of Water Resources Planning and Management*, 124(2), 69-78.

Fuglseth, A. M. and K. Gronhaug (1997). "IT-enable Redesign of Complex and Dynamic Business Processes: the Case of Bank Credit Evaluation", *Omega*, 25(1), 93-106.

Gameda, S. and J. Dumanski (1998). "Soilcrop - A Prototype Decision Support System for Soil Degradation - Crop Productivity Relationship", in: S. A. El-Swaify and D. S. Yakowitz (Ed.), *Multiple Objective Decision Making for Land, Water, and Environmental Management*, Lewis Publishers, 349-362.

Garza, O. F., A. J. Golub, et al. (1992). "Load Cubing at R J. Reynolds", *Production and Operations Management*, 1(2), 151-158.

Garza, J. M. d. l., D. R. Drew, et al. (1998). "Stimulating Highway Infrastructure Managment Policies", *Journal of Management in Engineering*, 14(5), 64-72.

Ghose, S. and D. L. Nazareth (1994). "Selecting Appropriate Support for Marketing Decisions", *Omega*, 22(5), 443-456.

Gonzalez-Andujar, J. L., J. L. Garcia-de Ceca and A. Fereres (1993). "Cereal Aphids Expert System (CAES): Identification and decision making", *Computers and Electronics in Agriculture*, 8(4), 293-300.

Goodrich P.R. (1998), "Smart Pitchfork", in: S.A. El-Swaify and D.S. Yakowitz (Ed.), *Multiple Objective Decision Making for Land, Water, and Environmental Management*, Lewis Publishers, 189-195.

Gough, J. D. and J. C. Ward (1996). "Environmental Decision-Making and Lake Management", *Journal of Environmental Management*, 48(1), 1-15.

Grabowski, M. and S. Sanborn (1992). "Knowledge Representation and Reasoning in a Real-time Operational Control System: The Shipboard Piloting Expert System (SPES)", Decision Sciences, 23(6), 1277-1296.

Grabski, S. V. and D. Mendez (1998). "Implementation of a knowledge-based agricultural geographic decision-support system in the Dominican Republic: a case study", *Information Technology & People*, 11(3), 174-193.

Green, P. E. and A. M. Krieger (1992). "An Application of a Product Positioning Model to Pharmaceutical Products", *Marketing Science*, 11(2), 117-132.

Grobler, D. C. and J. N. Rossouw (1991). "Applications of a decision support system to develop phosphorus control strategies for South African reservoirs", in *Water Quality Modeling*, 273-396

Guariso, G. and H. Werthner (1989). *Environmental Decision Support Systems*. New York, Halsted Press.

Hagland, M. (1998). "IT and Point-of-care Decision Support", *Health Management*, 19, 10-15.

Hanif, S. and Arief, S (1998). "A Tactical Decision Support System for Empty Railcar Management", *Transportation Science*, 32(4), 306-329

Hastak, M., D. W. Halpin, et al. (1996). "COMPASS-New Paradigm for Project Cost Control Strategy and Planning", *Journal of Construction Engineering and Management*, 122(3), 254-264.

Hernandez, M., P. Heilman, et al. (1998). "Use of a DSS for Evaluating Land Management System Effects on Tepetate Lands in Central Mexico", in: S. A. El-Swaify and D. S. Yakowitz (Ed.), *Multiple Objective Decision Making for Land, Water, and Environmental Management*, Lewis Publishers, 571-583.

Heylighen, F. (1992), "A Cognitive-systemic Reconstruction of Maslow's Theory of Self-actualization", *Behavioral Science*, 37, 39-58.

Hill, P.H. *et al.*, (1982), *Making Decisions. A Multidisciplinary Introduction*, Reading, MA: Addison-Wesley.

Hipel, K.W., D.M. Kilgour, et al. (1998). "Using the Decision Support System GMCR for Resolving Conflict in Resource Management", in: S. A. El-Swaify and D. S. Yakowitz (Ed.), *Multiple Objective Decision Making for Land, Water, and Environmental Management*, Lewis Publishers, 23-47.

Hoffman, T. (1993). "Union Pacific rail says, 'I know I can'", *Computerworld*, 27: 91- 93.

Hokkanen, J. and P. Salminen (1994). "The Choice of a Solid Waste Management System by Using the ELECTRE III decision-aid method", in: M. Paruccini (Ed.), *Applying Multiple Criteria Aid for Decisions to Environmental Management*, Kluwer Academic Publishers, 111-154.

Holsapple, C. W. and A. B. Whinston (1996). *Decision Support Systems. A Knowledge-based Perspective*, New York: West.

Howells, O. (1998). "Ecozone-II - A Decision-Support System For Aiding Environmental-Impact Assessments In Agriculture And Rural-Development Projects In Developing-Countries", *Computers And Electronics In Agriculture*, 20(2), 145-164.

Huff, S. L., S. Rivard, A. Grindlay and I. P. Suttie (1984). "An Empirical Study of Decision Support Systems", *INFOR*, 21(1), 21-39.

Huynh, T. and C. Lassez (1990). "Expert Decision Support System for Option-based Investment Strategies", *Computers and Mathematics with Applications*, 20(9-10), 1-14.

IIASA (Apr 1999). "IIASA - WATER RESOURCES PROJECT - DESERT MODEL", http://www.iiasa.ac.at/Research/WAT/docs/desert.html.

Islei, G., G. Lockett and M. Stratford (1990). "Resource Management and Strategic Decision Making in Industrial R&D Departments. Decision Support Using Judgmental Modelling in the Chemical Industry", *Engineering Costs and Production Economics*, 20(2), 219- 229.

Jacucci, G. (1996). "Developing Transportable Agricultural Decision-Support Systems: An Example", *Computers and Electronics in Agriculture*, 14(4), 301-315.

Johnson, M. L. (1996). "GIS in Business: Issues to Consider in Curriculum Decision-Making", *Journal of Geography*, 95(3), 98-106.

Johnston, M., Y. Wand and M. Curran (1993). "An Expert System to Support Site Preparation Decisions Related to Reforestation" *INFOR*, 31(3), 221-243.

Jones. A.J., R. Selley and L.N. Mielke (1998). "Multi-Objective Decision Support Strategies for Cropping Steep Land", in: S. A. El-Swaify and D. S. Yakowitz (Ed.), *Multiple Objective Decision Making for Land, Water, and Environmental Management*, Lewis Publishers, 93-103.

Kadas, R. (1995). "Healthcare Data Analysis Systems", *Health Management Technology*, 16(7), 12-14.

Kainuma, M., Y. Nakamori and T. Morita (1990). "Integrated Decision Support System for Environmental Planning", *IEEE Transactions on Systems, Man and Cybernetics*, 20(4), 777-790.

Kampke, T. (1988). "About Assessing and Evaluating Uncertain Inferences Within the Theory of Evidence", *Decision Support Systems*, 4(4), 433-439.

Keegan, A. (1995). "Decision-support Systems Aiding New Care Delivery", *Health Management Technology*, 16(7), 50-55.

Keen, P. G. W. and M. S. Scott-Morton (1989). Decision Support Systems: An Organizational Perspective. Reading, MA, Addison-Wesley.

Keeney, R.L. (1992). *Value-focussed Thinking. A Path to Creative Decision Making*, Cambridge, MA: Harvard University Press.

Kersten, G.E. (Apr 1999). "IS Resource Page", *Carleton University*, http://www.business.carleton.ca/~gregory/teaching/is_materials/.

Kersten, G.E., (1997). "Support for Group Decisions and Negotiations. An Overview", in J. Climaco (ed.), *Multicriteria Analysis*, Heilderberg: Springer Verlag, 332-346.

Kersten, G.E., S. MacDonald, et al. (1993). "Knowledge-based Simulation for Medical Education", *Modelling and Simulation: Proceedings of IASTED International Conference*, M.H. Hamza (Ed.), Anaheim, CA: IASTED, 630-633.

Kersten, G.E. and S. Szpakowicz (1995). "Forming Decision Making Skills with a Patient Simulator", *Control and Cybernetics*, 24(3), 301-326.

Kira, D. S., M. I. Kusy, et al. (1990). "A Specific Decision Support System to Develop an Optimal Project Portfolio Mix under Uncertainty", *IEEE Transactions on Engineering Management*, 37(3), 213-221.

Kleijnen, J. P. C. (1993). "Simulation and Optimization in Production Planning: A Case Study", *Decision Support Systems*, 9(3),: 269-280.

Kovacs, G.J., J.T. Ritchie and T. Nemeth (1998). "CERES Models in Multiple Objective Decision-Making Process", in: S.A. El-Swaify and D.S. Yakowitz (Ed.), *Multiple Objective Decision Making for Land, Water, and Environmental Management*, Lewis Publishers, 281-289.

Lawton, G. (1993). "Transforming Numbers into Graphics Yields Insight from Images", *Bank Management*, 69, A16-A18.

Leibovici, L., V. Gitelman, et al. (1997). "Improving Empirical Antibiotic-Treatment - Prospective, Nonintervention Testing Of A Decision-Support System", *Journal Of Internal Medicine,* 242(5), 395-400.

Levary, R. R. and M. D. Renfro (1991). "Application of Assembly Line Balancing Techniques to Installment Lending Operations of Commercial Banks", *Computers and Industrial Engineering,* 20(1), 104-109.

LGI (Apr 1999), "The Data Warehousing Information Center – Decision Analysis Tools", http://pwp.starnetinc.com/larryg/decision.html.

Liang, T.-P. and S.-Y. Hung (1997). "DSS and EIS applications in Taiwan", *Information Technology & People,* 10(4), 303-315.

Liberatore, M. J. and A. C. Stylianou (1993). "The Development Manager's Advisory Systems: A Knowledge-Based DSS Tool for Project Assessment", *Decision Sciences,* 24(5), 953-976.

Little, J. D. C. (1970), "Models and Managers: The Concept of a Decision Calculus", *Management Science,* 16(8), 35-43.

Lovejoy, Stephen B., J.G. Lee, et al. (1997). "Research Needs for Water Quality Management in the 21st Century: A Spatial Decision Support System", *Journal of Soil and Water Conservation,* 52(1), 18-22.

Lysgaard, J. (1992). "Dynamic Transportation Networks in Vehicle Routing and Scheduling", *Interfaces,* 22(3), 45-55.

Mak, H-Y, A. P. Mallard, T. Bui and R. Au, "Crisis Management Support Systems using Workflow", May 1999, *Decision Support Systems* (in print).

Mak H-Y. and T. Bui, "Modeling Experts Consensual Judgments for New Product Entry Timing", *IEEE Transactions on Systems, Man and Cybernetics,* 1996, 26(5), 659-667.

Mallach, E. G. (1994). *Understanding Decision Support Systems and Expert Systems,* Burr Ridge, IL: Irwin.

Manley, B. R. and J. A. Threadgill (1991). "LP Used for Valuation and Planning of New Zealand Plantation Forests", *Interfaces,* 21(6), 66-79.

Martin, C. H., D. C. Dent and J. C. Eckhart (1993). "Integrated Production, Distribution, and Inventory Planning at Libbey-Owens-Ford", *Interfaces,* 23(3), 68-78.

Marakas G. M. (1999). *Decision Support Systems in the 21st Centure,* Upper Saddle River, NJ: Prentice Hall.

Marathe, A., V. Ghate, P. Abhyankar and V. D. Vartak (1991). "Computer Aided Decision Making in Social Forestry", *Indian Forester,* 117(5), 403-410.

Martinsons, M. and R. Davison (1999). "The Balanced Scorecard: A Foundation For The Strategic Management Of Information Systems", *Decision Support Systems,* 25(1), 71-88.

Matthew, P.L. and B.A. Peasley (1998). "A Case Study in the Use of an Expert System as a Multiobjective Decision Support System (MODSS) – Boobera Lagoon Environmental Management Plan", in: S. A. El-Swaify and D. S. Yakowitz (Ed.), *Multiple Objective Decision Making for Land, Water, and Environmental Management,* Lewis Publishers, 655-666.

Meyer, J. E. (1992). "SCADA for Financial Planning and Management", *Management Quarterly,* 33(1), 20-30.

McNurlin, B. C. and Sprague, R. H., Jr. (1993). Information Systems Management in Practice. Engelwood Cliffs, NJ, Prentice Hall.

Migliarese, P. and E. Paolucci (1993). "A System for Group Production Planning in Manufacturing", *Interfaces,* 23(3), 29-40.

Mintzberg, H., D. Raisingham and A. Theoret (1976). "The Structure of 'Unstructured' Decision Processes", *Administrative Science Quarterly,* 21, 246-275.

Mira Da Silva, L., J. Park, and P.A. Pinto (1998). "Decision Support Systems for Irrigated Zones: An Integrated Approach to Land Use Planning and Management in Southern Europe", in: S.A. El-Swaify and D.S. Yakowitz (Ed.), *Multiple Objective Decision Making for Land, Water, and Environmental Management,* Lewis Publishers, 599-614.

Mishra, R. B. and S. Dandapat (1993). "A Knowledge-based Interpretation System for EMG Abnormalities", *Int. Journal of Clinical Monitoring and Computing*, 10(2), 131-142.

Missikoff, M. (1998). "An Object-Oriented Approach To An Information and Decision-Support System For Railway Traffic Control", *Engineering Applications Of Artificial Intelligence*, 11(1), 25-40.

Mohanty, R. and S. Deshmukh (1997). "Evolution Of A Decision-Support System For Human-Resource Planning In A Petroleum Company", *International Journal Of Production Economics*, 51(3), 251-261.

Moon, D.E., S.C. Jeck, et al. (1998). "Elements of a Decision Support System: Information, Model, and User Management", in: S.A. El-Swaify and D.S. Yakowitz (Ed.), *Multiple Objective Decision Making for Land, Water, and Environmental Management*, Lewis Publishers, 323-334.

Morgan, R. F. (1996). "An Intelligent Decision Support System for a Health Authority: Solving Information Overload", *Journal of the Operational Research Society*, 47(4), 570-582.

Moribayashi, M. and C. Y. Wu (1990). "Decision support system for capital budgeting and allocation", *Computers and Industrial Engineering*, 19(1-4), 524-528

Muth, R. M. and R. G. Lee (1986). Social Impact Assessment in Natural Resource Decision Making: Toward a Structural Paradigm, in H. A. Becker and A. L. Porter (Eds.), *Methods and Experiences in Impact Assessment*, Dordrecht, Reidel. 168-183.

Naesset, E. (1997). "Geographical Information-Systems In Long-Termforest Management And Planning With Special Reference To Preservation Of Biological Diversity - A Review", *Forest Ecology And Management*,93(1-2), 121-136.

Osmond, D. L., R. W. Gannon, et al. (1997). "WATERSHEDSS: A Decision Support System for Watershed-scale Nonpoint Source Water Quality Problems", *Journal of the American Water Resources Association*, 33(2), 327-342.

Ott, J. (1992). "Federal Express Develops C3I-Based Information System", *Aviation Week and Space Technology*, 137, 57-60.

Ozdamar, L., M. Bozyel, et al. (1998). "A Hierarchical Decision-Support System For Production Planning (With Case-Study)", *European Journal Of Operational Research*, 104(3), 403-422.

Paige, G.B., J.J. Stone, et al. (1998). "Overview of a Decision Support System for the Evaluation of Landfill Cover Designs", in: S. A. El-Swaify and D. S. Yakowitz (Ed.), *Multiple Objective Decision Making for Land, Water, and Environmental Management*, Lewis Publishers, 153-165.

Paruccini, M. (1994). *Applying Multiple Criteria Aid for Decision to Environmental Management*, Dordrcht: Kluwer.

Park, Y. H., E. Park and C. A. Ntuen (1990). "Investment Decisions. An Integrated Economic and Strategic Approach", *Computers and Industrial Engineering* 19(1-4), 534-538.

Payandeh, B. and D. Basham (1993). "'FIDME-PC': Forestry Investment Decisions Made Easy on Personal Computers", *International Journal of Modelling and Simulation*, 13(2), 72-6.

Pingry, D. E., T. L. Shaftel and K. E. Boles (1991). "Role for Decision Support Systems in Water-delivery Design", *Journal of Water Resources Planning and Management*, 117(6), 629-644.

Pinter, J., M. L. D. Fels, et al. (1995). "An Intelligent Decision Support System for Assisting Industrial Wastewater Management", *Annals of Operations Research*, 58(1), 455-477.

Power, D. and F. Quek (Apr 1999)."ISWorld Decision Support Systems Research Page", http://power.cba.uni.edu/isworld/dss.html.

Power, D. J. (Apr 1999)."Decision Support Systems (DSS) Resources - Home Page", http://dss.cba.uni.edu/DSShome.html.

Power, J. M. (1988). "Decision Support Systems for the Forest Insect and Disease Survey and for Pest management", *Forestry Chronicle*, 64, 132-135

Power, J. M. (1993). "Object-oriented Design of Decision Support Systems in Natural Resource Management", *Computers and Electronics in Agriculture*, 8(4), 301-24.

Proudlove, N.C., S. Vadera, et al. (1998). "Intelligent Management Systems in Operations: a Review", *Journal of the Operational Research Society*, 49(7), 682-699.

Quek, F. (Apr 1999). "Decision Support Systems Courses", http://is.lse.ac.uk/iswnet/dss/courses.htm

Rao, H. R., V. S. Jacob, F. Lin, D. Robey and G. P. Huber (1992). "Hemispheric Specialization, Cognitive Differences, and Their Implications for the Design of Decision Support Systems Responses" *MIS Quarterly*, 16(2), 145-154.

Rauscher, H. M. (1999). "Ecosystem Management Decision Support for Federal Forests in the United States: A Review", *Forest Ecology And Management,* 114(2-3), 173-197.

Rios, J. P. (1993). "Investment in New Production Capacity Under Uncertain Market Conditions", Industrial Engineering 25(6), 31-40.

Riverside Technology (Apr 1999). "Decision Support Software Systems for Water Resource Management", http://www.riverside.com/dss/dssshow/next/index.html.

Robotham, M.P. (1998). "AGFADOPT: A Decision Support System for Agroforestry Project Planning and Implementation", in: S. A. El-Swaify and D. S. Yakowitz (Ed.), *Multiple Objective Decision Making for Land, Water, and Environmental Management*, Lewis Publishers, 167-176.

Ross, S. and I. Hannam (1998), "Multiple Objective Decision Support Systems Used in Management of Temperate Forest Ecosystems in Southeast Australia", in: S. A. El-Swaify and D. S. Yakowitz (Ed.), *Multiple Objective Decision Making for Land, Water, and Environmental Management*, Lewis Publishers, 667-684.

Sandia National Laboratories (1999). "Environmental Decision Support & Resource Coalition", http://www.nwer/sandia.gov/coalition/.

Sankaran, J. K. and R. R. Ubgade (1994). "Routing Tankers for Dairy Milk Pickup", *Interfaces,* 24(5).

Santhanam, R. and J. Elam (1998). "A survey of knowledge-based systems research in decision sciences (1980-1995)", *Journal of the Operational Research Society,* 49(5), 445-457.

Sauter, V. (1997). *Decision Support Systems*, New York: Wiley.

Scot Morton, M. S. (1971). "Management Decision Systems. Computer-based Support for Decision Making", Division of Research, Harvard University, Cambridge, MA.

Shannon, P. W. and R. P. Minch (1992). "A Decision Support System for Motor Vehicle Taxation Evaluation", *Interfaces*, 22(2), 52-64.

Sharkey, P. D., M. J. DeHaemer, et al. (1993). "Assessing the Severity of Patients' Illnesses to Better Manage Health Care Resources", *Interfaces*, 23(4), 12-20.

Shaw, R., J. Doherty, et al. (1998). "The Use of Multiobjective Decision Making for Resolution of Resource Use and Envirnomental Management Conflicts at a Catchment Scale", in: S. A. El-Swaify and D. S. Yakowitz (Ed.), *Multiple Objective Decision Making for Land, Water, and Environmental Management*, Lewis Publishers, 697-716.

Shtienberg, D., A. Dinoor and A. Marani (1990). "Wheat Disease Control Advisory, a Decision Support System for Management of Foliar Diseases of Wheat in Israel", *Canadian Journal of Plant Pathology*, 12, 195-203.

Silvert, W. (Apr 1999)."Decision Support Systems for Aquaculture Regulation", http://www.maritimes.dfo.ca/science/mesd/he/staff/silvert/dss/dss.html

Simonovic, S. P. (1992). "Reservoir Systems Analysis: Closing Gap between Theory and Practice", *Journal of Water Resources Planning and Management*, 118(3), 262-280.

Sisodia, R. S. (1992). "Marketing Information and Decision Support Systems for Services", *Journal of Services Marketing*, 6(1), 51-64.

Sobanjo, J. O., G. Stukhart, et al. (1995). "Evaluation of Projects for Rehabilitation of Highway Bridges", *Journal of Structural Engineering,* 120(1), 81-99.

Sophocleous, M. and T. Ma (1998). "A Decision Support Model to Assess Vulnerability to Salt Water Intrusion in the Great Bend Prairie Aquifer of Kansas", *Ground Water*, 36(3), 476-483.

Sophocleous, M. A., J. K. Koelliker, et al. (1999). "Integrated numerical modeling for basinwide water management: The case ofthe Rattlesnake Creek basin south-central Kansas", *Journal Of Hydrology*, 214(1-4), 179-196.

Sorensen, C. G. (1998). "A decision support system for planning field operations", *Computers in Agriculture*, 446-454.

Specht, J. and Owls (1995). "Loggers Share Forest Wealth", *GIS World*, 8(10), 36-38.

Sprague, R. H., Jr. and E. D. Carlson (1982). Building Effective Decision Support Systems. Engelwood Cliffs, NJ, Prentice Hall.

Sprague, R. H. and H. J. Watson (1997). *Decision Support for Management*. Upper Saddle River, NJ: Prentice Hall.

Srihari, K. and M. Cala (1992). "Knowledge Based Decision Support for PCB Assembly Using SMT", *Computers and Industrial Engineering*, 23(1-4), 405-408.

Swetnam, R., J. Mountford, et al. (1998). "Spatial Relationships Between Site Hydrology And The Occurrence Of Grassland Of Conservation Importance - A Risk Assessment With GIS", *Journal Of Environmental Management*, 54(3), 189-203.

Stewart, T. J. (1991). "Multi-criteria Decision Support System for R&D Project Selection", *Journal of the Operational Research Society*, 42(1), 17-26.

Stodolak, F. and J. Carr (1992). "Systems Must Be Compatible with Quality Efforts", *Healthcare Financial Management*, 46(6), 72-77.

Stone, N.D., D. Faulkner, et al. (1998). "CROPS: A Constraint-Satisfaction System for Whole-Farm Planning", in: S. A. El-Swaify and D. S. Yakowitz (Ed.), *Multiple Objective Decision Making for Land, Water, and Environmental Management*, Lewis Publishers, 527-537.

Subramaniam, C. and S. Kerpedjiev (1998). "Dissemination of Weather Information to Emergency Managers: A Decision Support Tool", *IEEE Transactions on Engineering Management*, 45(2), 106-114.

Suresh, N. C. (1990). "Towards an Integrated Evaluation of Flexible Automation Investments", *International Journal of Production Research*, 28(9), 1657-1672.

Sutherland, F. R. and J. J. Lambourne (1991). "Development of a Decision Support Mapping Utility for Water Resources Planning" *Water*, 17(4), 281-288.

Tanzi, T. and R. Guiol (1998). "A System For Motorway Management Based On Risk Rate Estimation", *Safety Science*, 30(1-2), 9-23.

Tecle, A., B. Shrestha, et al. (1998). "A Multi-objective Decision-Support System For Multiresource Forest Management", *Group Decision And Negotiation*, 7(1), 23-40.

Tessmer, A. C., M. J. Shaw and J. A. Gentry (1993). "Inductive Learning for International Financial Analysis: A Layered Approach", *Journal of Management Information Systems*, 9(4), 17-36.

Thrall, G. I. and S. M. Elshaw-Thrall (1990). "Computer Assisted Decision Strategy for Evaluating New satellite Hub Sites for a Local Utility Provider", *Computers, Environment and Urban Systems*, 14(1), 37-48.

Turban, E. and Aronson, J.E. (1998). *Decision Support Systems and Intelligent Systems*. Upper Saddle River, NJ: Prentice-Hall, Fifth Edition.

Twery, M.J., S.L. Stout and D.L. Loftis (1998). "Using Desired Future Conditions to Integrate Multiple Resource Prescriptions: The Northeast Decision Model", ", in: S. A. El-Swaify and D. S. Yakowitz (Ed.), *Multiple Objective Decision Making for Land, Water, and Environmental Management*, Lewis Publishers, 197-203.

Tyagi, N. K., K. C. Tyagi and N. N. Pillai (1993). "Decision Support for Irrigation System Improvement in Saline Environment", *Agricultural Water Management*, 23, 285-301.

Vickner, S.S. and D.L. Hoag (1998). "Advances in Ration Formulation for Beef Cattle Through Multiple Objective Decision Support Systems", in: S.A. El-Swaify and D.S. Yakowitz (Ed.),

Multiple Objective Decision Making for Land, Water, and Environmental Management, Lewis Publishers, 291-298.

Vogel, D. R. and J. F. Nunamaker Jr. (1990). "Group Decision Support System Impact: Multi-Methodological Exploration", *Information & Management*, 18(1), 15-28.

Wadsworth, R. and M. Brown (1995). "A Spatial Decision-Support System to Allow the Investigation of the Impact of Emissions from Major Point Sources Under Different Operating Policies", *Water Air And Soil Pollution*, 85(4), 2649-2654.

Walker, D. and A. Johnson (1996). "NRM Tools - A Flexible Decision-Support Environment For Integrated Catchment Management", *Environmental Software*, 11(1-3), 19-24.

Wagner, P. and F. Kuhlmann (1991). "Concept and Implementation of an Integrated Decision Support System (IDSS) for Capital-intensive Farming", *Agricultural Economics*, 5, 287-310.

Williams, S. B. and D. R. Holtfrerich (1998). "A Knowledge-Based Reasoning Toolkit for Forest Resource Management", in: S. A. El-Swaify and D. S. Yakowitz (Ed.), *Multiple Objective Decision Making for Land, Water, and Environmental Management*, Lewis Publishers, 251-268.

Wilmes, G. J., D. L. Martin and R. J. Supalla (1990). "Decision Support Systems for Improved Irrigation Management", *Visions of Future*, American Society of Agricultural Engineers, 594-600.

Winsemius, P. and W. Hahn (1992). "Environmental Option Assessment", *Columbia Journal of World Business*, 27(3, 4), 248-266.

Wood, D. B. and S. M. Dewhurst (1998). "A Decision Support System for the Menominee Legacy Forest", *Journal of Forestry*, 96(11), 28-32.

Wybo, J.L. (1998). "FMIS: A Decision Support System for Forest Fire Prevention and Fighting", *IEEE Transactions on Engineering Management*, 45(2), 127-131.

Xiang, W. N. (1993). "A GIS method for Riparian Water Quality Buffer Generation" *International Journal of Geographical Information Systems*, 7(1), 57-70.

Yakowitz, D.S., B. Imam and L.J. Lane (1998a). "Effects of Optional Averaging Schemes on the Ranking of Alternatives by the Multiple Objective Component of a U.S. Department of Agriculture Decision Support System", in: S.A. El-Swaify and D.S. Yakowitz (Ed.), *Multiple Objective Decision Making for Land, Water, and Environmental Management*, Lewis Publishers, 217-232.

Yakowitz, D.S., J.J. Stone et al. (1998b). "Evaluation of a Prototype Decision Support System for Rangelands in the Southwest United States", in: S.A. El-Swaify and D.S. Yakowitz (Ed.), *Multiple Objective Decision Making for Land, Water, and Environmental Management*, Lewis Publishers, 363-378.

Yakowitz, D.S., J.J. Stone et al. (1992). "Evaluating Land Management Effects on Water Quality Using Multi-Objective Analysis within a Decision Support System", *American Water Resources Association 1st International Conference on Ground Water Ecology*, April, Tampa, Fl, 365-373.

Yakowitz, D.S., J.J. Stone et al. (1993). "Evaluating Land Management Practices with a Decision Support System: An Application to the MSEA Site Near Treynor, Iowa", *Soil and Water Conservation Society Conference on Agriculture Research to Protect Water Quality*, Minneapolis, MN, Feb, 404-406.

Yost, R. and J.Z.C. Li (1998). "Developing an Integrated Nutrient Management Decision Aid", in: S. A. El-Swaify and D. S. Yakowitz (Ed.), *Multiple Objective Decision Making for Land, Water, and Environmental Management*, Lewis Publishers, 177-185.

Zhang, X., Chen, S., et al. (1998). " Season No-Tillage Ridge Cropping System: A Multiple Objective Tillage System for Hilly Land Management in South China", in: S. A. El-Swaify and D. S. Yakowitz (Ed.), *Multiple Objective Decision Making for Land, Water, and Environmental Management*, Lewis Publishers, 685-696.

Zhu, X., R.G. Healey, et al. (1998). "A Knowledge-Based Systems Approach to Design of Spatial Decision Support Systems for Environmental Management", *Environmental Management,* 22(1), 35-48.

Zwietering, M. H., T. Wijtzes and J. C. De Wit (1992). "A Decision Support System for Prediction of the Microbial Spoilage in Foods", *Journal of Food Protection,* 55, 973-979.

Glossary

Ad-hoc query is a spontaneous or unplanned question or request to extract information from one or more databases.

Aggregate data results from applying a process to combine and summarize data elements.

Computer-mediated communication involves the use of computers to create, store, deliver, and process communication functions.

Computer supported cooperative work (CSCW) involves the use of computers to support work jointly conducted by multiple participants.

Corporate planning system is a DSS that holds and derives knowledge relevant to planning decisions that cut across organizational units and involve business functions.

Critical success factors are issues and areas of business activity in which results are necessary for a company to achieve its objectives.

Data binary (digital) representations of facts, text, graphics, bit-mapped images, sound, analog or digital video segments; it is processed to obtain information.

Data dictionary is database about data and database structures; it is a catalog of all data elements, containing their names, structures, and information about their usage.

Data-driven DSS or Data-oriented DSS allows for model construction from data with the use of data mining and learning methods and during the decision support process.

Data element is an elementary unit of data that can be identified and described in a dictionary or repository and cannot be further subdivided.

Data mining are statistical, learning and other methods for the construction of models from large databases and data warehouses.

Data warehouse is a large database comprising historical and current data and designed to support decision making in organizations. It is a subject-oriented, integrated, time-variant, and nonvolatile collection of data.

Data visualization is a set of methods and tools to easily and quickly view graphical displays of information from different perspectives.

Decision analysis methods include decision trees, multi-attribute utility, Bayesian models, multi-objective and multi-criteria methods.

Decision analysis tools help decision makers to decompose and structure problems.

Decision room is physical arrangement for a group DSS in which workstations are available to participants.

Decision systems are computer based programs and technologies intended to make routine decisions, monitor and control processes, and aid or assist decision makers in semi-structured and/or non-routine decision situations.

Decision support systems (DSS) are interactive computer-based systems intended to help decision makers utilize data and models to identify, analyze and solve problems and make decisions.

Declarative knowledge is passive knowledge expressed as statements of facts about the world.

Dependent variables

Descriptive model is physical, conceptual or mathematical models that describe situations as they are or as they actually appear.

Deterministic model is constructed under a condition of certainty under the assumption of only one possible result for each alternative decision.

Dialog generation and management system (DGMS) is a software management package in a DSS whose functions in the dialog subsystem is similar to that of a DBMS in a database.

Dialog system is the hardware and software that create and implement a user interface for a DSS. A DSS dialog system creates the human-computer interface.

Domain expert a person who has expertise in the domain in which a specific expert system is being developed.

Drill down/up - An analytical technique that lets a DSS user navigate among levels of data ranging from the most summarized (up) to the most detailed (down).

DSS generator is a computer software that provides tools and capabilities that help a developer quickly and easily build a specific DSS.

DSS development tools comprises software components (e.g., editors, code libraries, specific objects, and visual interfaces) that facilitate the development of a specific DSS.

Enterprise-wide DSS is a system that supports a large group of managers in a networked-centric environment.

Evolutionary (iterative) design process is a systematic process for system development that is recommended for use in creating DSS.

Executive information systems (EIS) is a system intended to provide current and appropriate information to support executive decision making for managers using a networked workstation. The emphasis is on graphical displays and an easy to use interface that present information from the corporate database.

Executive support system (ESS) is an executive information system that includes specific decision aiding and/or analysis capabilities.

Expert system is a man-machine system with specialized problem-solving expertise. The "expertise" consists of knowledge about a particular domain, understanding of problems within that domain, and "skill" at solving some of these problems.

Facilitator is a person who manages the use of a group decision support system from initial planning through actual operation.

Feasibility study is conducted to determine the technical and economic prospects for developing a system prior to actually committing resources to actually developing it.

Functional DSS is a system that holds and derives knowledge relevant for decisions about a business function in an organization performs.

Generator is a software packages that designed to expedite programming efforts that are required to build information systems, especially expert and decision support systems.

Goal-seeking is system's capability of providing values of both independent and dependent variables that they must assume in order to attain assumed values of the resulting dependent variables.

Geographic information systems (GIS) is a support system that represents data and relationships among them with the use of maps.

Graphical user interface (GUI) is a program interface that uses a computer's graphics capabilities to make the program easier to use. It comprises icons, graphic, menus, text boxes, etc.

Group decision support systems (GDSS) is an interactive, computer-based system that facilitates solution of problems by a group of people.

Groupware is software designed to support more than one person working on a shared task; it provides mechanisms to communicate, collaborate, coordinate and keep track of on-going projects and tasks.

Heuristics are informal, experiential and judgmental rules used in a given domain to evaluate, verify and solve problems of this domain.

Hypermedia is a combination of several types of media such as text, graphics, audio, and video.

Hypertext is an approach to handle information (text) by linking documents via hyperlinks. It allows users to move among topics and documents..

Icon is a visual, graphical representation of a document, program or concept.

Independent variables are used to determine values of dependent variables (decision variables); controlled by the environment.

Inference is a reasoning process of drawing a conclusion from observed or assumed facts about a problem.

Inference engine is a part of an expert system that actually performs the reasoning function.

Information is a meaningful and interpretable data; data that has been processed to add or create meaning..

Information system architecture is a specification of the business processes and rules, systems structure, technical framework, and product technologies for business information systems. .

Interdependent decisions is a set of decisions that are interrelated.

Internet is a collection of packet-switching networks and routers that uses the TCP/IP protocol suit and functions as a single, cooperative virtual network.

Intranet is an internal organizational network using TCP/IP with at least one web server that is only accessible by an organization's members or others who have specific authorization.

Knowledge is the body of facts and principles accumulated by human kind or the act, fact, or state of knowing; it is information that incorporates the relationships between concepts and ideas

Knowledge acquisition is the process of extraction and formal representation of knowledge.

Knowledge base is an organized collection of facts, rules, and procedures in such a way that can be processed by an inference engine.

Knowledge-based systems depend on a rich base of knowledge to perform difficult tasks and its main components are a knowledge base, inference engine and interface.

Knowledge engineer is responsible for the technical side of the development, testing and validation of a knowledge base an expert system.

Knowledge engineering (KE) - The engineering discipline that involves integrating knowledge into computer systems in order to solve complex problems normally requiring a high level of human expertise.

Knowledge management (KM) is the distribution, access and retrieval of unstructured information about "human experiences" between interdependent individuals or among members of a workgroup. Knowledge management involves identifying a group of people who have a need to share knowledge, developing technological support that enables knowledge sharing, and creating a process for transferring and disseminating knowledge.

Knowledge Management Software (KMS) is software that can store and manage unstructured information in a variety of electronic formats. The software may assist in knowledge capture, categorization, deployment, inquiry, discovery, or communication.

Metadata provides semantic information about the data in a database or a data warehouse and is stored in a data dictionary. It allows to locate the contents of the data warehouse and serves as a guide to the algorithms used for summarization of current detailed data.

Methodology is a system of principles, practices, and procedures applied to a specific branch of knowledge.

Middleware is software that allows applications to interact across hardware and network environments.

Model base is a collection of preprogrammed quantitative models (e.g., statistical, financial, logical, optimization and simulation) organized as a single unit.

Model-driven DSS emphasizes access to and manipulation of models in order to obtain values of decision (independent) variables. Model parameters are determined with the use of stored data and via interactions with users.

Modeling tools are specialized programs that help developers build models quickly.

Multiparticipant DSS is a system that supports multiple participants engaged in a decision-making task.

On-line analytical processing (OLAP) is software for manipulating multidimensional data from a variety of sources that has been stored in a data warehouse.

Operational or transaction database is a database-of-record for a transaction-update system. It contains detailed data used to run the day-to-day operations of the business.

Organizational DSS is a multiparticipant DSS designed to support a decision maker in a setting that has a more elaborate infrastructure than a group (i.e., involving specialized roles, restricted communication patterns, differing authority levels). **Procedural knowledge** is compiled knowledge related to the performance of some task.

Prototyping is an approach in system development in which a scaled down system or portion of a system is constructed in a short time, tested, and improved in several iterations.

Semistructured decision is one in which some aspects of the problem are structured and others are unstructured and unknown.

Sensitivity analysis allows to determine the range of model parameter values for which the solution or its structure remains the same (e.g., optimal).

Shell is an expert system development tool that consists of a rule set manager and an inference engine capable of reasoning.

Specific DSS is a computer-based system that actually helps a person accomplish a specific task.

Structured decision (also routine or programmed decision) is a standard or repetitive decision for which solution techniques are already available; the structural elements in the situation, e.g. alternatives, criteria, environmental conditions, are known, defined and understood.

Symbolic processing uses of symbols, rather than numbers, combined with rules and heuristics, in order to process information and solve problems.

Unstructured decisions - This type of decision situation is complex and no standard solutions exist for resolving the situation. Some or all of the structural elements of the decision situation are undefined, ill-defined or unknown.

User-friendly is used to evaluate of a DSS user interface and it indicates the ease of learn, understand, and use of the system.

User interface (dialog) is the component of a computerized support system that allows bi-directional communication between the system and its user. It is a set of commands or menus through which a user communicates with a program.

Web-based DSS is a system that resides on servers and delivers decision support tools to a client via Web browser.

"What If" analysis allows answer user's requests about effects that are caused by the user change of some of the input data or values of independent variables.

Contributors

Dave Abel is a chief research scientist with CSIRO Mathematical and Information Sciences. He is a graduate of University of Queensland (B.Sc., Diploma of Information Processing) and James Cook University (M.Sc. and Ph.D.). He has specialized in spatial information systems since 1980. His work has included investigations of access methods, data modeling, systems integration, decision support systems and distributed systems. His recent activities include consultancies for strategic review and planning in government and architectural design for national and regional spatial data infrastructures.

Hans-Dieter Bechstedt is with the International Board for Soil Research and Management in Bangkok, Thailand. He is a social scientist with advanced degrees in psychology and sociology. He specializes in rural sociology and social anthropology and conducts research and training in participatory rural appraisal for sustainable development

Lioubov V. Bourmistrova graduated from Moscow State University, 1996 (Applied Mathematics). As a Ph.D. student in applied mathematics at Moscow State University, her research interests include approximation theory, multiple criteria decision and negotiation support, integrated assessment of environmental problems.

Tung X. Bui is Matson Navigation Company Distinguished Professor of Global Business at the University of Hawaii College of Business Administration at Manoa, Honolulu. He earned a doctorate in Managerial Economics from the University of Fribourg, Switzerland and a Ph..D. in Information Systems from New York University. Prior to joining the University of Hawaii, he held faculty positions in Asia, Europe and North America. Bui has conducted research in information technology management, group decision and negotiation, decision support systems, and economic development in South East Asia. He has taught such courses as economic evaluation of information systems, management of information technology, decision support and expert systems, business process engineering, and electronic commerce. Professor Bui has published seven books and over 100 papers. He has also been a regular consultant and adviser to governmental agencies and companies.

Frada V. Burstein is a senior lecturer at the School of Information Management and Systems (SIMS), Monash University, Melbourne, Australia. She received her Masters of Sciences in Computer Sciences in 1978 from Tbilisi State University and Doctoral degree in decision support systems in 1984 from the Institute of Cybernetics Georgian Academy of Sciences. She has been a staff member of Monash University since 1992. Dr Burstein is a leader of the formal research group in knowledge management at the

SIMS. Her research interests include intelligent decision support, organizational memory, knowledge modeling.

Vladimir A. Bushenkov graduated from the Moscow Institute for Physics and Technology, 1979 and received his Ph.D. in Mathematical Cybernetics from Moscow Institute for Physics and Technology, 1983. Since 1987 he has been with Computing Center of Russian Academy of Sciences as senior researcher. Since 1995, he has been part time teaching at Moscow Peoples Friendship University. He is an author of two books and about 30 papers. Current fields of research: numeric analysis, stability theory, and systems for decision and negotiation support.

Muthu Chandrashekar has obtained his B.Tech. in Mechanical Engineering from the Indian Institute of Technology, Kanpur, India, in 1969 and M.A.Sc. and Ph.D. in Systems Design Engineering from the University of Waterloo in 1970 and 1973 respectively. His research interests were in system thinking, modeling, analysis and design. He has mastered interdisciplinary approaches for applications in engineering, energy, and environment. He has also worked extensively in the area of rural sustainability, international development, and technology transfer.

Dr. Muthu Chandrashekar has passed away due to a heart attack in the first week of January 1998.

Victor S. Chabanyuk has received his Ph.D in Mathematical Cybernetics from the Kyiv State University, Kyiv, Ukraine, in 1988. He develops and uses environmental management information systems based on geo-information technologies, beginning from the accident at the Chornobyl Nuclear Power Plant in 1986. Since 1991 he has been working with private companies in the field of environmental and land management information systems. He is now a director in the Intelligence Systems GEO, Ltd., Ukraine. His research interests include the similarity of information systems and semantic data models, hybrid semantic data models in GIS, methodologies of information systems analysis and design, object-oriented databases and GIS.

Émerson C. Corrêa graduated in Mechanical Engineering in 1995 and received his M.Sc. in Systems Engineering in 1996 at Federal University of Santa Catarina, Brazil. His research areas of interest are multicriteria decision aids, problem structuring methods and strategic management. He is mainly interested in the introduction of these disciplines into practice. He is a lecturer at Universidade Lusiada and a project leader at Cised Consultores in Lisbon. He is a member of the International Society on Multiple Criteria Decision Making and of the European Working Group on Multicriteria Decision Aid.

Carlos Bana Costa received his Ph.D. in Systems Engineering from IST in 1993. His research areas of interest are management and decisions sciences, in particular multicriteria decision analysis, problem structuring methods, and strategic and engineering management. He is also a decision-aiding consultant in many private and public organizations. Professor Bana e Costa is author of more than 80 articles, the editor of *Readings in Multiple Criteria Decision Analysis*, and co-author of the MACBETH Approach. He is member of the International Executive Committee of the International Society on Multiple Criteria Decision Making.

David Cray is an associate professor of International Business in the School of Business at Carleton University. He received his Ph.D. from the University of Wisconsin in 1981 for his work on control systems in multinational corporations. He is currently analyzing the interaction of gender and culture in cross-cultural negotiations. In his most recent book, *Making Sense of Managing Culture* published in 1998 with Geoff Mallory, he outlines a cognitive approach to international management.

Leonardo Ensslin received his Ph.D. in Industrial and Systems Engineering from USC Los Angeles, USA in 1973. He is a professor of multicriteria decision aid and decision making at the EPS-UFSC in Florianopolis, Brazil, and a private consultant in the South of Brazil, working with public and private groups. Before joining UFSC he worked in agribusiness in Uruguay. He specialises in structuring decision maker's decision contexts in order to improve understanding and to generate alternatives to improve the overall performance and has consulted widely in this area.

Benoit Gaily, received his MBA from INSEAD in 1997 and a Ph. D. in applied mathematics from the Université Catholique de Louvain, Belgium in 1994. Since 1995 he has been a consultant with McKinsey & Company, initially in Belgium and now in Ireland.

Eduardo Gamboa is a senior developer with the Procyon Group, a Web consultant in San Francisco, USA, specializing in Web database applications using Java technology. He has been developing software for over twenty-five years, during most of which he served as a database consultant in various World Bank assisted development projects in the Philippines. He was an instructor at the School of Economics, University of the Philippines, a lecturer at the Ateneo Computer Technology Center, and a Fellow of the Social Weather Stations. EcoKnowMICS started as his graduate thesis proposal at the London School of Economics. He has a B. S. in Mechanical Engineering and a M. A. in Economics.

Samuel Gameda is a research scientist with the Land Resources Program, Eastern Cereal and Oilseed Research Centre, Agriculture and Agri-Food Canada, Ottawa. He graduated in agricultural engineering in 1978, and obtained M. Sc. with specialization in soil water management in 1981. He carried out several years of investigation as a research associate on soil-machinery interactions under intensive agricultural production, following which he completed his Ph.D. in 1993 at McGill University in Montreal, Canada. He has also developed an expert system on soil degradation and crop productivity relationships. His interests are in the integration of local knowledge and scientific research for land management. His research is focused on the application of information technology for land resource management and sustainable development.

Patrick A.V. Hall, is professor of computer science at the Open University, United Kingdom. He graduated in Electrical Engineering from Cape Town University, and then in Mathematics from Cambridge University before taking his doctorate in machine learning from London University. He has spent about half his career in academic life, and half in industry. During the 1970s he spent two and a half years in Saudi Arabia concerned with the Arabisation of computing, and recently spent 9 months in Nepal looking at cultural aspects of computing and the Internet in South Asia. In the past he has researched in relational databases, computer aided design, software testing, software reuse, multimedia, and distance education. His current research interests are software architectures and cross-cultural computing, and more generally in the interplay between the way software in constructed and how it is used.

To Bao Ho is currently a professor at Graduate School of Knowledge Science, Japan Advanced Institute of Science and Technology, and a senior researcher at the Institute of Information Technology in Hanoi. He received a B. Eng. degree from Hanoi University of Technology, and M.Sc. and Ph.D. degrees from Paris 6 University, and a Habilitation from Paris Dauphine University. His research interests include machine learning, knowledge-based systems, decision support systems, and knowledge discovery in databases.

Michel Installé is a professor, since 1971, at the Center for Systems Engineering and Appied Mechanics (CESAME) of the University of Louvain, Louvain la Neuve, Belgium. He graduated as an Electrical and Mechanical Engineer in 1963, got a Master degree in Automatic Control in 1968 and a Ph.D. degree in Electrical Engineering in 1971 at Stanford University, USA. He worked for many years on adaptive and learning systems and mathematical modeling techniques. His current interests are in applying his background to the solution of problems related to the issue of sustainable development and more particularly to the fields of the development of decision support systems and their application to the search of suitable solutions to environmental and socio-economic problems.

Gregory (Grzegorz) E. Kersten is a professor of decision sciences and information systems at the Carleton University School of Business, the Director of the Centre for Computer Assisted Management, the principal investigator of the InterNeg Project and a member of the Ottawa-Carleton Computer Science Institute. He was a visiting professor at the Naval Postgraduate School, Monterey, CA and Hong Kong University of Science and Technology, and a senior research scholar at the International Institute for Applied System Science. He is a Vice-Chairperson of the INFORMS Group Decision and Negotiation College, a departmental editor of the *Group Decision and Negotiation Journal* and member of the editorial boards of *Journal of Decision Systems*, *INFOR*, and *Control & Cybernetic Journal*. Gregory received M. Sc. in Econometrics and Ph.D. in Operations Research from the Warsaw School of Economics, Poland. He has been a consultant for projects of the International Development Research Centre, Association of Universities and Colleges of Canada, Department of National Defense, and the Co-ordinator of the "Canadian management training and development program for Poland", Department of External Affairs Canada. His research interests include individual and group decision-making, negotiations, artificial intelligence, decision support, web-based systems and electronic commerce. He has written over hundred refereed papers, co-authored three books and has published articles in some twenty journals.

Margaret J. Kersten is a teacher of English as a second language at Carleton University. She received her M.A. in Linguistics from the University of Warsaw. She is a co-author of the text-book *Look and Speak* (with Z. Sznuk) and a number of teaching materials. She has published articles in several journals and presented papers at international conferences. Her main areas of interest include language and culture, and cross-cultural communication.

Anantha Mahadevan is a graduate student at the School of Business, Carleton University and the Data Manager in the InterNeg Project. His masters thesis is on constructing descriptive models of newly researched social phenomenon using data mining techniques.

Zbigniew Mikolajuk is a senior program specialist at the International Development Research Centre in Canada responsible for research project development in the areas of information and telecommunication technologies. He is an adjunct professor at the Carleton University School of Business. He received Ph.D. in computer science from the Warsaw Technical University. From 1970 to 1981, he was an assistant professor at Warsaw Technical University, postgraduate researcher at Kyoto Industrial University, and lecturer at Warsaw School of Economics. From 1981 to 1993 he held research positions in industry working for Shell, Phillips, and Gandalf. His current research interests include decision support systems, knowledge management systems, Internet applications and software engineering. He is the author and co-author of 3 textbooks and over 25 papers published in journals and presented at conferences.

T. R. Madanmohan received his B. Tech in Civil Engineering from Gulbarga University, in 1985 and his Ph. D from the Management Studies Department of the Indian Institute of Science Bangalore, in 1992. He is currently an assistant professor in Technology and Operations at the Indian Institute of Management, Bangalore, India. His research interests include service warranty, product platforms, technology strategy in pharmaceutical industry, technological indicators, emergence of new service designs, Web based inter-cultural negotiations, project management in knowledge industry and academic-industry interaction. He is a member of IEEE and ORSI

Sunil Noronha, is a research staff member at the IBM T. J. Watson Research Center in New York. He received his B. Tech in Electrical Engineering from the Indian Institute of Technology, Madras, in 1987, and his Ph.D. from the Computer Science and Automation Department of the Indian Institute of Science, Bangalore, in 1993. In 1994 he joined Carleton University's School of Business as a postdoctoral fellow and worked on the Negoplan and InterNeg projects on negotiation support. Since June 1997 he has been with the Electronic Commerce Systems group at IBM, leading a research effort in personalization. His other interests include several topics in artificial intelligence, decision support systems, data mining and information economies. Dr. Noronha is a member of the IEEE, ACM and INFORMS.

Xia Li received his Ph.D. from the Centre of Urban Planning and Environmental Management, University of Hong Kong. He continued his post-doctoral research in the Centre during 1997-98. He is now a professor and deputy director of the academic committee of Guangzhou Institute of Geography. He has carried out many research projects in the Pearl River Delta in China related to monitoring land use changes and modelling for sustainable development. He has published over 40 articles in remote sensing and GIS, of which many appeared in top international journals and top Chinese journals. Dr. Li has received 8 awards for his outstanding scientific and technological contributions from the Guangdong Government and Guangdong Academy of Sciences since 1988. He is the recipient of the Outstanding Young Scholars Award of the Chinese Geographical Society (1998). He has also been the deputy secretary of Guangdong Association of Remote Sensing and is present on the editorial board of the Chinese Journal, *Tropical Geography*.

Gordon Lo is a graduate student at the School of Business, Carleton University and the Negotiation Manager in the InterNeg Project. His thesis is on constructing negotiation software agents for electronic commerce.

Alexander V. Lotov graduated Moscow Institute for Physics and Technology, 1969 (Flight Dynamics and Control). Received Ph.D. in Computational Mathematics received from the Computing Center of Russian Academy of Sciences (CCAS), 1973. He is an author of four books and about 70 other publications. He is a member of the Council and of the Governing Body of Russian OR Society, of the Editorial Advisory Board of the International Society for Multiple Criteria Decision Making and the Deputy Head of the Scientific Council "Decision Making with Multiple Criteria" of the Russian Academy of Sciences. His current fields of research include systems analysis, multiple criteria decision making, decision and negotiation support (including on computer networks), integrated assessment of environmental problems, approximation theory and dynamic systems.

Olexandr V. Obvintsev received his Ph.D. in Mathematical Modelling from Cybernetics Institute, Kyiv, Ukraine in 1996. Since 1985 he has been teaching at the Mathematics and Mechanics Faculty of Kyiv State University, now as an Associate Professor. Since 1997, he has been a involved with the Intelligence Systems GEO as a team leader on a part time basis. His research interests include pattern recognition, object-oriented modeling and design, and distributed systems.

Shaligram Pokharel obtained his B.E. in Mechanical Engineering, with Honors from the University of Kashmir, India in 1985, and M.Sc. and Ph.D. in Systems Design Engineering from University of Waterloo, Canada in 1992 and 1997 respectively. Shaligram's current research is in the area of energy and environment planning and policy formulation, geographical information systems, decision support systems, mathematical programming and engineering, and environmental education. Shaligram has been working with the Water and Energy Commission Secretariat, Ministry of Water Resources in Nepal for about 12 years. At the time of writing this paper, Shaligram was a research fellow in Systems Design Engineering at the University of Waterloo.

Kumudini Ponnudurai graduated in 1999 from the School of Business, Carleton University, Canada. She is the Manager of the Web resources in the InterNeg Project (http://www.interneg.org). Her research interests include Web-based decision support and knowledge discovery from databases.

Ahmed Rafea is a professor in Computer Science, the Vice Dean of the Faculty of Computers and Information, Cairo University and the Director of the Central Laboratory for Agricultural Expert Systems, Egyptian Ministry of Agriculture and Land Reclamation. Dr. Rafea got his Ph.D. from University Paul Sabatier, Toulouse, France in 1980. He served in the American University in Cairo, San Diego State University, and National University in the United States. Dr. Rafea is the founder of the Central Laboratory for Agricultural Expert Systems that was established in 1991. He was the principal investigator and manger for several projects for developing expert systems in agriculture. His publication record exceeds 80 papers in scientific journals, conference procedures, and edited books.

Mohammad Rais is a research scientist with the National Institute of Science, Technology, and Development Studies (NISTADS), CSIR, New Delhi, India. He is a graduate of G.B. Pant University of Agriculture and Technology, Pantnagar (B.Sc., M. Sc.) and Indian Institute of Technology, Kanpur (Ph.D.). His areas of interest lie in the use of geographic information systems, expert systems and decision support systems for land use planning, environmental management, and sustainability assessment. He has organized several World Bank assisted training programs at NISTADS on pollution information systems for middle level officers. He was a fellow at the International Institute for Software Technology of the United Nations University (IIST/UNU), Macau, during 1995-96. During 1996-98, he worked as project leader of the IDRC sponsored project on "Decision support systems for evaluating sustainable land management in South-east Asia" at the International Board for Research Management, Bangkok.

Adisak Sajjapongse is a network coordinator for the ASAILAND Management of Sloping Lands Network, IBSRAM, Bangkok. He specialized in soil science, with a BSc from Kasetsart University, Thailand (1967), and received an MS (1971) and PhD (1973) from the University of Minnesota, USA. He has also served as an associate crop management specialist, AVRDC, Taiwan (1979-1988), a vegetable specialist at CARDI, St. Lucia, West Indies (1984-1986), Chief, Soil Division, TISTR (Thailand Institute of Scientific and Technological Research), Bangkok, Thailand (1976-1979), and postdoctoral fellow at IITA, Ibadan, Nigeria (1974-1976).

Ramita Sharma received a bachelor degree in Information Technology from the Universiti Utara Malaysia and completed her Master of Computing in 1996 in Monash University (Melbourne). She currently holds a position as Lecturer in the School of Information Technology, Universiti Utara Malaysia.

Helen Smith is a senior lecturer in the School of Information Management and Systems (SIMS), Monash University, Melbourne, Australia. She has a bachelor of science from Melbourne University and has been a member of the staff of the School and its predecessors since 1971. Ms. Smith's research interests include intelligent decision support, organizational memory, knowledge acquisition and modeling.

Dayo Sowunmi completed his bachelors degree (Computer Engineering) in 1991 from the University of Ife (Nigeria) and Master of Computing degree in 1996 from Monash University (Melbourne). He is currently the general manager at a communications agency based in Melbourne where he is involved in marketing-database analysis, design and development. His interests include strategic planning, intelligent decision support, knowledge management and the integration of database and web-based technologies.

Kerry Taylor has worked as a computer science researcher, practitioner and teacher. She has developed information systems for application in scientific research, financial research, library management, financial management and publishing. Since joining the CSIRO in 1995 her research has focussed on environmental information systems, particularly on supporting integrated access to heterogeneous information sources in databases, files and models. Kerry holds a Ph.D. in Computer Science and Information Technology from the Australian National University for her research in artificial intelligence, and a B.Sc. from the University of New South Wales.

D. Roland Thomas is a professor of Business Statistics and the Supervisor of Graduate Studies at the School of Business, Carleton University, Ottawa, Canada. He received a Ph.D. in Aeronautical Engineering in 1968 from the Imperial College of Science and Technology, England, and later (1980) completed an M.Sc. in Mathematics at the Department of Mathematics and Statistics, Carleton University. In a varied career, he has been a research associate in mathematics, a computer analyst, a senior consultant in statistics and statistical computing, and a manager of the academic division of a university computer center. Since 1986, he has been a faculty member at Carleton University, and also spent 1993/94 as a Visiting Professor at the Department of Statistics, University of Auckland, New Zealand. He is an active researcher and has held an operating research grant from the statistics committee of the Natural Sciences and Engineering Research Council of Canada since 1987. He has published widely in a variety of formats, including refereed articles in over fifteen different academic journals.

Jean-Claude Vansnick received, in 1973, his Ph.D. in Mathematics from the Free University of Brussels and, in 1974, the Royal Academy of Belgium award (Section "Mathematical Sciences"). Since then, he has extended his areas of interest to measurement theory and decision aid. His paper "Strength of Preference: Theoretical and Practical Aspects" was selected as National Contribution of Belgium for the Tenth Triennal Conference of the International Federation of Operational Research Societies (IFORS'84). He is member of several groups on Multicriteria Analysis. He is also co-author of the MACBETH approach currently used by many business and other organisations.

Gavin Walker has worked at CSIRO since 1991, applying object oriented techniques to spatial decision support interfaces and architectures. His current work is in intelligent transport systems. Gavin studied at James Cook University completing a BSc.

Graham Williams is a Senior Research Scientist with CSIRO Mathematical and Information Sciences. He received his Ph.D. in Computer Science from the Australian National University. His research was in artificial intelligence with a focus on expert systems, knowledge representation, and machine learning. He was involved in the development of one of the first expert systems for geographic domains and has been a Consultant, Research and Development Manager, and then Marketing Manager with BBJ Computers. In CSIRO he has implemented intelligent spatial planning systems and has developed ideas for representing and reasoning with expectations in spatial information systems. His current interests are in data mining.

Anthony Gar-On Ye is the Chair Professor and Assistant Director of the Centre of Urban Planning and Environmental Management, Associate Dean of the Graduate School and Director of the GIS Research Centre of the University of Hong Kong. He is a Fellow of the Hong Kong Institute of Planners, Royal Town Planning Institute, and Royal Australian Planning Institute and a Member of the Chartered Institute of Transport and British Computer Society. He has published over 20 books and monographs, and over 100 international journal articles and book chapters related to urban development and planning in Hong Kong and China on the use of information technology in urban planning. He has been a consultant for projects of the Hong Kong Government, the World Bank, Canadian International Development Agency, Urban Management Program, and the Asian Development Bank. He has been Chairman of the Hong Kong Geographical Association, Vice-President of the Hong Kong Institute of the Commonwealth Association of Planners, Vice-President of the Hong Kong Institute of Planners, and Founding President of the Hong Kong Geographic Information System Association. At present, he is the Secretary-General of the Asian Planning Schools Association (APSA) and Chairman of the Geographic Information Science Study Group of the International Geographical Union (IGU). He is also on the editorial board of international journals, including *Environment and Planning B, Computers, Environment and Urban System, Transactions in GIS, Progress in Planning,* and *International Planning Studies.*